John King

Urological Dictionary

Containing an Explanation of Numerous Technical Terms

John King

Urological Dictionary
Containing an Explanation of Numerous Technical Terms

ISBN/EAN: 9783744670081

Printed in Europe, USA, Canada, Australia, Japan

Cover: Foto ©Andreas Hilbeck / pixelio.de

More available books at **www.hansebooks.com**

Fig. 39.

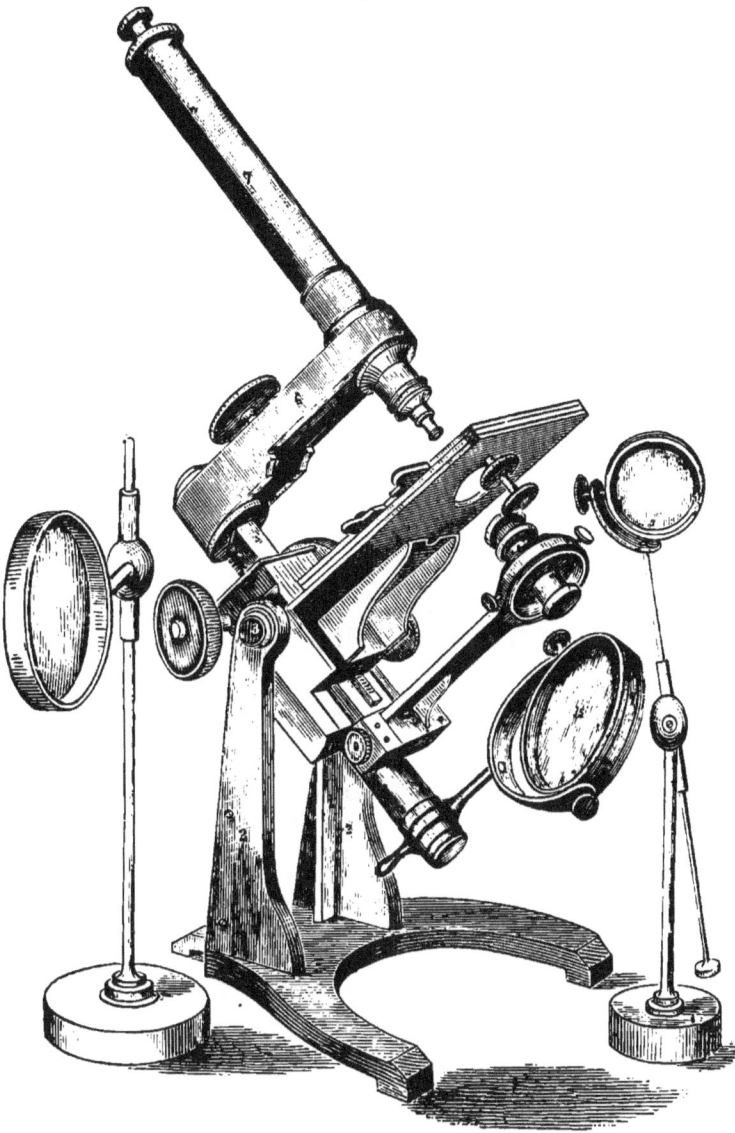

Large Microscope.

UROLOGICAL DICTIONARY:

CONTAINING

An Explanation of Numerous Technical Terms; the
Qualitative and Quantitative Methods Employed
in Urinary Investigations; the Chemical
Characters, and Microscopical Appear-
ances of the Normal and Abnormal
Elements of Urine, and their
Clinical Indications:

BY

JOHN KING. M. D.

With Twenty-seven Useful Tables and Thirty-nine Wood Cuts.

CINCINNATI:
WILSTACH, BALDWIN & CO.,
Nos. 141 and 143 RACE STREET.
1878.

PREFACE.

THE capability of examining urine to assist in the diagnosis, prognosis, and therapeutics of disease, as manifested by the numerous works that have already appeared upon the subject, is recognized at this day as an important and necessary acquisition to medical practice,—one which no physican who values his professional standing can afford to ignore or neglect. The present Dictionary, while it lays no claim to erudition, nor to originality, is designed as a *multum in parvo* to the physician and the chemist, affording information concerning urinary technicalities and urinary investigations together with their clinical importance, that is not to be found in any one book yet published, and which, it is hoped, will prove instructive and useful.

In the determination or analyses of the several normal and abnormal elements encountered in the renal excretion, a number of tests and reactions are given that have been collected from the best sources, fully posting the reader, in these matters, up to the present time; and while many of them are tedious, frequently requiring a more thorough knowledge of chemistry than is generally had by the mass of practitioners, as well as a greater amount of time than can ordinarily be spared during an active practice, yet, in every instance where it was deemed necessary, the more simple and rapid processes have been described and pointed out by the indicator, ‡, as explained hereafter; so that examinations, sufficiently exact for clinical purposes, may be undertaken by any one without the embarrassing and irksome complication of extensive chemical reagents, instrumental apparatus, repeated filtrations, drying precipitates, minute and exact weighings, etc., without loss of time, and without difficulty. To the medical man especially, these are important considerations.

An exact knowledge of the quantity of any element, even to the fraction of a grain, that may be present in a given urine, as essential as it may be to science, is not so urgent for the medical practitioner, who can almost always satisfy all the requirements of practice, by simply determining the approximate amount of such or such important constituent, and then carefully

PREFACE.

observing its daily variations,—whether it remains stationary, diminishes, or augments, under the influence of the treatment pursued. In the Preliminary Remarks much valuable information will be found, leading to the result just named.

In the Appendix several Tables, etc., are given, to aid the investigator in effecting rapid calculations during his observations; and I embrace this opportunity to return my thanks to my young friends Mr. John U. Lloyd and Mr. Nathaniel W. Lord, for their kindness in furnishing the Table of Symbols. I must likewise acknowledge my indebtedness to the host of eminent authors from whose writings I have compiled much important and precious matter, in many instances, without special credit. Nor can I allow this occasion to pass, without expressing my gratification at the elegant style in which my publishers have issued the work.

J. K.

CINCINNATI, February, 1878.

PRELIMINARY REMARKS.

N. B. For information concerning any urinary constituent, reagent, testing apparatus, or technicality, the name of which is *italicised* throughout the pages of this work, the reader will refer to it under its proper alphabetical heading.—It should likewise be stated that, with any given urine, the physician can separately perform such investigation as he may judge proper, qualitative or quantitative, without any preliminary operations (except for determining the presence of albumen), by simply turning to the article treating upon the urinary constituent in which he is interested, and pursuing the method (or one of the methods) therein named. A regular methodic analysis will require him to pursue the course indicated in the Table for Qualitative Analysis (page 10), and for Quantitative (page 11). For the advantage of practitioners who have not the time for lengthy and expensive processes, the more simple and rapid methods, with the ordinary normal and abnormal urinary elements, will, under their respective heads, be prefixed with the indicator ‡.

Explanation of Abbreviations.

c. c. Cubic centimetres.
cgrms. Centigrammes.
fl. oz. Fluid ounces.
grm. or grms. Gramme, or grammes; as 0grm. .02,=two hundredths of of a gramme; or 0.02 grm.
mgrms. Milligrammes.
min. Minims.
mm. Millimetre : as 0mm. .006. = six thousandths of a millimetre; or 0.006 mm.
sp. gr. Specific gravity,
✕. Diameters : as ✕ 500 = 500 diameters.

The processes usually described for the examination of urine, as, repeated filtrations, drying and weighing of precipitates, volumetric operations, etc., together with the expensive and complicated instruments required, as, chemical balance, saccharometer, drying ovens, graduated burettes, etc., as desirable and valuable as they may be in the laboratory, or in the hospital, are embarrassing and discouraging to the general practitioner, who, not having the time and many other material conditions, requires simple and rapid processes for his investigations. The important point for him is, to ascertain whether the urine contains elements not found in it when in a normal condition, and whether its normal elements are deficient or in excess ; bearing in mind that albumen, biliary coloring matter, and sugar, are the three most important morbid soluble substances met with in this fluid, indicating more or less danger, the same as a deficiency of urea. It is not indispensable that he should know to the sixth part of a grain, the absolute quantity of such or such element present, but it is highly necessary that he should carefully watch its daily variations. And if, in testing the urine of a patient from time to time, he will always place himself under the same operative circumstances, and surround himself with the same precautions, he will be enabled to arrive at constant and sufficiently accurate results.

In most instances, the physician has but little time to devote to quantitative analyses, and he will find the method, named hereafter in the succeeding paragraphs, ordinarily sufficient to determine not only the presence of albumen, phosphates, urates, etc., in the specimens of urine under examination, but likewise their daily variations,—whether they increase or diminish in quantity, or remain stationary,—and thus be enabled to appreciate the value of the treatment pursued. True, this method is not an absolutely exact one, but it is sufficiently so for practical purposes, and what is lost in exactness is gained in the saving of time and in instrumental simplification.

For the more extended urinary investigations the medical man should be fully provided with every convenience for prompt and accurate analyses. Among which may be named the possession of a small laboratory room furnished with shelves to securely hold chemicals and apparatus, so that they may always be within easy reach ; a stout, well-made table, four or five feet by two, and furnished with drawers for the reception of infrequently employed materials; sufficient apparatus and chemicals, that there may be no delay nor trouble in the examinations ; a good light (a northern is preferable) for both the chemical and microscopical researches; various accurately computed tables for expediting the determination of certain reactions and results ; and in addition to which a blank book will be found decidedly useful for recording the various results of the investigations. The liquids and other chemicals should be pure, and, together with the apparatus, should be kept constantly clean when not in use, in their proper places, and easy of access for prompt and thorough examinations, so as to occasion the least possible delay or inconvenience. Apparatus kept in an uncleanly or disorderly condition, impure or carelessly arranged chemicals, as well as an insufficiency of apparatus and chemicals, are sources of great annoyance and inconvenience, interfere with precision and accuracy of results, and tend to render the practitioner disgusted with, and ultimately opposed to, all urinary examinations.

In the examination of urine, the different specimens passed at various periods of the day, should be tested both singly and combined, especially in maladies presenting obstinate, severe, or obscure symptoms; and that recently voided should be examined at the time, and likewise after it has stood for 12 or 24 hours. And however difficult or troublesome it may be to accurately collect all the urine passed by a patient within 24 hours, it must be borne in mind that, unless this quantity be correctly known, the examination will be of no value, whatever may be the method of analysis pursued. It may be observed here, that, if the number of cubic centimetres of urine passed be multiplied by its sp. gr. the result will approximatively give the weight of the urine in grammes. .

For qualitative analysis, 200 or 300 c. c. (6 or 10 fl. oz.) of the urine are sufficient; but when it is desired to follow the qualitative by the quantitative analysis, the urine to be investigated should be divided into two large parts, say 500 to 1,000 c. c. (16½ to 33½ fl. oz.), each of which may be subdivided into as many portions as may be necessary. It will prove a great economy of time and money, beside giving much clearer results, to operate upon small quantities only, at a time, say 5 or 10 c. c. (81 or 162 min).

The quantity of a precipitable substance in urine can be approximatively ascertained by placing a given volume of the urine in a graduated jar, or test tube, allowing it to stand a sufficient time, and then observing the height of the precipitate that occurs; by employing the same volume of urine in this manner, daily, and observing the height of the precipitate in each specimen at the end of 24 hours, or at the visit of the next day, closely comparable results may be obtained. In this manner may be followed the daily quantitative variations of such or such element, and the curve they represent may be traced in a book kept for this purpose. Substances, which it is not desirable to test in this manner, may be removed from the urine, by agents that will rapidly precipitate them,—the precipitate being removed by filtration; or, they may be prevented from precipitation by the addition of an agent or agents that will hold them in solution.

In disease, whether acute or chronic, it is frequently required that the urine passed for several consecutive days must be examined daily, or oftener, before the average condition of this fluid can be correctly arrived at. And, in doing this, the several portions of urine passed during each 24 hours should be collected in separate vessels, so that samples of each one can be examined previous to combining them.

The specimen of urine to be examined should be voided into a well cleansed bottle, and a small portion of it be at once tested, to determine its sp. gr., its acidity or alkalinity, its color, odor, general appearance, and whether albumen or sugar be present. This done, the remainder, or a portion of it, should be placed in a clean cylindrical vessel, and covered, so as to keep out dust and foreign matters. It is then to be set aside until a sediment is formed at the bottom of the fluid; in summer this requires about 12 hours, and about 24 hours in winter. Small portions of this sediment are to be removed, by means of a small *pipette*, for microscopical or clinical examination, as may be required. Ordinarily, the urine first passed early in the morning, or the last half of it, is that which is examined; this will usually answer to give a general idea of the condition of this fluid, but where completeness and accuracy is required, samples of the whole of the urine passed in 24 hours must be analyzed, as heretofore observed. It must not be forgotten that, even in a state of health, the urine will be found to vary more or less daily, in its specific gravity, its acidity, its color, and the amount of its constituents. Too great a variation is the evidence of some morbid condition of the system.

In most cases the principal characters of a urinary deposit may be determined at the bedside, by means of a *Coddington* or *Stanhope lens*; or, still better, by a portable pocket microscope. But, if the deposit can not be examined at once, and has likewise to be carried some distance for this purpose, it will be useful to have several small well-closed sample tubes, in each of which a few drops of the deposit may be placed; these should be wrapped in paper, upon which the patient's name is written. To prevent any change or putrefaction, a layer of oil will serve to protect the deposit from atmospheric contact, provided the tube be exactly filled before closing it. A small fragment of camphor, or a little carbolic water (carbolic acid 1, water 1,000), will likewise answer.

Fig. 1.

Coddington Lens.

In all instances, when a specimen of urine, or of a urinary deposit, is presented for investigation, it is necessary to ascertain how long since the fluid was voided, in order to determine concerning any changes, decomposition, etc., that may have occurred. And, in cases where several days may pass before a specimen of urine can be examined, it may be preserved by the above named articles, or by the addition of *chloral*, 10 grains to each fluidounce of urine; or, of *salicylic acid*, 3 grains to each fluidounce of urine. However, chloral is not suitable for urine that is alkaline, or that contains sugar, or urates, as it seriously interferes with the tests ordinarily employed for the estimation of these substances. Larjorrois states that 1–40,000th its weight of fuchsine, or aniline violet, will thoroughly preserve urine from putrefaction. The addition of a little borax will likewise preserve urine for several days, but should not be used where the determination of soda salts is required.

Albumen, it should be remembered, embarrasses or prevents both qualitative and quantitative analyses; hence, when present, it should be promptly removed by one of the processes hereafter indicated. It will also be proper to refer to the article on *Sugar* before proceeding to the other examinations.

In order to effect a systematic examination of a specimen of urine, the following order will be found the most convenient, always ascertaining the quantity of this fluid passed each day, for several successive days:

1. Observe the *color* and *appearance* of the urine.

2. Observe its *odor*.

3. Examine its *reaction*, whether acid, neutral, or alkaline.

4. Ascertain its *specific gravity*.

5. Test for *albumen*. If albuminous, examine under the microscope for *Renal Casts*, *Pus Corpuscles*, and *Red Blood Corpuscles*. If the urine contains albumen, remove it, and then test the filtered urine for other substances, if indicated. But if neither albumen or sugar be present, unless there be some special indication, no farther examination will be required.

6. Test for *sugar*.

7. Test for *coloring matters, normal* or *abnormal*.

8. Any sediment found in urine, which has been allowed to stand for 12 or 24 hours, must be examined microscopically, and often chemically. The more common sediments are:

A. Pink or reddish, and dissolved on heating it in the test tube,—*urates*.

B. White crystalline, dissolved when acetic acid is added,—*phosphates*. Insoluble in acetic acid, but soluble when nitric acid is added,—*oxalate of lime*.

c. White amorphous flocculent, becoming ropy when liquor potassa, or ammonia, is added,—*pus*.

D. Brownish-red crystalline,—*uric acid*.

E. Red amorphous,—*blood*.

Other substances may be present in the urine, as *spermatozoids, fungi, epithelial cells, cancerous fragments, mucus, biliary pigments*, or *acids, leucin, tyrosin, cystin, carbonate of lime, hippuric acid, extraneous substances*, etc.

Or, the order given in the following table (on page 10), which slightly varies from the preceding, may be pursued, and will give similar results; for the details refer to the articles in the body of the book.

I. General Table of the Course to Pursue for Qualitative Analysis.

1. Color of the urine.
- Normal colors.
  - Urine pale; colorless to straw yellow.
  - do normal; golden yellow to amber.
  - do highly colored; reddish yellow to red.
  - do dark colored; brownish; dark, beer color.
- Abnormal colors.
  - Essential, arising in the interior of the organism.
    - Coloring matters of the blood.
    - Biliary pigments.
    - Uroxanthine; indican.
    - Uroerythrin.
  - Accidental, derived from without, and only passing through the organism.
    - Various coloring matters.

2. Odor.
- Essential.
  - Normal, sui generis.
  - Abnormal.
    - Urinous, due to Carbonate of Ammonia.
    - Sulphureted ? (Beale).
- Accidental, from odoriferous substances introduced into the organism.
  - Very varied by
    - Asparagus.
    - Oil of Turpentine.
    - Cubebs; Copaiba.
    - Sandal, etc.

3. Aspect.
- Urine clear.
- do turbid (slightly).—Cloudy or Flocculent.
- See *Mucus, Epithelia.*
- Urine sedimentary.—See *Examination of Sediments.*

4. Chemical reaction with litmus.
- Normal.—Acid.
- Abnormal.
  - Neutral. { By Carbonate of Ammonia.
  - Alkaline. { By fixed Alkalies.

5. Specific gravity.
- 1015 to 1020.—Normal.
- A persistent sp. gr. below 1015.—Test for albumen.
- 1028 to 1050.—Test for sugar; excess of urea.
- 1001 to 1008.—Test for deficient urates.
- 1005 to 1030.—Test for phosphates. See sp. gr. of urine, with each urinary constituent.

6. Detection of the abnormal elements in the urine. (Divide the urine into several portions, in which we successively search for the following principles:
- Albumen.
- Sugar.
- Bile. { Coloring matters. { Biliary acids.
- Blood.
- Fibrin.
- Fat.

7. Examination of the sediments. { Microscopically and Chemically.

In a methodical investigation of urine, *quantitative analysis* must perfect the information obtained from the qualitative. *But as regards medical practice exclusively, it is rarely necessary to perform a complete and exact analysis;* however, the medical man may be frequently desirous of effecting the quantitative analysis of such or such urinary element, the variations of which he is interested in ascertaining. Therefore, the following brief table defining the order to be systematically pursued in this analysis, is given; the detection and determination of the urinary constituents will be found in the body of the work under their respective heads.—

II. General Quantitative Analysis.

1. Quantity of urine excreted in 24 hours.	Averaging 1,500 grammes (about 60 fl. ounces.)			
	Variations.	According to the following physiological condition:	*a.* The more or less copiousness of water in the blood.	
			b. The excretory activity of the kidneys.	
		According to the pathological condition.	Diminution of urine.	Acute diseases. Approach of death.
			Increase of urine.	The increase temporary. — Hydruria. Dropsy.
				The increase permanent. — Polyuria.
2. Density of the urine.	Normal sp. gr. about 1.020.			
3. Quantity of urinary pigment.				*Urohematin. Bile Pigment Blood. Accidental.*
4. Quantitative analysis of some of the constituent elements of the urine.				*Urea. Uric acid. Phosphates. Chloride of Sodium. Sulphates.*

Instruments and Reagents.

These will vary according to the character of the analysis, and the extent to which the physician wishes to carry his investigations. He should always have a book in which to sketch, as far as possible, the various appearances met with in the urine in its original state, as well as when acted upon by reagents; he should likewise possess another book in which is kept the date, name of the person, age, occupation, etc., appearance and character of the urine, symptoms present, curves of variation of important urinary characters, when these can be examined daily, reference to drawings in the drawing-book, as well as to the treatment, and its result.

‡1. Bedside Case.

A neat pocket case, holding the articles named below, carefully and compactly fitted, and, at the same time, easy of access, will be found very useful for urinary investigation at the bedside :—

Urinometer.

3 test tubes, $\frac{1}{2}$ fl. oz. each.

1 small alcohol lamp.

1 watch glass with wire support and handle.

1 small vial Acetic Acid, $\frac{1}{2}$ fl. oz.

1 " " Liquor of Potassa at $\frac{1}{10}$th $\frac{1}{2}$ fl. oz.

3 glass slides and a few thin glass covers.

1 graduated small pipette, 6 in. long.   1 small brass forceps.
1 small thermometer.   2 small vials for extra reagents when
1 wire holder for test tubes.   required.
Strips of blue and red litmus paper.   Small plate of platinum.
1 glass jar of 2 fl. oz., graduated into half fl. drachms; or of 8 grammes
    graduated into c. c.
1 Coddington lens; or a small microscope of about 200 diameters (Nachet).

More extensive investigations in the medical laboratory will require cases
containing the instruments and reagents as follows :

The fluids named should be in quantity from 2 to 4 fluid ounces each, with
the exception of alcohol and water, which should be in the quantity of from
6 to 8 fluid ounces each.

II. Instruments for Qualitative Analysis.

6 or 12 test tubes.
4 conical glasses, or champagne
    glasses.

Fig. 2.   *Fig. 3.*

Conical Test Glasses.

2 or 4 porcelain capsules.
Alcohol lamp.
2 small dropping pipettes.

1 pipette, graduated into 5, 10, and
20 c. c.
Achromatic microscope of ✕ 300 to
    ✕ 500.
Slides and thin glass covers.
Urinometer and containing glass.
Small thermometer.
1 burette.
Filtering paper.
Blue litmus paper.
Reddened litmus paper.
1 graduated cylindrical jar of 1,000
    c. c.
1 glass funnel.
Swabs for cleaning test tubes.
3 glass stirring rods.
Coddington lens. *Fig. 1.*
Small brass forceps.

These should always be well cleansed immediately after use, so as to be in
readiness for the next investigation.

Reagents for Qualitative Analysis.

Pure Nitric Acid.
    " Hydrochloric Acid.
    " Acetic Acid.
    " Sulphuric Acid.
Nitroso-nitric Acid.
Chromic Acid, 4 grs. to 1 fl. oz.
Alcohol.
German Yeast.

Distilled water.
Solution of Acetate of Lead, 1 part
    to 4 parts of distilled water.
Liquor Ammonia.
Liquor Potassa.
Pavy's Solution in separate parts.
Tincture of Iodine, 1 part to 4 parts
    of distilled water,

Besides the other reagents named throughout the work.

If pharmacists would keep pure chemicals on hand, they could dispense the reagents whenever the physician's prescription therefor was received. It seems to us that, in examinations made at the bedside, the patient should bear the small expense of investigations pursued in his interest; and, it is very frequently the case that, where daily examinations are desired, the physician can instruct some intelligent person around the patient how to conduct the processes, and observe and report the results.

Figs. 5 and 6.

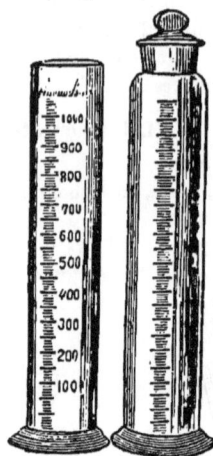

Fig. 4.

Wash Bottle.

Graduated Glass Jars.

III. Instruments for Quantitative Analysis.

Urinometer and containing glass.
Apparatus for determining urea.
Nest of beakers. *Fig.* 8.
2 precipitating jars.
3 conical glasses. *Figs.* 2, 3.
12 test tubes, and rack.
4 porcelain capsules, various sizes.
Small platinum capsule.
Hot water drying oven. *Fig.* 9.
3 glass stirring rods.
Sp. gr. bottle, 50 or 100 grms.
2 funnels; large and small.
Blue and reddened litmus paper.
Wash bottle for washing precipitates on filter. *Fig.* 4.
2 graduated burettes, and holder. *Fig.* 32.

Copper water bath.
Portable sand bath.
Thermometer.
12 glass slides.
24 thin glass covers.
2 brass forceps.
Agate mortar.
2 dropping pipettes.
3 pipettes of 5, 10, and 20 c. c. *Figs.* 11 and 27.
Pipette of 50 c. c., graduated into $\frac{1}{10}$ths. *Fig.* 28.
Small retort stand.
Litre flask. *Fig.* 38.
Chemical balance to weigh at least $\frac{1}{50}$th grain. *Fig.* 7.
Sett of gramme, and grain weights.

2 graduated glass jars of 100 c. c. and 500 c. c. *Figs.* 5 and 6.

Fig 7.

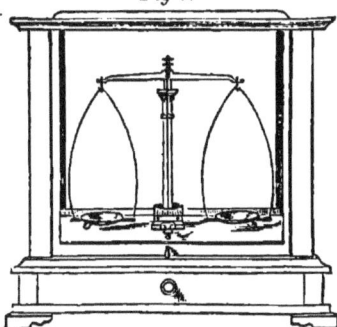

Chemical Balance.

2 test tube holders.
6 ground glass covers.
Graduated test glass.
Microscope, of × 100, 200, and 500 or thereabouts.
Platinum foil.
Platinum wire.
Spirit lamp.
Blowpipe.
2 small tripods.
6 watch glasses.
Bibulous paper.
Filtering paper.
Cylindrical jar of 2,000 c. c. graduated into $\frac{1}{2}$ths.

Reagents for Quantitative Analysis.

Absolute Alcohol.
Alcohol, 95 per cent.
Sulphuric Acid.
Hydrochloric Acid.
Nitric Acid.
Oxalic Acid.
Lime Water.
Baryta Water.
Chloroform.
Ether.
Pavy's Solution, in separate parts.
Perchloride of Iron.
Bichloride of Platinum.

Distilled Water.
Ammonia.
Oxalate of Ammonia.
Liquor Potassa.
Solution of Soda.
Solution of Hypochlorite of Soda.
Carbonate of Soda.
Phosphate of Soda.
Chloride of Ammonium.
Chloride of Calcium.
Chloride of Barium.
Nitrate of Silver.
Sulphate of Copper.

Volumetric Solutions,

Besides the other instruments and reagents referred to throughout the work.

Fig. 8.

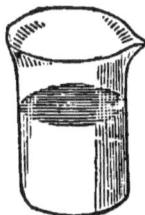

Beaker Glasses.

The *Beaker Glasses, Fig.* 8, should consist of thin, well annealed glass; they are used for containing hot fluids, as well as for heating fluids when an elevated temperature is required. *Precipitating jars* are for collecting precipitate in quantity. *Conical glasses, Figs. 2 and 3,* are for containing urine to determine sp. gr., and to collect small amounts of precipitate from any given specimen; the bottom of the cavity of these glasses should be pointed, not rounded. *Porcelain capsules* are for heating substances that can not well be placed into the beakers; in observing reactions, they are seen best on a white ground; also for the evaporation of fluids. *Platinum capsules, foil* and *wire*, are for holding matters requiring an elevated temperature, for calcination, fusion, or decomposition, and for ascertaining the effects of intense heat upon minute fragments of calculi, sedimentary substances, etc. *Blowpipe*, to direct the flame of the spirit lamp upon bodies to be calcined or fused. *Sand bath*, a shallow sheet-iron vessel, filled with hot sand, upon which is placed a glass vessel, or porcelain capsule, watch glass, etc.; the iron vessel is exposed over a spirit lamp, or other source of heat, and the temperature raised to the degree required. *Water bath* (seen to the left of *Fig.* 10), is a copper or porcelain vessel containing water, and in which fluid another similar vessel, capsule, or glass dish, holding a fluid to be heated or evaporated, is placed, and the whole is then exposed to a source of heat, as required. *Hot water drying oven, Fig.* 9, is a copper vessel, in which substances are placed to be dried that require for this purpose a temperature higher than that of boiling water; when a less elevated temperature is required for dessication, a tin water bath is employed. *Tripods, Retort Stands* (seen to the right of *Fig.* 10), are for holding vessels under exposure to lamp heat, and for supporting filters, dishes, retorts, etc.

Fig. 9.

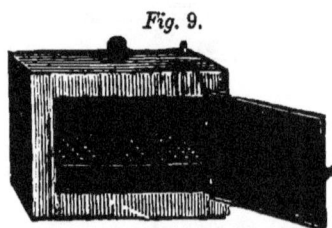

Hot Water Drying Oven.

Dilute solutions of the different reagents are made by the addition of 1 part of the chemical agent to 1, 2, 4, 8, or more, parts of distilled water, as may be required. In all cases the reagents employed should be *strictly pure*, else there will constantly be a risk of having an investigation interfered with or rendered useless by some unforeseen and inexplicable accident. As quantitative or volumetric analyses are generally conducted in a laboratory, the practitioner undertaking them may keep on hand a quantity of each reagent, say 1 or 2 ounces of solid chemicals, and 6 to 8 ounces of liquids, or even more, the bottles containing which should be closely stopped. The small bottles, with capillary orifices (Beale's), for holding the principal fluid reagents, will be found exceedingly useful in micro-chemical investigations.

All the apparatus, reagents, and volumetric solutions may be had of Messrs. Bullock and Crenshaw, No. 528 Arch Street, Philadelphia, Pa.; of Mr. E. B. Benjamin, No. 10 Barclay Street, New York City; of Mr. W. J. Rohrbeck, No. 4 Murray Street, New York City, etc. Dr. W. H. Pile, of Philadelphia, furnishes accurate urinometers, minim, and other graduated pipettes, specific gravity bottles, glass tubes, jars, etc., graduated as may be ordered. Practical chemists, most philosophical instrument makers, and first-class surgical instrument makers, can supply many of the articles referred to.

Fig. 10.

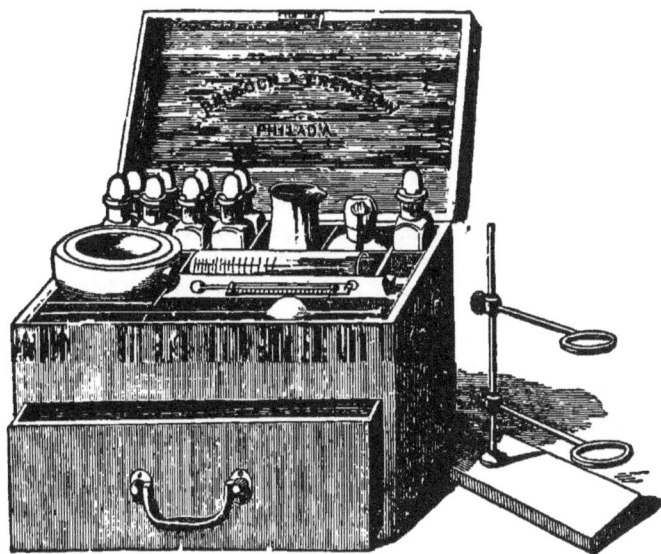

Chest of Apparatus and Chemicals for Urinary Analysis.

UROLOGICAL DICTIONARY.

A.

Abnormal Constituents of Urine, when met with, indicate the existence of more or less serious disease, and afford the practitioner highly important information. Each one of these constituents is significant of a particular disease or morbid condition, that has occasioned its presence, the detection of which enlightens the diagnosis in cases of doubt and obscurity, especially in the earlier stages of such disease. The more common and important abnormal constituents are, in point of order: *albumen, sugar, bile, blood, fibrin, fat, leucin, tyrosin,* and some likewise include *urinary sediments,* considered abnormal not in point of composition, but in point of place. Those less frequently encountered, and the indications of which are imperfectly understood, are: *alkapton, inosite, benzoic acid, lactic acid, butyric acid, allantoin,* and *sulphureted hydrogen.*

Abnormal Deposit in Urine. Any visible deposit in urine which has stood from the time of voiding for twelve or twenty-four hours. See *Urinary Sediments.*

Accidental Abnormal Constituents of the Urine, are those substances which are derived from food, drinks, drugs, etc., and which pass into the urine either changed or unchanged. Their presence is not always a pathological indication, though it is frequently useful to determine such presence, especially in cases of poisoning, when the toxic agent has passed into the urine. The processes to pursue for such determination, are the same as are employed when such substances are found mixed with organic matters, as in cases of death from poison. These accidental constituents are *alkaline carbonates, antimony, arsenic, benzoic acid,* bismuth, *bromine, bromide of potassium, camphoric acid, chlorate of potassa, chloroform,* cobalt, gold, *iodine, iron,* lead, *mercury, nickel, phosphorus, quinine, salicin, santonin, senna, strychnia, succinic acid,* sulphide of potassium, *tannin, tar,* zinc. See *Coloring Matters; Urinary Sediments.*

Acetate of Lead. *Lead* or *Plumbic Acetate.* There are two acetates of lead employed in urinary investigations: 1. *Neutral Acetate of Lead, Neutral Plumbic Acetate, Superacetate of Lead, Sugar of Lead, Lead Acetate,* and which is determined from the subacetate of this metal by not causing a copious precipitate with carbonic acid, which only partially decomposes it, and, likewise, by not forming a precipitate with solution of gum Arabic; with sulphuric acid, acetate of lead forms a white precipitate of sulphate of lead, at the same time evolving vapors of a vinegar odor; it also forms precipitates with lime water, sulphate of lime, the alkalies, and salt. In solution, it is used to precipitate coloring matters of highly colored urine, when these interfere with the

2

examination; it does not precipitate the sugar in diabetic urine. It is also employed as a reagent. The solution consists of 1 part sugar of lead to 4 or 6 parts of distilled water. Label: "*Solution of Acetate of Lead for Decolorizing Urine.*" 2. *Basic Acetate of Lead, Basic Lead, or Plumbic Acetate, Subacetate of Lead, Diacetate of Lead, Tribasic Acetate of Lead.* This salt, when added to gummy, mucilaginous, or saccharine solutions, causes a white flocculent precipitate, which is an intimate mixture of the organic substance and the oxide of lead. Tannin, albumen, and many animal substances are also precipitated by it. The organic substance may be eliminated by the addition of sulphuric acid to the precipitate, which forms an insoluble sulphate of lead with the oxide. Subacetate of lead also forms precipitates with the solutions of the alkalies or their carbonates, with sulphuric and hydrochloric acids free or combined, with the soluble iodides and chlorides, and with solutions of all the neutral salts. With carbonic acid, it forms a copious precipitate of carbonate of lead. In solution, it forms a precipitate with most vegetable colors. When added to urine, it precipitates nearly all the coloring matter, as well as considerable saccharine matter should this be present, and hence should not be employed to render diabetic urine more transparent. Its solution should be clear and colorless, and is made by dissolving 1 part of the basic salt in 4 parts of distilled water; this solution has an alkaline reaction, is readily decomposed by the carbonic acid in the atmosphere, and, consequently, should be kept in well-stopped bottles, and labelled: "*Solution of Basic Acetate of Lead.*"

Acetic Acid. This acid is employed pure, and also diluted, 1 part of pure glacial acetic acid in 5 parts of water. Acetic acid dissolves the earthy and triple phosphates; the urates with subsequent formation of uric acid crystals; and carbonate of lime with effervescence. It dissolves part of the constituents of pus and blood corpuscles, causes these cells to swell to nearly twice their natural size, and renders them transparent, showing a thin, clear outline, and one or more internal nuclei. It dissolves the fibres of muscular tissue and leaves intact the dartoic fibres. It does not precipitate albumen, but precipitates cystin from its ammoniacal solution. The dilute acid coagulates mucus, forming an opaque corrugated membrane, precipitates *mucin* in the form of fine threads, and renders the nuclei of the mucus corpuscles more distinct.

Acetone. A colorless thin liquid, of sp. gr. 0.814 at 32° F., and boiling at 132° 8′ F. It is soluble in water, alcohol, and ether. Acetone is found in the urine and blood of diabetic patients, and imparts to the former its peculiar odor.

Acidity of Urine. From 1.29 to 1.95 grammes pass in health during each 24 hours. This acidity is due to a cause or causes yet imperfectly understood. It may be owing to the presence of an acid phosphate of soda, or, to free lactic or hippuric acids, to acid urates, or, to some organic acid or its acid salt. Bence Jones supposed that the acidity of the urine may, to a certain extent, be taken as a measure of the gastric acidity. Vogel ascertained that the maximum of acidity in the urine was during the night; the minimum during the forenoon; and the medium in the afternoon subsequent to the dinner meal. The urine is generally acid when just voided, as may be determined by its reddening blue litmus paper; in health this acidity remains for several days, becoming more intense for a time, but should the acidity be lost within 24 hours after the voiding of the urine, it indicates some disease requiring prompt attention. The amount of the acidity should be determined as soon as possible after the urine has been voided, as it becomes more intense upon standing, except in a few instances, or when decomposition occurs. This is done by a volumetric process, carefully neutralizing the acidity by a solution of caustic soda, every c. c. of which indicates exactly 10 milligrammes of a solution of one gramme of pure, dry oxalic acid in exactly 100 c. c. of distilled water. 10 c. c. of this solution of oxalic acid are placed in a small beaker, to which 6 or 8 drops of tincture of litmus are added, or enough to produce a distinct red color. Let this rest upon some white paper, and carefully add the soda solution until the original blue color of the fluid has been restored. Suppose this reaction has required 6 c. c. of the soda solution indicating 100 milligrammes of oxalic acid; we add to 600 c. c. of the soda solution 400 c. c. of distilled water, and thus obtain 1,000 c. c. of a standard solution, 1 c. c. of which neutralizes 10 milligrammes of oxalic acid, and which should be tested to ascertain its correctness. Now to 50 or 100 c. c. of the acid urine, gradually add this standard solution, testing each time with litmus paper, until it no longer develops any redness, when from the amount of standard solution employed we can determine the amount of the unknown free acid of the urine equivalent to so much oxalic acid. In this process, it will be better to place a drop of the urine upon the litmus paper, from time to time, as the standard solution is added, and, also, to be careful not have an excess of alkali in the urine.—To lessen alkalinity of the blood and increase the acidity of the urine, diminish vegetable food and augment the animal; or use the stronger mineral acids, in proper doses, diluted, avoiding the vegetable.—The so-called "*Acid Fermentation*" of acid urine, resulting in a decomposition of this fluid into ammonia, uric acid, fungus spores, etc., is supposed to be due to the presence of mucus and other organic substances, which, acting as a ferment, develop lactic and acetic acids during the progress of decomposition of these substances, and thus increase, temporarily, the intensity of the acidity of the urine. This, however, has not been satisfactorily demonstrated. See *Chemical Reaction.* Sulphydric acid (sulphureted hydrogen) in urine turns paper, washed with solution of acetate of lead, black.

Adiposuria. Urine containing considerable fatty matter.

Agents *that exert an influence upon certain conditions of the urine,* are briefly as follows:—1. The *solids are increased* by digitalis, belladonna, colchicum carbonate of potassa, white Rhine wine, etc.—2. The *solids are diminished* by citrate of quinia and iron, ammonio-citrate of iron with quassia, alcohol, beer, coca, tea, coffee, and Paraguay tea; also, by opium, morphia, conium, calabar bean, hyoscyamus, cannabis, etc., which, however, from their action upon the nervous system, may sometimes be contraindicated.—3. The *fluids* *are increased* by sweet spirits of nitre, beer, gin, turpentine, whisky, coffee without sugar or milk, large draughts of soft water, infusion of verbesina virginica, of triglochin maritimum, etc.—4. The *fluids are diminished* by conia, citrate of quinia and iron; iron, copper, and ammonio-citrate of iron with quassia; arsenic and cantharides may almost wholly arrest it.—5. *Urea is increased* by water, salt, coffee without sugar and milk, cubebs, cantharides, atropia, tea, bicarbonate of potassa, liquor potassa, eggs, gelatin, milk, animal soups, jellies, and all nitrogenized foods.—6. *Urea is diminished* by chlorate of potassa, sugar, starch, fat, cream, cod-liver oil, beer, coca, coffee, tea, Paraguay tea, all non-nitrogenized foods, citrate of quinia and iron, calabar bean, benzoic acid, acetate and phosphate of soda, colchicum, acetate of potassa, quinia, alcoholic drinks.—7. *Uric acid is increased* by phosphate of soda, bicarbonate of potassa, liquor potassa, alcoholic drinks, beer, port wine, catawba wine.—8. *Uric acid is diminished* by tea, coffee, quinia, colchicum, atropia, acetate of potassa, cod-liver oil, alcohol, matico and epigea repens.— 9. *Alkalinity is increased* by alkalis, alkaline carbonates, alkaline mineral waters, apples, lemon or orange juice, citrates, tartrates, and acetates; cod-liver oil lessens acidity.—10. *Alkalinity is diminished* by acid phosphates, chloral hydrate, benzoic acid, Campeachy wood, salicylic acid, alcoholic drinks, etc.

Albumen. In all cases the first and most important step, after observing the color, appearance, odor, and acid or alkaline reaction of the urine, is to determine the presence or absence of albumen, before proceeding to test for any other substance, and this should be done as soon after the urine has been voided as possible. It is not the occasional appearance of albumen in the urine that renders the prognosis unfavorable, but only when it remains persistently, and especially when it is attended with *renal casts, pus, or blood globules.* A low specific gravity of the urine would lead to a suspicion of albumen, yet this substance often exists in the fluid when its sp. gr. is high. The following table will give the best course to pursue for the QUALITATIVE ANALYSIS:

III. Table.

‡ Preliminary Steps.

1. { a. The urine is turbid or sedimentary. See 2.
    { b. The urine is clear and transparent. See 2b.

2. Filter or Decant.

    a. The sediment must be examined under the microscope.

    b. The filtered urine is

    x. Acid or Neutral. — See *Detection.*

    y. Alkaline. — Neutralize with nitric acid until a slightly acid reaction is given, then see *Detection.*

Remarks.—The neutralization advised above (2 y) may appear useless, because in "Detection," below, it has been stated that the addition of 1-10th nitric acid will neutralize the urine. But it may happen then that no precipitate of the albumen takes place, even on heating, because the proportion of acid, a part of which will have been neutralized, will be found insufficient; for it must not be forgotten that the addition of a few drops of nitric acid prevents the albumen from coagulating by heat, a fact due to the setting free the phosphoric acid of the urinary phosphates, in which albumen is freely soluble. But, if an excess of nitric acid be added, it displaces or predominates over the phosphoric acid, and the albumen is precipitated; hence, before proceeding to the Detection, the urine must always be brought to unity, that is, to a neutral or slightly acid reaction. It may be proper to observe, however, that when the quantity of albumen in the urine is quite small, nitric acid is apt to retain it in solution, for which reason many prefer acidulating the urine with acetic instead of nitric acid. The proportion recommended for acidulating the urine is 2 drops of acetic, or 0.5 c. c. of nitric acid, to 15 c. c. of urine. When the deposit (precipitated or not by heat) in the urine, consists only of phosphates, the addition of acetic or nitric acids renders the fluid clear by dissolving the phosphatic salts.

IV. Table.

† Detection.

1. Into a test tube containing 10 c. c. of urine, pour 1 c. c. of nitric acid.	*a.* An evident coagulation is produced	*Albumen.*
	b. There is doubt; the liquid is only turbid, or there is a slight precipitate..	See 2.
2. Heat the liquor.	*c.* The whole is redissolved	*Uric acid, Nitrate of Urea.*
	d. The turbidity or the precipitate remains........................	Albumen, or see 2x
2x. The turbidity remains.	Add Alcohol in small quantity.	*e.* The turbidity disappears.. — Resins.
		f. It persists.... — Albumen.

The step 2x, above, is necessary, when there is a certainty or suspicion that the patient has been taking cubebs, copaiba, turpentine, sandal, etc.

Observations.—The preceding process is the clearest and most exact of any. It is useless for the physician to attempt processes for recognizing minute traces of albumen in the urine, as it would only embarrass him, and be of no advantage, from the fact that small quantities of albumen may be present in this fluid, under the influence of diet. It may be observed, relative to the above table, (referring to its letters and figures) that:—1. The employment of nitric acid is always disagreeable; but, in testing for albumen, it is preferable to acetic acid, which has the inconvenience of precipitating *mucin.* Again, it would be improper to conclude that, because cold nitric acid gives a white precipitate, an acid urine is albuminous, it is important that this precipitate be insoluble by heat. One-tenth of nitric acid is the most advantageous; but, in doubtful or difficult cases, the testing must be repeated by adding more, as well as less, acid, and it should invariably be added drop by drop.—*b.* Any doubts may be removed by examining a drop of the precipitate under a magnifying power of 300 diameters; either some amorphous, granulated masses of albumen will be recognized, or the crystalline forms of uric acid, etc. It is usually prudent to wait a few minutes to afford time for these crystalline productions.— 2, *c.* If the fluid be again allowed to cool, the precipitate will be reproduced, as these salts are almost insoluble in the cold liquid.—2 *x, e.* This cause of error (precipitate of resinous substances in the urine) has not been confirmed by any author since Maly made it known. It should be stated that when the urine contains much coloring matter, the albuminous deposit, instead of being white, is colored, greenish if biliary matters, brownish red if blood, and violet if indican, be present, etc.

Other Methods. — The above method of qualitatively determining the presence of albumen will be found fully sufficient for all practical purposes; other methods, however, have been described, and it may prove useful to make a few of them known.—1. Fill a test tube one-third or one-half of its depth with the suspected urine, and then allow 20 or 30 drops of pure colorless nitric acid to slowly run along the side of the inclined tube down to the bottom of the urine; if albumen be present, it will appear in the form of a white disc, with a more or less distinct outline above the layer of acid, varying in thickness according to the proportion of albumen. At times this disc is so transparent, that it has to be held in a certain manner in the light to become apparent; and, again, a few minutes may elapse before it will be seen. If the urine holds a large proportion of urates, uric acid will eventually be separated, and, as a rule, will collect near the free surface of the fluid, being separated from the albuminous disc by a larger or smaller clear interval, and will show the crystalline forms of this acid under the microscope. Should it unfortunately happen that these crystals become mixed with the albumen, the experiment will have to be repeated. This is a delicate test, detecting 1 grain of albumen in 1,000 of urine, and is especially useful when the albumen is so small in amount as not to become recognizable by heat. Should carbonic acid or carbonates be present in the urine, to such an amount as to occasion effervescence, this test will prove unavailable. *Heller.*—2. Dissolve 10 grammes of tannin in 200 grammes of distilled water, and then add 10 grammes of ether to preserve the solution. This precipitates gelatin, modified albumen, and other organic substances from urine, and should be used for a long time consecutively, as, soups rich in gelatin, etc., cause the urine of persons eating them to give an abundant precipitate with this solution. *Bouchardat.* Two parts of tannin in 100 of

alcohol, when added to urine in the proportion of $\frac{1}{6}$th its volume, will detect $\frac{1}{2}$ grain in 1,000 grains of urine; should urates be precipitated, heat
will cause them to be dissolved. *Almen.*—3. Place a portion of the suspected urine in a test tube, and carefully pour on it an equal quantity of
ordinary alcohol, without allowing the two fluids to mix. At the junction
of the two, a milky haze will be seen, and if the albumen be in considerable
quantity, small prolongations of it will form in the alcohol. This will frequently show albumen where it is not suspected. *Betz.*—4. Wash the albumen, precipitated by heat or nitric acid, with water, to remove all traces of
urine which interferes with the reaction; then add a little liquor potassa to
dissolve the albumen, and let a drop or two of the liquor of Barreswill (or
of Fehling's solution) fall into the solution. Immediately a very fine, rich
violet color demonstrates the presence of the albumen. If the albumen is in
minute quantity, the solution must be heated. This color will be produced
if $\frac{1}{150}$ part of albumen be present. *Boulaud.*—5. Place a few c. c. of a
saturated aqueous deep yellow solution of picric acid into a test tube, at the
surrounding temperature. Let a few drops of the suspected urine fall upon
this, and if albumen be present, in passing through the solution, a characteristic whitish furrow or line will be produced. Although a good test, and
one that may be applied in the presence of phosphates or urates, it is only
useful when albumen is present to an amount greater than is required for
most of the other tests; on heating the fluid, the albumen collects in a lump
and floats. *Galippe.*—6. Albumen in the urine, or in solution, under the action of a polarizing apparatus (*albuminimetre*) turns the plane of polarization
towards the left. Each minute of the instrument corresponds to 0.180
grammes of albumen, each degree to 10.800 grammes. *Becquerel.*—7. This
process is sometimes employed to confirm the result of action of heat and
nitric acid; to a fresh portion of the urine add acetic acid, filter to remove
mucus, and then add solution of potassium ferrocyanide. A white precipitate indicates albumen.—8. Albumen is very soluble in phosphoric acid;
hence, in examining urine for it, first, precipitate the phosphoric acid by adding a few drops of solution of chloride of lime, and then a little ammonia;
let it stand for 24 hours, filter to remove the phosphates, and then test the
filtered fluid for albumen.—‡ 9. To 100 grammes of urine add 4 or 5 drops
of acetic acid, 2 c. c. of non-concentrated nitric acid, and 10 c. c. of a solution composed of 1 part, each, of crystallized phenic acid and acetic acid,
and 2 parts of alcohol at 90°.—Shake the mixture, throw it upon a filter,
that has been carefully dried and weighed previously, and wash the contents
of the filter with water holding 1 part of phenic acid to 100 of water in
solution. Dry the filter, and weigh it. The difference between the two
weighings of the filter will be the weight of the albumen present. *Méhu.*

QUANTITATIVE ANALYSIS (or Estimation) of Albumen in the Urine.
The amount of albumen passing with the urine may vary from a mere
cloudiness to an ounce during the 24 hours, though this maximum quantity
is uncommon; the average amount usually met with varies between 1$\frac{1}{2}$ and

3 drachms in the 24 hours. The methods usually employed for the exact estimation of albumen in any given urine, are difficult to perform, and require the use of accurate balances and weights. For ordinary clinical purposes an approximative process only will be necessary, one based upon the comparative estimation of the amount of albumen precipitated in a given volume of urine.—When the presence of albumen has been recognized by the process above indicated (in the table under *Detection*), it will suffice to repeat the operation in a graduated tube, agitating it strongly, and then allowing it to rest for twenty-four hours. (Tubes holding from 20 to 50 c. c., and perfectly graduated in fifths of a c. c., can be had for a dollar or two. *Fig.* 11.) The flocculent precipitate at first rises to the surface of the fluid,

Fig 11.

Pipette graduated into 5ths.

and afterwards falls to the bottom. When there is but little albumen, the precipitate may be hastened by gently heating the urine. This experiment must be repeated every day with fresh quantities of the urine, voided day after day, and by reading the height of the precipitate in the graduated tube on each occasion, a curve of the variations of the albumen may be traced and preserved.

This is a most excellent approximative method, giving correct comparative results, and may be employed for the estimation of several other precipitates. Thus, in diluting a fluid containing a precipitate of known height, with water one-half, this height will diminish one-half; whatever may be the diameter of the tube employed, the precipitate will occupy the *same volume*, although the heights will be naturally greater as the diameter of the tube is smaller.—If, in the performance of the above named process, the urine be filtered, so as to remove from it any epithelia, urates, or foreign substances, the result will be still more exact. When the amount of albumen is quite small, so that its depth in the tube is hardly appreciable, its proportion may be expressed by the terms, "cloudiness," or "opalescence;" in larger amounts by the terms, one-fourth, one-six, or one-tenth, etc., according to the height of the precipitate.

Other Methods.—1. A very exact method, rarely required however, is to add acetic acid to a measured amount of urine, and boil it in a test tube. Collect the precipitate on a previously dried and weighed filter, wash well, dry, weigh in an accurate balance, and then calculate the proportion of albumen.—2. Fill a wide-mouthed burette to a multiple of 100 with albuminous urine; add a slight amount of a solution of 4 grains chromic acid in 1 fluidounce of distilled water; shake the mixture and set it aside for 24 hours; read off the amount of precipitate, and calculate the amount of albumen for the whole of the urine passed, by multiplying. *J. Dougall.*—3. To the albuminous urine add one-half its volume of a solution of common salt, (2 parts to 10 of water), and shake the mixture. Now add solution of tannic acid until the albumen is completely precipitated. Collect the precipitate upon a weighed filter, and wash it with distilled water until all the salt is removed; after which, treat it with boiling alcohol until the filtrate passes without giving indications of tannic acid. Dry the filter and the precipitate; weigh it, and the weight, minus that of the filter, will give the amount of albumen present in the urine operated upon. This is stated to be a superior method. *L. Girgensohn.*

Microscopic Examination of the Sediment of an Albuminous Urine. This sediment may contain : 1. Crystallized salts, as *uric acid, ammonio-magnesian phosphates,* and *oxalate of lime.* Uric acid will be frequently met with in very pale, feebly acid urine of patients having albuminaria of long standing. Though Beale has stated that the presence of uric acid in albuminous urine would lead to the supposition that the case was acute and of short duration. —2. Amorphous salts, as *urates, phosphate of lime, carbonate of lime.*—3. Figurate, or amorphous elements of various characters, as, *a. epithelial cells* from the *urethra,* the *vagina,* the *bladder,* or the *ureter,* etc.; *b. leucocytes, blood globules; c.* elements coming from the uriniferous tubules (*renal casts*), and which appear in the form of cylinders of varied aspect.—All these sedimentary elements are referred to under their proper heads, *which see.*

Albumen, precipitated from urine, when examined under the microscope with a power of 300 or 400 diameters, is in the form of small, irregular, granular, amorphous masses, not changed by acetic acid. For purposes of examination, it may be prepared by adding a small quantity of white of egg to some clear, fresh urine, shaking the mixture thoroughly ; precipitate the albumen by the method above named. Sometimes filamentous threads, debris of the lax cellular tissue enclosed in the albumen of the egg, will be present, and which may be removed from the urine by filtering previous to precipitating the urine. An aqueous solution of albumen treated with sulphocyanate of iron gives a blood-red color, which is removed when a few drops of ammonia are added.—E. Reichardt states that he has found albuminous urine to contain a substance very similar to dextrine.

Clinical Import. Urine of sp. gr. 1010 to 1012. An albuminous urine may be considered an important objective sign of disease. It is met with in renal congestion, in many acute febrile and inflammatory diseases, in emphysema, heart disease, cirrhosis, poisoning by lead, abdominal tumors, etc., in which affections its presence may be owing to impediments of the circulation of the blood, to a specific poison, etc. Sometimes the urine may be albuminous for several consecutive days, and then disappear, without any attributable cause. It may also be present when an albuminous diet is continued for some time, as, for instance, the eating of eggs. Albumen is of more serious import when its presence is permanent, and the urine contains pus, blood, or casts. *Pus* and *blood* indicate disease or abscesses of one or more of the urinary organs. *Blood* may also be present in the urine in purpura, scurvy, and other debilitating maladies. *Renal casts* should always be carefully sought for when albumen is detected in the urine ; their presence, being dependent on structural changes in the kidney, affords evidence that the albumen is owing to these changes, and that an organic renal disease exists, of a more or less serious character. The greater the amount of the albumen, and the more abundant the casts and renal epithelium, or fatty cells, the stronger the indication of the existence of kidney disease (acute or chronic Bright's disease). See *Fig.* 30. In the latter months of pregnancy and during the puerperal state the urine should be examined from time to time, to ascertain if albumen be present; for, whether it be due to passive renal congestion or to actual Bright's disease, its presence indicates the existence of one of the probable causes of puerperal eclampsia.

Albuminaria. A term applied to urine in which albumen is present, as is the case in Bright's disease of the kidneys, renal congestion, etc.—Arterial tension ascertained by sphygmographic traces of the pulse in scarlatina previous to the appearance of albumen in the urine, while at the same time the characteristic crystalloid substances of the blood, especially hemoglobine, were found in it, has led M. Mohamed to consider these as the signs of the prealbuminuric stage. He has found them in certain conditions predisposing to albuminaria, as, at the commencement of alcoholic and lead poisoning, and during pregnancy. The presence of the crystalloid substances of the blood in the urine may be recognized by the blue coloration it assumes when two drops of ozonic ether and one of tincture of guaiac are added to this fluid. This coloration fails as soon as the albumen becomes abundant.

Albumininimetre. A polarizing apparatus devised by Becquerel for the detection and estimation of albumen in the urine; it is a modification of Mitscherlich's polarizer for the detection and estimation of sugar. The analyser is not a Nichol's prism, but a birefracting prism, cut in such a manner that one image only is in the field of vision. To this analysing prism a lens is adapted, by means of which the examiner is the better enabled to observe the effects produced. It is seldom used on account of its expensiveness, as well as its inapplicability to urine containing small quantities of albumen.

Albuminuric. A term applied to a person whose urine is albuminous; likewise applied by some to the urine itself, or to the condition present occasioning the albuminaria.

Alcohol coagulates or precipitates from their solutions, albumen, gum, dextrine, liquid starch, in white flakes and clots; the alcohol should be strong. It dissolves resins, volatile oils, and a certain quantity of fat oils. Physiologists generally suppose alcohol to be consumed during respiration, but Lieben has ascertained its presence in urine by the following process: Place the suspected liquid in a test tube with a few centigrammes of iodine, and a few drops of solution of caustic soda. Gently heat the mixture and a very characteristic precipitate of iodoform ensues. The $\frac{1}{50000}$th part of alcohol may be detected by this process.

Algæ. Very fine filaments arranged in straight or waved clusters variously crossing each other, and often forming a felt-like net-work so thick as to intercept the light and prevent their filamentous structure from being distinguished, except along the margins of this mass, consisting of *Leptothrix*. With these filaments are constantly observed, included in their midst, or swimming very freely in the urine, a multitude of minute rods or wands (vibriones, bacteria, etc.). They may be studied in the coat or fur upon the tongue, especially in the morning, and in which they are found mingled with granular epithelial scales. In somewhat old deposits of alkaline urine, especially when they contain mucus, they may also be found more or less masked by phosphates and urates, which a drop or two of acetic acid will cause to disappear. Algæ never exist in fresh urine. Other allied species of algæ are likewise found in putrefied urine, particularly *Leptomitus*. See *Vegetable Organisms*.

Alkaline Phosphates. See *Phosphates*.

Alkaline Urine. This is commonly due to alkaline fermentation. Urine when just discharged is usually acid; it is rarely voided of an alkaline character, unless as the result of certain remedies taken inwardly, or when the urine is retained for a long time in the bladder. Healthy urine requires 24 hours or longer to become alkaline, and which is due to fermentation; but when the alkalinity occurs soon after it is passed, or within 24 hours afterward, some unfavorable condition exists which hastens the fermentation. During the fermentative process, the *urea* is decomposed, carbonate of ammonia is formed, as well as sulphide of ammonium, and the urine becomes of a fetid ammoniacal odor, and loses its transparency from the presence of bacteria, torula, and deposits of phosphates and urates. If acid phosphates are present in the fermenting urine, they become converted into neutral, and then into basic, phosphatic salts. The ferment is supposed to be decomposing mucus, pus, or other organic substances which act by catalysis upon the urea. This form of alkaline urine indicates some chronic affection of the bladder and urinary organs, or some malady of the spinal cord. Mucus, and other elements that by decomposition may act as catalytic bodies, may be removed from fresh urine by repeated filtrations, and then the urine will keep unchanged for a long time. However it is sometimes the case, when no morbid condition exists, that the urine is voided of a neutral or alkaline character, due to the use of acid fruits, certain foods, and medicine, or to the presence of an alkaline earth, or a fixed base of potash or soda; the alkali existing in the blood to excess, which excess is separated and removed by the kidneys. This condition of the blood is the result of a long use of the alkalies, either alone or combined with carbonic acid, or with vegetable acids; it may also coincide with a general condition of weakness, defective nutrition, chlorosis, anemia, etc.

An alkaline condition of the urine is determined by reddened litmus paper, which is restored to its original blue color. If, on drying the paper, or on gently heating it, the blue color disappears, and the red returns, the alkalinity of the urine is due to an ammoniacal salt; if it remains permanent, to a fixed alkali. See *Chemical Reaction. Ammoniacal Urine.* —Ammoniacal alkalescence changes the reddish-yellow color of logwood paper to a dark violet. If a glass rod, moistened with hydrochloric acid, be placed in ammoniacal urine, white vapors are given off. Carbonate of ammonia, in urine, gives effervescence when a dilute acid is added.

Alkaloids. By agitating urine, which has been rendered alkaline, with chloroform, many alkaloids will be taken up by it. Agitate the chloroform solution with acidulated water, which removes the alkaloid. In this way may be detected aconitia, atropia, caffeina, cinchonia, codeia, conia, emetina, hyoscyamia, narcotina, nicotina, physostigmina, quinia, quinidia, strychnia, thebaina, theobromina, and veratria. Brucia, colchicia, or papaverina, dissolve more slowly; narceina is taken up in only small quantity; sabadillia requires heat; picrotoxin should be removed from an acid solution; and morphia and solania are not removed at all by the chloroform. *J. Nowak.*— Phosphotungstic acid, as well as phosphomolybdic acid, precipitates alkaloids from their acid solutions; filter and wash the precipitate, which may then be still further investigated. *Schering.*

Alkapton. A peculiar substance found in urine by Bœdecker, which becomes of a brown color when acted upon by an alkali; it reduces the copper salts, but not those of bismuth, and does not undergo fermentation with yeast.

Allantoin A substance met with in the urine of calves, and also stated to have been found in the urine of man, and that of dogs. It may be prepared by placing 100 grammes of uric acid in one or two litres of distilled water, which has been acidified with acetic acid; to this is added the binoxide of lead, obtained from 1,500 grammes of minium. Expose the mixture for a time to the light without heating it; lastly, heat to boiling, wash the residue on the filter with boiling water, and evaporate the filtered liquid. From 30 to 32 grammes of allantoin are obtained. At first dialuric acid and urea are formed, then allantoin and oxalic acid. *Mulder.* Allantoin separates in colorless, prismatic, glass-like crystals, which are soluble in 160 parts of boiling water, in 30 of cold water, in boiling alcohol, but not in ether. It forms combinations with the oxides of lead, copper, and zinc; and, under the influence of yeast, at 86° F., it is decomposed into urea, oxalate, and carbonate of ammonia, and a new syrupy acid. Mulder has combined it with nitric, sulphuric, and other acids.

Alloxan. When 1 part uric acid is treated with concentrated nitric acid, sp. gr. 1.42, 4 parts, effervescence ensues, nitrogen and carbonic acid are set free, and a crystalline mass is formed, consisting of urea and alloxan; the latter being in colorless rhombic octahedra, which are disintegrated when exposed to atmospheric influence. The crystals are soluble in tepid water, have an acid reaction with litmus, and stain the skin purple. By evaporation the crystals may be obtained from their aqueous solution, in a state of greater purity.—When uric acid is dissolved in diluted warm nitric acid, *alloxantine* is formed in the solution; on evaporating the solution almost to dryness, and then adding nitric acid, alloxan is obtained, whch gives a deep purple color (*murexide*) with ammonia.

Amanita Muscaria. A fungus plant of the Agaric or Mushroom tribe, which is used by the Tartars to produce intoxication. Its intoxicating principle passes unchanged into the urine, imparting its intoxicating properties to this fluid.

Ammonæmia. A term applied to that condition of the blood in which an excess of ammonia is present. In this condition the breath is ammoniacal, as well as the perspiration; and the urine is ammoniacal at the time of voiding it. This condition is due to a reabsorption into the blood of the decomposed products of urea, when urine is retained within the bladder for a long time. Ammonæmia is often curable by catheterization. See *Uræmia.*

Ammonia. This alkali is employed in the same cases as soda and potassa, its action being a little less energetic. It is used to neutralize acids introduced into a fluid, to aid in the formation of certain ammoniacal salts, to distinguish leucocytes from epithelia, the former being rapidly dissolved by it, while the latter are not affected, or only become paler, and to dissolve carmine. It renders vibriones and bacteria motionless without dissolving them, while it dissolves monads and spermatozoids. It is frequently employed in solution of full strength; but, for most purposes, a mixture of 1 part of the strongest liquor ammonia with 3 parts of distilled water, will be found of sufficient strength; it should be kept in a well stopped bottle. In addition to this, the following ammoniacal preparations are likewise employed:

1. *Solution of Hydrochlorate of Ammonia.* Take of pure hydrochlorate of ammonia 1 part, distilled water 10 parts (by weight). Mix, dissolve, filter, and label, "*Hydrochlorate of Ammonia. Detection of Lime in Presence of Magnesia.*"

2. *Solution of Molybdate of Ammonia.* Take of pure molybdic acid 1 part; pulverize and place it in a porcelain capsule, or a glass balloon; sprinkle and moisten it with pure liquor ammonia 5 parts. Heat gently, and agitate until the solution is completed. Then add, previously mixed together, the following: Pure nitric acid, 36°, 8¾ parts, distilled water 3¾ parts (all by weight). Agitate the whole together, and allow it to rest for some time in sunlight, or in a warm place, until a light lemon-colored precipitate is deposited. Decant the colorless liquor, and preserve it in a well closed ground-stopper bottle; label, "*Molybdate of Ammonia. Detection of Phosphoric Acid.*"— A quantity of this reagent must be employed, at least equal to that of the fluid in which the presence of phosphoric acid is sought, and should the precipitate be tardy in appearing, gently heat it (104° F.). This is a very sensitive reagent; the only cause of error will be the presence of *arsenic acid* in the fluid, which precipitates moreover by the other reagents of phosphoric acid.

3. *Solution of Oxalate of Ammonia.* Take of pure oxalate of ammonia 1 part, distilled water 24 parts (by weight). Dissolve, filter, and label, "*Oxa-*

late of Ammonia. Detection of Lime." N. B.—The oxalate used must leave no residue when evaporated on a platinum spatula.

4. *Solution of Picro-Carminate of Ammonia.* Take of *Ammoniacal Solution of Carmine*, q. s., and add it, drop by drop, to a saturated and filtered solution of picric acid in distilled water, until this solution becomes neutralized, or assumes the tint of gooseberry syrup. Any slight precipitate resulting from a commencing acidity may be removed by filtration. This solution colors the nuclei a delicate rose, and the rest of the cellule yellow. It likewise gives a yellow color to fibrin, mucus, and hyaline urinary casts. The coloring only remains while the elements are in the picro-carminate; if water or acetic acid be added, the yellow color disappears.

Ammonia in urine is passed in quantities varying from 5 to 13 grains in 24 hours, equal to 19 or 36 grains of chloride of ammonium. To detect it, place 20 c. c. of fresh urine, filtered, in a glass or porcelain dish. Place a glass triangle across this dish, to support a smaller flat dish containing 10 c. c. of the *standard sulphuric acid* for this analysis. Place this on a ground-glass plate quickly, cover with a receiver, and hermetically close with tallow, having just previously added 10 c. c. of milk of lime to the urine in the lower dish. Set this aside for 48 hours. Now gradually add the standard solution of caustic soda for this analysis to the acid in the upper vessel (which has absorbed all the ammonia liberated), until this is neutralized. Every c. c. of this soda solution, less than its graduated standard, corresponds to 0.00715 gramme of ammonia. That is, if it requires 30 c. c. of the soda solution to exactly neutralize the 10 c. c. of sulphuric acid previous to the operation; and, subsequently, it is found that only 26 c. c. of the soda solution have been used to neutralize,—the difference between the quantity required in the operation, 26 c. c., and that previous to it, 30 c. c. equals 4 c. c., which corresponds with the amount of ammonia evolved and absorbed. And, as 1 c. c. of soda solution equals 0.00715 gramme of ammonia, so 4 c. c. equals 0.0286 gramme of ammonia in the 20 c. c. of urine employed in the process. 1,000 grammes of this urine would then contain 1.43 grammes of ammonia.—See *Chloride of Ammonium ; Oxalate of Ammonia ; Purpurate of Ammonia ; Thionurate of Ammonia ; Urate of Ammonia ; Sulphureted Hydrogen. Nessler's Test.*

Ammoniacal Urine. Freshly voided normal urine contains a small amount of ammoniacal salts; but urine is not considered ammoniacal unless it becomes neutral or alkaline within a short time after its discharge from the bladder, or while it remains in this organ before being passed. This is due to some morbid condition, as healthy urine usually requires about 24 hours after its discharge before it becomes alkaline. See *Alkaline Urine.* Ammoniacal urine is fetid, pale, and turbid, and when permanent, is indicative of disease, as calculus in the bladder, catarrh of the bladder, paralysis of the bladder, ulceration of the bladder, prostatic enlargement, urethral stricture, disease of the spinal cord, etc. Pus is frequently present in this character of urine. Very minute quantities of ammonia may be detected by a dilute solution of sulphate of copper, which occasions a greenish turbidity and precipitate of basic sulphate of copper. Solution of silicate of soda at the 100th, injected daily into the bladder, has removed the ammoniacal condition of the urine, and restored it to its normal state. See *Nessler's Test*.

Ammonio-Magnesian Phosphates. See *Phosphates*.

Ammonio-Oxide of Copper. *Ammoniacal Solution of Copper.* Dissolve recently precipitated oxide of copper in liquor ammonia. This solution dissolves cellulose of *cotton, flax*, and *hemp*, but has no action on *silk* or *wool*.

Amygdalin. When taken into the system, emulsin being absent, amygdalin does not produce hydrocyanic acid. Formic acid can be found in the urine after large does of amygdalin.

Analyses of Urinary Sediments. The first step consists in completely separating the deposit it is desired to examine by filtration. The sediment remaining upon the filter, whatever be its character, must then be washed with several drops or more of distilled water; a small quantity of it is then placed upon a platinum capsule, and heated in the flame of a spirit lamp. The following results may happen:—1. The mass remains intact, and is not changed at a red heat.—2. The mass turns black, swells up, and finally there remains a white residue permanent at a red heat.—3. The mass chars, burns, and completely disappears.

In the first instance, the sediment will be composed solely of inorganic substances unalterable at a red heat; indicating the presence of *phosphate of lime*, or *magnesia*, or *carbonate of lime*, or *magnesia*. Indeed, these are the only immediate principles that are insoluble in water, and unchangeable at a red heat.

In the second instance, the deposit, which has only been partly decomposed, may consist of *urate of soda, potash, lime*, or *magnesia, oxalate of lime*, or *ammonio-magnesian phosphate*, the only principles that may thus be precipitated and be partially decomposed at a red heat.

In the third instance, a red heat causes the disappearance of the entire sediment, which may consist of *uric acid, urate of ammonia, cholesterin, cystin, margaric*, or *stearic acid*, principles that are completely decomposed at a red heat.—Suppose the deposit belongs to one of these three groups, the analysis may now be proceeded with, granting, for greater simplicity, that it shall only contain substances belonging to the same order.

The deposit remains unchanged at a red heat. Place a little of it, in a watch-glass or on a glass slide, under the microscope, and then moisten it with a drop of acetic acid; if a large quantity of air bubbles are disengaged, visible through the microscope, *carbonate of lime*, or *of magnesia*, is present; if no air bubbles are set free, it will then be *phosphate of lime*, or *of magnesia*. On adding to the solution of the carbonate or phosphate in the acid a few drops of oxalate of ammonia, there will be a precipitate if it is *lime*, and no precipitate if it is *magnesia*.

The deposit is partly decomposed by heat. Place the precipitate in a watch-glass, add some distilled water to it, and boil it upon a water bath. If the precipitate dissolves, slightly concentrate the liquor, and place a drop of it on a glass slide, under the microscope. On cooling, crystals will be observed to form of, *urate of potassa, soda, lime,* or *magnesia.* As this examination of the crystals may not probably be sufficient, an ulterior examination may be had recourse to; thus, to be certain that it is really a urate, some of the precipitate is placed into a platinum capsule and moistened with nitric acid,— this is then dried over a spirit lamp, being careful not to burn the mass. When dried, let a drop or two of ammonia fall upon it, and if it be a urate under analysis, the whole will assume a magnificent purple color.

To determine what urate it is, burn or calcine the mass a little, and then add a few drops of water to it; if it remains insoluble, it is magnesia,—urate of magnesia. If it dissolves, it is either soda or potassa. Place a drop of the solution upon a glass slide in the field of the microscope, and add to it a drop of chloride of platinum solution. If a yellow precipitate is formed,. composed of octohedral crystals, it is potassa,—urate of potassa; if no precipitate occurs, it is soda,—urate of soda.—If the precipitate (first obtained by partial decomposition under heat) does not dissolve when boiled in water, it is *oxalate of lime,* or *ammonio-magnesian phosphate.* Dissolve it in a drop of hydrochloric acid, then add a few drops of ammonia to it, to neutralize the acid added, if a precipitate reappears, examine it under the microscope, to determine from the form of the crystals, which of these two salts it is.

The deposit completely disappears by the action of heat. Treat some of it with a little water, and then boil it; it is partially dissolved, for, on cooling, crystals of *uric acid* are precipitated, recognizable under the microscope. The experiment heretofore named for the determination of urates, may also be repeated, to-wit, the production of the purple color by the action of nitric acid and ammonia. If the deposit remains insoluble in the boiling water, but is soluble in boiling alcohol, forming crystals on cooling, it is *cholesterin, margaric,* or *stearic acid,* substances easy to determine under the microscope. See *Fats; Urinary Sediments.*

Anazoturia. A deficient amount of urea in the urine.

Animalcula Seminalia. See *Spermatozoa.*

Antimony, Detection of in Urine. The process given for the detection of *arsenic* in urine is the one to be pursued for the detection of antimony. The brilliant tint of antimony differs from that of arsenic, being of a more decided violet color, and the metal is less readily sublimed by heat, and the spots, instead of presenting a bright metallic lustre like those of arsenic, are dull. The deposit of antimony is with difficulty dissipated by the flame of a spirit lamp; those of arsenic are readily driven off. The antimony stains are scarcely, if at all, removed when touched with a solution of hypochlorite of lime or of soda, while the stains of arsenic disappear at once. Again, the sublimate formed by antimony is amorphous or granular, and requires a more elevated temperature for its formation.

Apparatus for Examining Urine. See page 16.

Appearance of the Urine. In *color, transparency, turbidity,* and *sediments.* Which *See.*

Araneosa Urina. Urine containing cobweb-like filaments.

Ardor Urinæ. A sensation of heat or scalding along the urethra during micturition, which is due to irritation or inflammation of the mucous membrane of the part, or to an unhealthy condition of the urine.

Arenosa Urina. Urine containing sandy or gravelly deposits.

Arsenic, Detection of in Urine. Evaporate the urine to a small bulk; strongly acidulate it with one-seventh or one-eighth its volume of pure hydrochloric acid, and then boil it. While boiling introduce into it a small piece of freshly brightened copper foil or fine copper gauze, and continue the boiling for 15 or 20 minutes longer. If the copper becomes colored a metallic grey, fresh pieces should be added from time to time until the last one added presents no sensible change of color. Now remove the pieces of copper covered with the grey deposit, wash them, and then dry them between folds of bibulous paper. The deposit readily adheres to the copper unless it be very thin, and presents a very distinct metallic brilliancy, of a steel-grey color, or, unless it forms a very thin layer, of a somewhat bluish tint. Under the influence of heat it completely disappears, and the copper resumes its ordinary aspect. Or, the pieces of tarnished copper foil or gauze may be placed in a perfectly dry glass tube, of as narrow diameter as possible, and which should be closed at one end, and loosely covered at the other. Upon applying heat to the part where the foil lies, there will be deposited at the cold part of the glass tube, a grey or dark metallic ring of arsenic, beyond which will be observed a white sublimate of arsenious acid, showing, under the microscope, cubic and octohedral crystals.—If no deposit forms on the copper, the urine (without removing the copper) should be still more concentrated by boiling, and if no metallic tarnish be then observed, neither arsenic or *antimony* are present.

This is a very simple process (Reinsch's), and consists in precipitating metallic arsenic from its solutions by means of pure clean copper. As arsenic, by preference as it were, accumulates in the liver, from which it is not readily eliminated, in small quantities, it is always proper to concentrate the urine previous to testing it.

2. Place pure zinc in a small flask containing distilled water acidulated with sulphuric acid, then add some of the urine to be tested, and loosely stop the neck with cotton, so as to prevent any drops of the mixture from being thrown upon the test paper, which is to be held directly over the neck. The test paper is ordinary tissue paper moistened with a solution of bichloride of mercury, and used while moist. If arsenic be present a lemon-yellow spot will be formed on the paper, from the arseniureted hydrogen set free; if antimony be present, a brown spot. Pure hydrogen has no effect upon the paper. *Mayençon and Bergeret.*

Arseniureted Hydrogen. This substance inhaled, causes the urine to become highly albuminous and black, from rapidly dissolved blood pigment.

Asparagus. This article when eaten gives rise to a few crystals of oxa-late of lime, and imparts a peculiar unpleasant odor to the urine. In large quantity it is apt to occasion temporary glucosuria.

Aspergillus Glaucus. *Mucor glaucus.* The blue mold on cheese, lard, and butter, and also found in urine. It has also been detected in the lungs and air cavities of birds. It may be known by its non-partitioned filaments, bearing at their extremities chains or rays of spores arising from a globular head at the apex of each fertile filament. This fungus may be studied upon preserved fruits and certain kinds of cheese. See *Vegetable Organisms.*

Assafœtida. This article occasions a most disgusting odor from the urine.

Atropia. A strong aqueous solution of bromohydric acid saturated with free bromine, produces a more or less bright yellow amorphous precipitate when added to a solution of atropia, which in a short time becomes converted into granules and short needle-like opaque crystals; this precipitate is hardly soluble in caustic alkalies and the ordinary mineral acids, and is insoluble in acetic acid. Should the crystals become redissolved, a little more of the reagent will reproduce them. This will detect the $\frac{1}{25000}$th part of atropia. To detect this alkaloid in urine, first remove albumen, then to 10 c. c. of the filtered urine, add 4 or 5 drops of sulphuric acid, and 12 c. c. of strong alco-hol; agitate the mixture thoroughly, gently heat it for about 15 minutes, and when cold strain through fine muslin; wash the residue on the muslin with alcohol, and strongly express it. Concentrate the strained liquid on a water bath, at a moderate temperature; again strain; evaporate to a small quan-tity; filter; add liquor potassa to render the filtrate slightly alkaline; and then add twice its volume of chloroform to dissolve the liberated atropia. Allow the chloroform to evaporate spontaneously. To the residue add sev-eral drops of water containing a trace of sulphuric acid, and examine the solution by the above bromine reagent. If the residue is not sufficiently pure for testing, or in case no crystals are produced, again dissolve it in chloro-form, and proceed as before. The yellow precipitate, the peculiar crystals formed, and the dilatation of the pupils occasioned by a drop of the solution placed on the eye-ball, will be satisfactory evidence of the presence of atropia. Daturia, which is identical with atropia, gives similar reactions. In some cases, where the quantity of atropia present is very small, a larger volume of urine, say 20 c. c., will give more satisfactory results.

Azoospermia. Partial or complete loss of the vitality of spermatozoids.

Azoturia. Excess of urea and other nitrogenous principles in the urine. As any functional lesion in which the organic combustions are executed too rapidly, from whence, general emaciation, and dryness, and roughness of the skin. *Bouchard.*

B.

Bacteria. Minute vegetable organisms observed in decomposing urine, consisting of minute staff-shaped bodies, having a stiff, vacillating non-undulatory motion, and a length of from .002 to .005 millimetres. See *Vegetable Organisms.*

Balsam of Peru, as well as Balsam of Tolu, increase the amount of hippuric acid in the urine, or, occasion its presence, owing to the benzoic or cinnamic acid existing in them.

Barium. Ba. Is the metallic basis of the earth *baryta,* or *oxide of barium.* This agent, or its salts, are chiefly employed in the detection of carbonic and sulphuric acids. The different forms in which it is employed in urinary investigations are the following:—1. *Baryta Water.* Take of crystals of hydrate of baryta, 1 part; distilled water, 20 parts; mix, dissolve with the aid of heat, filter, and bottle. This water is used to precipitate magnesia, as well as to detect or remove acids which form insoluble compounds with it; as, carbonic and sulphuric. Carbonate of baryta is entirely soluble in nitric acid; sulphate of baryta is insoluble in this acid.—2. *Chloride of Barium Solution.* (*Barium Chloride.*) For volumetric analysis of sulphuric acid in urine; this is of two strengths, *a.* and *b.* *a.* Finely powder and then air-dry, crystals of chloride of barium, 30.5 grammes, and then add distilled water to make the solution amount to one litre. Of this solution 1 c. c. will precipitate 10 milligrammes of anhydrous sulphuric acid. Bottle and label, "*Strong Chloride of Barium. Detection of Sulphuric Acid.*" *b.* Of the preceding solution add 1 c. c. to 9 c. c. of distilled water. Of this solution, 1 c. c. will precipitate 1 milligramme of sulphuric acid. Bottle and label, "*Dilute Chloride of Barium. Detection of Sulphuric Acid.*"—3. *Chloride of Barium Solution,* for detecting sulphates. Dissolve 1 part of Chloride of Barium in 10 parts of distilled water, by weight. Filter, bottle, and label, "*Chloride of Barium. Detection of Sulphates.*"—4. *Baryta Solution.* Mix together 2 volumes of a saturated baryta water and 1 volume of a saturated solution of nitrate of baryta, at the ordinary atmospheric temperature. This is used to precipitate the phosphates from urine, when examining it for chlorides, etc. Bottle and label, "*Baryta Solution. Precipitation of Phosphates.*" The addition of an equal amount of distilled water to this solution forms the *Solution of Baryta for the analysis of urea by Liebig's process.*

Barreswill's Liquor. This fluid is often used in the examination of urine for sugar, but on account of the difficulty of preserving it, requiring it to be freshly prepared each time it is to be used, Fehling's and Pavy's modifications of it are preferred. It is made as follows, the quantities in parenthesis being from Beale: Take of cream of tartar, pure, 5 grammes (96 grains), crystallized carbonate of soda 4 grammes (96 grains), pure sulphate of copper 3 grammes (32 grains), caustic potash 4 grammes (64 grains), distilled water, a sufficient quantity to complete the volume of 100 c. c. (2 fluidounces). Dissolve the cream of tartar in some of the water by heat; pour this boiling solution into a solution of the sulphate of copper; then add the potash and the soda, and, lastly, enough distilled water to make 100 c. c. Of this reagent, 1 c. c. is decolorized by 1 centigramme of glucose. To use the fluid as made by Beale's formula, place equal bulks of urine and the fluid in a test tube, boil the mixture, and if sugar be present, a red precipitate of suboxide of copper ensues at once.

Baruria. Urine having great specific gravity.

Bas Fond. The most inferior portion of the bladder.

Benzamide. When hippuric acid is acted upon by oxidizing agents, as, peroxide of lead in water, it is decomposed into water, carbonic acid, and benzamide, which is soluble in alcohol.

Benzoate of Lithia. *Lithiac or Lithium Benzoate. Lithium Benzoicum.* Lithium forms a soluble salt with uric acid; and benzoic acid being transformed into hippuric acid in the organism, forms a soluble hippurate of lime (calcium hippuricum). Benzoate of lithia forms a light, white powder, soluble in water, and which has been of service in the removal of urate of lime in gout.

Benzoglycolic Acid. When hippuric acid is acted upon by hyponitrous acid, or when a solution of hippuric acid in nitric acid is exposed to the influence of a current of nitric oxide, nitrogen is evolved, and benzoglycolic acid crystallizes in the form of colorless prisms, which are readily soluble in alcohol and ether, sparingly soluble in water, and form neutral salts with one base.

Benzoic Acid, When taken internally, this acid is eliminated as *hippuric acid* in the urine. It has consequently been administered in ammoniacal urine with success, as follows: Take of benzoic acid 1 to 3 parts, neutral glycerin 3 to 4 parts, mucilage 150 parts; mix. Dose of the mixture 1 gramme daily, quickly carried to 4, 5, or 6 grammes per day. In this mixture the benzoic acid is held in suspension. It forms a hippurate of ammonia which is less poisonous than the carbonate; increases the acidity of normal urine; retards the decomposition of the urine and the consequent formation of the ammoniacal carbonate; and prevents the formation of insoluble phosphates, which give rise to cystitis, and calculus.

Bile in Urine. In certain pathological conditions of the body, bile is detected in the urine in the form of *bile pigments*, or *bile acids*, each of which must be searched for separately. Urine in which the biliary pigments or coloring matters are present, has a peculiar greenish-yellow, or dark-brown tint, and stains linen yellow. If the urine be shaken the froth will be yellow, and if the fluid contain anatomical elements, as, epithelia, pus, urinary cylinders, etc., these will likewise have an intense yellow color. Phosphatic sediments are frequently encountered in urine containing bile, which is often alkaline at the time it is voided. Sometimes, but not invariably, the crystals of ammonio-magnesian phosphate, are colored yellow. As the bile acids are seldom found in the urine, testing for the pigments is usually deemed sufficient. See *Coloring Matters of Urine.*

Bile Acids. These acids, in the urine, are, taurochloric, glycocholic, and their derivatives; they are rarely detected in the urine of icteric patients, but have been observed in this fluid in acute yellow atrophy of the liver, and in pneumonia, without the coexistence of the bile pigments. The cause of the presence of the bile acids in urine is but illy understood.

Detection of Bile Acids. For this purpose Pettenkofer's test is the one generally recommended, but it is tedious, troublesome, and, for the general practitioner, impracticable. Beside which it may lead one into error, because albumen, as well as several essential oils, give an analogous reaction; therefore, when albumen is present it must first be removed by coagulation and filtration.

1. *Hoppe's* modification of it renders the process very reliable, but does not diminish its tediousness. It is as follows : Caustic ammonia is to be added to the urine until it is faintly ammoniacal, and then treat with diacetate of lead as long as any precipitate occurs. Wash the precipitate collected on a filter, with distilled water; boil it with alcohol over a water bath, and filter while hot. Add a few drops of potash or soda to the filtrate, and evaporate the solution, over a water bath, to dryness. Again boil the residue, over a water bath, with absolute alcohol, continuing the evaporation until only a small quantity remains. Shake this with ether in a stoppered bottle, and after some time the alkaline salts of the bile acids will separate in crystals. These may now be verified by Pettenkofer's test, thus : Dissolve these crystals in a little distilled water, and place the solution in a porcelain capsule; a drop or two of saturated solution of cane sugar must now be added, and then treat the mixture by dropping strong sulphuric acid into it, in excess of the amount of bile acid present, or about a volume equal to that of its solution. Apply moderate heat, a beautiful cherry red, succeeded by a deep purple color, is formed, the latter being the characteristic reaction when bile acids are present. See *Cholic and Cholinic Acids.*

‡2. *Neukomm's* modification of Pettenkofer's test is very simple and exact. Prepare a solution of pure sulphuric acid 1 part, distilled water 4 parts; also a solution of cane sugar 1 part, distilled water 4 parts. Place a small portion of the urine in a porcelain capsule, or watch glass, and slowly and gently heat it over a water bath or spirit lamp, until but a few drops remain, to which add a drop or two of the solution of sulphuric acid, and then a drop of the saccharine solution. If biliary acids are present, the beautiful purple-violet color characteristic of Pettenkofer's test will appear at the border of the evaporated fluid. Bischoff states that this reaction does not occur with albumen or fat, but only in presence of the bile acids. Care must be had, in the above process, not to evaporate too rapidly, as it will give rise to the formation of black matters, masking the reaction. If a yellow color ap-

pears, the acid is acting on the sugar, and the desired reaction will not take place.

3. *M. Strasburg*, of Bremen, has given another modification of the preceding test, but it is imperfect and unreliable.

4. *M. Medin* has given a process by which both the biliary acids and pigments may be determined in one operation : Fifteen c. c. of the suspected urine are rendered alkaline by the addition of two drops of caustic soda solution, to which chloride of barium solution is added as long as a precipitate occurs. Then, the liquid is boiled on the precipitate, and filtered while boiling hot. The almost colorless filtered liquid is carefully neutralized by hydrochloric acid so as to leave it still slightly alkaline, when it is to be agitated with 4 or 5 c. c. of amylic alcohol. If the liquid presents an emulsive aspect, as often happens, the addition of a small amount of ordinary alcohol will remedy it. The amylic alcohol is now to be removed by means of a pipette, and evaporated to dryness. The light-colored residue thus obtained gives a beautiful purple-violet color, when heated with a trace of sugar and sulphuric acid, if bile acids were in the urine. Care must be had to *completely* volatilize the amylic alcohol before using the other reagents, as it alone will give a fine red color with sulphuric acid, which becomes still more beautiful when a trace of cane sugar is added to it. Urine containing a millionth part of bile will, by this process, give distinctly the reaction with Pettenkofer's test.

By washing the barytic precipitate with distilled water, then treating it by sulphuric acid and alcohol, with heat, a fine green color will be formed if bile pigment be present.

‡5. A very decisive and characteristic reagent for the detection of bile or bile acids, and which is not influenced (as to color) by the presence of sugar or albumen, is Methylaniline or Paris violet, 1 part dissolved in 500 parts of distilled water. A few drops of this solution added to normal urine renders it dichromatic ; blue by reflected light, and violet by transmitted ; but if bile be present, the urine assumes a blood-red color, with an abundant precipitate. The bile pigments, or cholic acid give, each, a precipitate, but no change of color. As chrysophanic acid is the only substance giving a like reaction with this reagent, the practitioner must ascertain whether the patient has taken rhubarb or senna, within 12 or 15 hours previous to passing the urine examined, before ascribing the change of color to bile. *M. Yvon.*

Bile Pigments. These coloring matters are : 1, *cholepyrrhine, brown pigment*, also termed bilifulvine, biliphæine, bilirubin, and cholophæine ; 2, *biliverdin, green pigment*, cholochloine, cholochlorine. The first named is the one more commonly present in the urine, and which is often converted into the latter. These pigments may be temporarily present in the urine in small quantities, during hot seasons (the same as we observe at this period of the year, a slight yellowish coloring of the sclerotica of many persons); but they are in greater amount and more marked in the urine of jaundiced patients. They may be present with or without the coexistence of the bile acids or their salts. The detection of the coloring matter of bile in the urine is seldom useful, except in doubtful cases, as jaundice is generally recognized by other symptoms. The coloring of the teguments is shown shortly after the appearance of the bile pigments in the urine; but it persists when these have disappeared. In jaundice, biliverdine and *biliprasine* predominate in the urine; the latter is biliverdine plus one atom of water. It is probable that this transformation occurs in the blood itself, because this substance is not a product of the normal secretion of the liver. Hardy states that if cholepyrrhin be injected into the veins, jaundice is produced, and biliprasine is found in the urine.—If the urine to be examined for bile pigment be of too dark a color, it should be rendered lighter by the addition of sufficient distilled water, previous to applying the tests.

Detection of Bile Pigments. ‡1. Into the clear urine pour some ordinary acetic acid; if these pigments are present, *persistent* green color appears which is darker in proportion to the greater amount of biliary coloring matter in the urine.—This is a very distinct reaction, and should be preferred to that with nitric acid. Indeed it is quite as sensitive, gives no play of colors, and remains for a long time, while the coloring determined by nitric acid is very fugaceous, especially when there is but little bile pigment. Besides, nitric acid has the inconvenience of giving analogous reactions with many organic compounds that may accidentally exist in the urine, indican, etc., also alcohol, essential oils, and many alkaloids.—‡2. Spread a thin layer of urine over a white porcelain or china plate, and let one drop of fuming nitric acid (nitroso-nitric acid) fall upon the center of this layer, a series of colors will be alternately and rapidly produced, and which will extend towards the circumference of the urine in the following order: Green, blue or violet, ruby-red or yellowish-red. (If the urine be albuminous, the same result may be obtained by letting the urine fall drop by drop upon some hydrochloric acid.) The green color is the characteristic one, the others may be obtained from the pigments common to the urine, and the whole of them do not always appear. Indican in urine will give a blue or violet, and even green, color with nitric acid, as is often the case in melanotic cancers, in which the urine is frequently a dark brown. *Gmelin.* Biliverdin in urine does not give the same reaction as cholepyrrhine, but gives the green color immediately with hydrochloric, sulphuric, or acetic acid, without the changes produced in cholepyrrhin by nitroso-nitric acid.—A. Hilger observes, however, contrary to what has generally been stated, that he has found all the biliary coloring matters to give the characteristic color changes, —green, violet, blue,—upon the addition of nitroso-nitric acid.—M. Cazeneuve states, relative to Gmelin's process, that there may exist an error in it if the suspected urine contains any alcohol. In a mixture of nitric acid and alcohol, the nitric acid is converted, by the alchohol, into nitrous acid and an immediate formation of highly volatile nitrous ether, no coloration being effected. But if the acid be allowed to flow slowly down the side of a test tube containing alcohol, it will pass to the bottom, and the reaction between the two fluids will slowly take place. In about a minute or so, a splendid greenish-blue zone will appear, having a yellowish-green zone below it, and, in a few seconds, a reddish zone above it. When the alcohol has become etherized, all coloration disappears.—3. Into a small beaker glass pour about 6 c. c. of pure hydrochloric acid, and add to it, by drops, just enough of the urine to slightly but distinctly color it. Mix the two, and then by means of a slender glass tube, funnel shaped at its upper extremity, drop in some pure nitric acid so as to form a layer beneath the fluid, or, this acid may be allowed to trickle down the side of the inclined glass to its bottom ; the play of colors, above named, will take place at the point where the acid is in contact with the mixture. If the nitric acid be now agitated with a glass rod, the colors will appear by the side of each other. When the urine is added to the hydrochloric acid, it will be colored green if biliverdin be present, and reddish-yellow if it be cholepyrrhin. *Heller.*—‡ 4. A most sensitive reagent is found in ordinary tincture of iodine. Two or three drops of this tincture let fall upon a small quantity (ʒi) of icteric urine in a test tube produces a magnificent emerald-green color, which remains about half an hour or longer, then becomes red, and finally yellow. When the urine is alkaline a few drops more will be required. This green color has been produced in a mixture of a few drops of icteric urine with distilled water 60 grammes (2 fl oz.). In case of a very feebly bilious urine, we should also act comparatively upon a normal urine at the same time, to render the difference of the tints more evident. The reaction will be more striking if an albuminous water be added to it, which will give a green precipitate; or, a solution of subacetate of lead, which will give a yellowish one. *Maréchal.* In this process the iodine tincture must not be allowed to mix with the urine, but to trickle down along the side of the test tube held nearly horizontally. The green color will be seen beneath the red layer of tincture of iodine, which must not be too strong lest the test fail. A very dark colored urine should be diluted with an equal volume of distilled water previous to applying the test. If the above rules be observed, the test will not fail in a single instance of bilious urine, and is fully as delicate as any other test now known.—5. Biliverdin is not separated from urine by chloroform, being insoluble in this agent, though its presence may, however, be revealed by the processes with the acids. Cholepyrrhine is soluble in this agent, the least quantity in the urine being separated from it; and this reaction gives a special process for detecting the latter pigment, when there are merely slight traces of it in the urine. Successively agitate, with some chloroform, several large quantities of the urine, being careful each time to allow the chloroform to settle, and then to draw off the supernatant urine. The chloroform sinks to the lower strata, and assumes a yellowish color from the presence of the pigment in solution, and a milky aspect due to the agitation. When this yellowish tint is sufficiently marked, the urine is drawn off for the last time. Now place a drop of the chloroform solution upon a glass slide, and allow it to dry. If cholepyrrhine be present, the residue will, when examined under the microscope, be found to consist of reddish-yellow crystals, which, moistened, under the microscope, with fuming nitric acid, will give changes of color, green, blue, red, violet, and yellow. The objective should be protected, or a çhemical microscope be employed, as the nitric acid vapors injuriously affect the brass of the objective. Alkalies dissolve these crystals.—We may likewise simply pour some fuming nitric acid upon the chloroform solution, and observe if the required colors are produced. *Valentiner.*—Ritter states that he has discovered a blue coloring matter in bile.

Bilharzia Hæmatobia. *Distoma Hæmatobium.* A parasitic entozoon common in Egypt, Cape of Good Hope, and some other hot countries, and first discovered by Bilharz, in Egypt. It is found in the intestines, in the bladder, in the ureter and pelvis of the kidney, producing very serious results. The frequency of fatal dysentery, calculous affections, hæmaturia, etc., to which the inhabitants of these countries are subject, are supposed to be mostly due to the presence of these worms in the urinary passages.

Bilifulvin. See *Bile Pigments.*

Biliprasin. See *Bile Pigments.*

Bilirubin. See *Bile Pigments.*

Biliverdin. See *Bile Pigments.*

Blood in Urine. (See *Coloring Matters of Urine.*) Urine may be mixed with blood corpuscles, which are known by their elements and distinctive characters; or, it may simply be tinged by the coloring matters of the blood, —the blood may be discharged from any part of the urinary-renal tract. It must be remembered that blood may likewise be found in the urine of women, of uterine or vaginal origin; and that it is likewise mixed with urine occasionally by the subjects themselves for purposes of deception. If from the kidneys, in small quantity, the urine being acid, the blood is diffused through this fluid, imparting to it a dark brown or smoky appearance. (See *Hematin.*) If the urine has an alkaline reaction, it will present a bright red color, like blood. Large quantities of blood in the urine may be known by the blood red color of this fluid, the coagula formed, and the brownish-red deposit; *albumen* will likewise be present in proportion to the amount of blood existing in the urine, and which is derived from the serum of the blood. The microscope affords the most ready and certain method of detecting blood in the urine, especially if this examination be made soon after the liquid has been passed. In some cases the corpuscles may be found perfect for several days, as when the urine is acid; but they sometimes decompose very speedily, especially if the urine be of low specific gravity, or ammoniacal, when they swell, and finally rupture from endosmosis; when the sp. gr. is high, they frequently contract, shrivel, and become distorted in various ways (crenated, etc.), from exosmosis. As the urine containing blood readily becomes alkaline, if it be desired to test it for *albumen*, it must first be rendered acid by the addition of acetic acid. ‡ Under the microscope, blood corpuscles will be observed non-nucleated, or entire, smooth, flat, non-granular, yellow or reddish-brown, and either detached or adhering in nummular form; the discs being biconcave, a change in focussing will give a dark circumference and a light center, or the reverse of this; and, if the fluid containing them be set in motion, so as to move the corpuscles, their biconcave form will be at once observed. The diameter of human blood corpuscles is constant, averaging about 0 mm .0075; that of the dog, 0 mm .00716; of the hog, 0 mm .006; of the horse and ox, 0 mm .0056; of sheep, 0 mm .005; of the rabbit, 0 mm .0069, etc. (See *Fig.* 22, page 129, Microscopic Appearances of Blood.) It may be proper to state that Prof. Fairfield, of the New York College of Veterinary Surgeons, who has been making careful examinations of blood corpuscles, has recently expressed doubts as to their disc-like form and biconcavity, which appearances he considers the result of the under correction of the microscopic lenses of the more powerful microscopes. He defines these corpuscles as being perfectly spherical bodies, containing internally a colorless protoplasm, and externally a thin coat slightly stained with coloring matter.

V. Table.

‡ Detection.

A. On cooling, the urine has a blood-red color.	I. It is transparent.	a. Add a few drops of hydrochloric acid.	1. The color becomes darker.	*Coloring matters of the blood.* See B.
			2. The color becomes clearer.	*Foreign coloring matters.* Rhubarb, etc.
	II. It is slightly turbid.	b. Allow it to rest until a precipitate forms, then examine this precipitate.... ,......................		According to III.
	III. There is a red sediment.	c. Observe under the microscope whether this precipitate is crystalline or amorphous; on heating it in a test tube it becomes dissolved.		*Urates; Uric Acid.*
		d. The microscope shows blood corpuscles; the precipitate is not soluble by heat.		*Blood.*
B. On cooling, the urine is reddish-brown, black-brown, or ink-black.	IV. It is not sedimentary, and the microscope shows no blood corpuscles in it. Boil the urine alone, or with a little acetic acid.	e. Formation of a reddish-brown coagulum.		*Coloring matters of the blood.* (Hemato-globulin and products of decomposition.

If we represent the average diameter of the blood corpuscles of man by 7, the globules of blood of other animals may be represented as follows:—Elephant, 9; dog, 7; rabbit, 6; horse, 5; sheep, 5; goat, 4; guinea pig, 2. The blood corpuscles of all mammiferous animals present circular disks under the microscope, with the exception of the llama and camel family, whose globules are elliptical, but always without a nucleus. The blood corpuscles of birds, are generally twice larger than those of man, and may be represented by 15; they are elliptical, biconvex, presenting a more or less distinct nucleus. The globules of reptiles and of amphibious animals are still larger (represented by 20 for those of the frog); they are still more elliptical, more convex, and with a very distinct granulous nucleus. The blood corpuscles of fishes may attain very large proportions; they usually present the same characters, with a few unimportant exceptions, as far, at least, as concerns the present subject.—White globules in human blood are spherical, about a third larger than the red, represented by $9\frac{1}{3}$, and present when in a neutral vehicle, a pale, granulous appearance, irregular outline, and a characteristic silver-white color. The addition of water causes them to swell, to present a smooth outline, with Brownien movements among the internal granules, and single, double, or multiple nuclei, which are rendered more distinct by the addition of a little acetic acid. In the healthy adult, there is about 1 white globule to every 350 red ones, the blood being taken from the skin; but food, exercise, disease, etc., may greatly augment the number of the white corpuscles. A very excellent instrument has been devised for counting the blood globules under the microscope. It is termed *Hematimetre*, of Hayem and Nachet, and is manufactured by Nachet. optician, Paris, France. A quadrillated ocular is necessary when using this instrument. It has been ascertained that the number of corpuscles in a drop of blood are temporarily increased when the serum of this fluid is subtracted by any morbid or therapeutical cause, as by purgation, diarrhea, obstinate vomiting, excessive perspiration, etc.; consequently, when counting the number of corpuscles by the Hematimetre, the proportion of the serum in which they are found should be taken into consideration, otherwise, the count may give a plethoric indication when there is only feebleness and exhaustion.—The presence of blood in urine or other fluids is readily determined by spectral analysis; but the positive determination of human blood in spots or stains has not been satisfactorily demonstrated. See *Micro-Spectroscopy.*

A. The preceding table will at least be found practical, though probably not perfect. When blood imparts a blood-red color to urine, a greater or lesser quantity of blood globules will be found present. If the color be brown or smoky, the method B may be required, because the globules are apt to be more or less dissolved or destroyed; the commencement of this decomposition occurring either in the bladder or after the urine has been voided.—A. I. Urine colored blood red by dissolved *hematoglobulin*, contains no blood corpuscles, consequently, in addition to the test by hydrochloric acid, that according to B must also be pursued.—The foreign articles which color urine red, so as to lead to a suspicion of the presence of blood, are numerous and often unexpected.—A. II. When the quantity of blood is quite small, the urine will present only a slight turbidity; in such cases, the urine should be allowed to rest for some time in a conical-shaped test glass, until the corpuscles become deposited at the bottom of the fluid. These, especially in warm weather, are liable to rapid decomposition, consequently the deposit must be examined as soon as it begins to form.—A. III. The more or less deep-red sediment, accompanying a blood-red urine is not rare in chronic affections of the liver (cirrhosis), and in the last stage of cardiac affections (affections of the mitral orifice). This sediment composed of urates and a little uric acid, is clearly determined by its microscopic characters, its solubility by heat, or by the addition of an alkaline solution. Should earthy phosphates be present, they will not dissolve by heat, nor in alkaline solution, and are readily distinguished under the microscope. As to the urine itself, should it contain albumen, albuminaria may be present.—*d*. When this red sediment consists of blood, the microscope will reveal the characteristic corpuscles, and which examination should be made at as early a period as possible.—B. This peculiar coloring may be present in the urine when it contains blood, either in a natural state, when a commencing decomposition has occurred, or, when it contains only the coloring matter of blood, hæmatoglobin, which decomposes, and gives, upon boiling it with acetic acid, hæmatin and albuminoid substances which coagulate. This reddish-brown coagulum, when dried and powdered and boiled in alcohol containing sulphuric acid, renders the fluid red or reddish-brown, if hæmatin be present. From this experiment it may be concluded that the urine contains the coloring matters of the blood in solution.

Reactions. *Water* added to blood corpuscles causes them to swell and assume a spherical form, and if the action of the water is prolonged, their coloring matter is abstracted, and they become pale, softened, completely vesiculous and transparent, and soon disappear under ordinary magnifying powers. Now, if a drop or two of a concentrated solution of sulphate of soda be added, the corpuscles again become visible, but are deformed, more or less angular, dentated, etc.—A solution of an *alkali* at $\frac{1}{10}$th rapidly dissolves the globules. A concentrated solution does not dissolve them, but causes them to greatly diminish in size.—*Acetic acid* at $\frac{1}{10}$th ordinarily renders blood corpuscles pale, and so faint as to be barely perceptible.—*Nitric acid* shrinks the corpuscles, and colors them greenish.— *Coloring reagents*, as well as *coloring matters*, are without action upon them. *Bile* dissolves them.

‡ Prof. Almen gives the following process [Prof. Van Deen's] for detecting blood in urine : Mix tincture of guaiac 2 c. c. with an equal volume of oil of turpentine, in a test tube, and agitate until an emulsion is formed. To this carefully add a little of the urine to be tested, so that it may sink to the bottom. As the two fluids come in contact the guaiac resin separates and forms a white, dirty yellow, or green precipitate ; but if blood be present, the resin will have a more or less intense indigo-blue color. This is a simple and reliable method, and is so extremely delicate as to produce the reaction with one twenty-thousandth part of a milligramme of iron, according to T. Schiellerup, of Copenhagen ; and hence requires great care and judgment as a test in legal cases.

Clinical Import. Urine which contains blood corpuscles, likewise contains fibrin and albumen, these substances being integrant parts of the blood. When a quantity of *fibrin*, or *albumen*, is found in the urine, not in proportionate relation with the amount of blood in this fluid, it may be decided that there has been an exudation of fibrin, or of albumen. Blood contained in the urine indicates that a hemorrhage has occurred at some point of the urinary apparatus. It must, however, be borne in mind, that with women, blood is frequently found in their urine during menstruation, and, when metrorrhagia is present.—If, in addition to the blood corpuscles, urinary casts or cylinders are also present, renal hemorrhage may be suspected. (See *Renal Tube Casts. Blood Casts.*) Small clots of blood are sometimes observed in a slightly-colored urine, which may be due to small intermittent hemorrhages from the surfaces of the vesical mucus membrane, or from the urethral surface, or they may be formed in the ureters, from which they are discharged with more or less severe pain. Decomposed blood is sometimes of serious import, for a reference to which, see *Hæmatoglobulin.*

The diseases in which blood is found in the urine, are, acute Bright's disease; congestion of kidney; cancer of kidney; tubercle of kidney; external injury to the kidney; calculus in pelvis and ureter; bilharzia hæmatobia; pyelitis; cancer of the pelvis and ureter; diseases of the bladder, and of the urethra; certain constitutional diseases, as, scurvy, purpura, etc.; stimulating diuretics; mechanical violences, etc. See *Table* XXII, which may aid in the diagnosis of the source of the blood met with in urine.

Bodo Urinarius. An animalcular organism sometimes observed in the urine, and first pointed out by Hassall. This animalcule is oval and round, often broader at one end, furnished with one or more cilia, about $\frac{1}{1800}$th of an inch in length and $\frac{1}{3000}$th of an inch broad, granular, and resembling a leucocyte or mucus corpuscle. It increases by division. It is found in albuminous urine. I have seen it in the urine of excessive drinkers of lager beer.

Bradysuria. Painful micturition, with incessant urinary tenesmus.

Bright's Disease. A term applied to several forms of kidney disease, characterized by the presence, in the urine, of albumen and urinary casts. The disease was first described by Dr. Richard Bright, of London; there are two forms of it, the acute and chronic, which are subdivided into several varieties. It is seldom curable, more commonly occasioning death.

Bromide of Potassium in the Urine. It is very probable that nearly, if not quite, all of this bromide taken into the stomach is eliminated through the urine. The quantity of bromide may be determined by means of a titrated solution of hypochlorite of soda placed in a burette. Acidulate the urine by citric acid, and sulphuret of carbon will take up the bromine abstracted by the nascent chlorine. *Coigniet.*—A method of detecting a bromide in the urine has been given as follows: Acidulate the urine with nitric acid, and then add argentic nitrate to precipitate the mixed chlorides and bromides. Wash this precipitate, and when dry fuse it with a mixture of chemically pure carbonate of soda and potassa—that prepared by igniting Rochelle salts is better than the commercial carbonate. Dissolve the mass in a little water, add enough hydrochloric acid to neutralize the solution, and filter into a clear test tube of white glass. Another similar test tube is filled to the same height with a weak solution of chloride of sodium. Into each liquid let fall one drop of a solution of chloride of gold, and agitate. If any bromide be present, on comparing the two fluids side by side, the first one will present a yellowish tint throughout, even if only slight traces of bromide exist. If an iodide be also present, it will interfere with the reaction, and should be separated, first, by palladium, and the palladium by sulphydric acid. *J. H. Bill, M. D.*

Bromine, and Iodine in the Urine. Cut a leaf of brief paper into small strips, and introduce one of them into the test tube or beaker glass containing the urine, letting its inferior extremity touch the bottom of the vessel. Then drop, so as to flow along the side of the glass, three or four drops of nitric acid. In a few minutes, upon removing the paper, a more or less dark, fine blue color will be observed at the inferior extremity of the paper, if *iodine* be present; or, a well marked orange-yellow, if *bromine.* Ordinarily, iodine is more readily detected in the urine in two or three hours after its administration. After having dropped the acid, at least ten minutes must be allowed for the blue tint to become manifest. Care must be taken to so drop the acid that it may run along the side of the glass vessel and collect at the bottom. An error must always be suspected, caused by the transparency of the paper impregnated with the transparent fluid, manifesting itself by a blueish tint.

Brownian Movements. *Brunonian Movements. Molecular Motion.* These terms are synonymously applied to a perpetual, more or less active, agitation among the molecular granulations floating in liquids, as observed under the microscope. These granulations vary in size, but are usually less than the five or six thousandth of a millimetre in thickness. Robert Brown, a botanist, first made known, in 1827, that finely powdered metals, stones, and even charcoal, when treated by acids and heat, presented this agitation, as long as they remained floating in the liquid. These movements are especially seen in the fovilla-grains of pollen, in certain diatoms and desmids, in the granules resulting from the decomposition of leucocytes and infusoria, in the *organites* (elementary bodies) so termed, of certain contagious maladies, etc. They generally present vibrating, oscillatory, or top-like spinning movements, and may become displaced four or five times their diameter in one direction, and then in another, without making any progression. Previous to Brown's experiments they were erroneously considered as characteristic of animality. They apparently play a large part in the infection from virus, and especially from bacteria and miasms of vegetable origin. They are supposed to be owing to some obscure chemical action, when under the influence of heat, and are of no known importance.

Butyric Acid. This acid exists in butter, and is evolved when this becomes rancid, occasioning an unpleasant odor. It is rarely found in normal urine; more frequently in abnormal. It also appears in diabetic urine upon standing for a time, as well as, occasionally, in the urine of pregnant women. The cause of its presence in urine is not satisfactorily known, though supposed to be one of the results of the metamorphosis of leucin. Anhydrous butyric acid is a very mobile liquid, colorless, of a powerful odor, and strongly refracting light. As a hydrate, and united with bases to form salts, it is exceedingly repulsive, having the odor of rancid butter. When its salts are heated with sulphuric acid, it is set free and emits its peculiar, offensive odor; and as it rarely exists in the free state in urine, if at all, the odor evolved by this reaction may determine its presence.

C.

Cacospermia. *Cacospermasia.* Defective, or ill-conditioned semen.

Calculi. Concretions found in various parts of the body, but more commonly in the tonsils, lungs, articulations, bile ducts, prostate gland, and urinary bladder; in the latter, a calculus is popularly termed "stone in the bladder." They differ in composition and size, and apparently result from the deposition and subsequent cohesion of an excess of normal or abnormal matters in the fluids of the parts wherein they are found. The substances which form a vesical calculus, either alone or in combination, are: *uric acid; urate of ammonia; urate of lime; urate of potassa; urate of soda; urate of magnesia; xanthic oxide; ammonio-magnesian phosphate; phosphate of lime; cystine; oxalate of lime; carbonate of lime; carbonate of magnesia; silica; oxide of iron; benzoate of ammonia; oxalate of ammonia; phosphate of iron; mica; hydrochlorate of ammonia; urea; animal matters.* Calculi are divided into organic, inorganic, and mixed or alternating. When a calculus is sawn through, and the cut surface polished and varnished, it will, in most instances, present a surface formed of concentric rings or layers, of different colors, of varying degrees of hardness, and of different chemical composition; and, to determine their character, a small portion of each layer, about the size of a pin's head, should be subjected to a separate examination. It is rarely the case that a calculus consists of only one substance. The nucleus around which urinary deposits form to constitute a calculus, are of various kinds; a very common nucleus is oxalate of lime; uric acid is by no means an infrequent nucleus; the phosphates are very rarely found forming the nucleus of a calculus. Other nuclei have likewise been encountered, as blood-clot, mucus, fibrine, and other foreign bodies forming within the bladder, as well as substances that have been introduced into the bladder accidentally, from without. Sometimes the calculus will have a cavity in its center instead of a nucleus.—When a calculus is moist, it should be gradually dried by means of a water bath, before proceeding to analyze it. The more common kinds of urinary calculi requiring analysis, are those containing uric acid, oxalate of lime, and mixed phosphates.

Constituents of the Various Kinds of Calculi.

When consisting of—1. *Uric Acid.* They are yellowish or reddish-yellow, lateritious, especially when moist; yield a wood sawdust-like powder when sawn into, and which emits a hydrocyanic odor when slowly heated in an open vessel, and, when the temperature is elevated to a red heat, it burns without residue. It forms unctuous compounds when triturated with caustic alkalies; readily dissolves in diluted alkalies, from which solution it is precipitated, by acetic or hydrochloric acid, in crystalline form. It is soluble in nitric acid; the solution, when evaporated and brought into contact with ammoniacal vapor, becoming of a rich purple color, which changes to violet on the addition of a little caustic potassa.—2. *Urate of Ammonia*, in calculi, is almost always combined with other substances, and when in the form of layer, is of an ash-grey color. It burns without residue, and evolves a strong odor of ammonia with alkaline solutions. It is more soluble in water than the urates of the fixed alkalies, and, with nitric acid, acts in a manner similar to that of uric acid.—3. *Urate of Lime*. Is less soluble in water than the other urates, and when present is in combination with oxalate of lime, phosphate of lime, or the other urates, but in very small quantity.—4. *Urate of Potassa*. This salt is soluble in about 400 times its weight of cold water, and in much less hot water. Its solubility prevents it from being present in a calculus, except in very small amount.—5. *Urate of Soda*. This salt is sometimes met with in calculi but seldom in any great quantity, though it forms the chief constituent in gouty concretions. In many respects it acts similar to the urate of potassa.—6. *Urate of Magnesia* is much less soluble in water than urate of ammonia, and is often met with in urinary calculi. It is generally found combined with uric acid, ammonio-magnesian phosphate, urate of ammonia, or phosphate of lime; sometimes it forms an entire layer in a calculus, but is seldom found as the only constituent of the calculus.

7. *Xanthic* or *Uric Oxide*, detected by M. Marcet, and others, in calculi; it forms a permanent yellow mass on evaporation from its solution in nitric acid, without becoming red by ammonia. It is very seldom met with. Stromeyer found it once in a large calculus, occurring as a white powder, which when dried, formed yellowish, hard wax-like masses.—8. *Ammonio-Magnesian Phosphate* forms a white, crystalline, translucent layer, or calculus, which gives a vitreous globule under the blowpipe, by a red heat. When triturated with alkalies, the calculus is not dissolved, but emits an ammoniacal odor. This phosphate is readily soluble in acetic, hydrochloric, and sulphuric acids.— 9. *Phosphate of Lime* is white, opaque, non-crystalline, non-vitrifiable, loses hardly anything by calcination, is insoluble in alkalies, but forms a thick magma when triturated with them, at the same time evolving heat, but no ammoniacal odor. They are less readily soluble in acids than the preceding double phosphate. These phosphates (8 and 9) sometimes form the entire calculus, at other times they will be found forming an internal or external layer of the stone. The calculi containing them are termed *Mixed Phosphate* or *Fusible Calculi*, because, when acted upon by flame with the blowpipe, they fuse and form a hard vitreous globule.—10. *Cystine* is very rarely found in calculi. It appears to be a feeble organic alkali, of a light-straw color, crystalline, inodorous, soluble in ammonia, but not in water or alcohol. Thrown upon burning charcoal it emits a garlicky odor.—11. *Oxalate of Lime*. These

calculi are grey, but more frequently deep brown, probably from accompanying animal matters. They are almost always disposed in undulating layers, with surface studded with tubercles, like that of mulberries; calcination decomposes them, giving a white slate-colored residue of carbonate of lime,—the lime being known by its acrid taste, and the carbonate by its effervescence with acids. When this residue consists of lime only, from a prolonged exposure to heat, it amounts to nearly one-third of the weight of the calculus. Oxalate of lime calculi are often met with; on account of their rough, warty surface, they are apt to occasion much irritation, with pain, inflammation and hemorrhage.—12. *Carbonate of Lime* rarely forms calculi by itself; they are whitish-grey, presenting an earthy chalk-like appearance, and, in a few cases, dark yellow or brown. When present in a calculus it is more apt to be mixed with oxalate of lime, earthy phosphates, etc. It is not fusible, but by prolonged exposure to a strong heat becomes converted into caustic lime.

13. *Carbonate of Magnesia* has rarely been found in human urine, or in calculi, though more frequently met with in those of animals. 14. *Silica,* is still more rarely found in calculi, though present in various fluids and solids of the body.—15. *Oxide of Iron* in calculi is probably due to the presence of *uroerythrin,* and is found in the ash obtained from a portion of the calculus.—16. *Benzoate of Ammonia, Oxalate of Ammonia,* etc., have been occasionally detected among the constituents of calculus. The *Animal Matters,* beside uric acid, cystine, and fibrinous matter, probably consist of altered mucus, bladder epithelia, blood, and urine pigment.

Analysis of Urinary Calculi. The calculus to be analyzed must first be divided into three portions, A, B, and C, each one of which must be acted upon separately. From A, is learned in a general way, the nature of the stone, organic, inorganic, or mixed, as well as of the bases that may be combined with the organic substances. B, is designed for the separation of matters soluble in acids. C, is to assist in ascertaining the principles soluble in water only, without being decomposed.

Portion A. This portion must be reduced, in an agate mortar, to an exceedingly fine powder. A part of it is weighed, placed in a platinum capsule and submitted to the action of an elevated heat (by blowpipe or otherwise). If the whole becomes dissipated, the calculus consisted of organic substances, as, uric acid, urate of ammonia, creatine, etc. If a residuum is left, it must be weighed to ascertain the percentage of loss by the heat, after which, its character must be determined by a chemical examination. To determine whether the organic matter of the calculus is uric acid or a urate, a small portion of the original powder (A) may be heated in a watch glass, porcelain capsule, or platinum spatula, with a little nitric acid; this solution effected, add a few drops of ammonia to neutralize the acid, and then evaporate to dryness. The presence of a more or less dark red or purple color indicates the existence of uric acid or of a urate. Any further examination of A will be unnecessary.

Portion B. This portion must also be reduced an exceedingly fine powder.

4

Place some of this in a platinum capsule (or a glass test tube), add a little concentrated hydrochloric acid to it, and boil it, because oxalate of lime, should it be present, is soluble only in this acid, boiling. If the specimen contained carbonate of lime there will be effervescence from the disengagement of carbonic acid, and the lime will exist in the solution in the form of chloride. If the solution be of a dark-brown color, oxalate of lime may be present; if it be a phosphate, the solution will hardly be discolored. In order to determine the substances held in solution in the acid liquor, we must verify them by a chemical or microscopical examination. For this purpose the solution must be filtered, to remove any organic substances present, and then proceed to examine the filtered liquor as follows:—1. Gradually drop ammonia into a portion of the solution, being careful to cease as soon as the acid is neutralized, which may be known by a cloudy appearance in the fluid, occasioned by the formation of crystals, which will soon be deposited; or, a piece of reddened litmus paper may be placed in the solution, which will be restored to its blue color as soon as the liquor becomes slightly alkaline. As soon as the precipitate has occurred, we add an excess of acetic acid which will dissolve the *phosphate of lime* and the *ammonio-magnesian phosphate*, but not the oxalate of lime, which, if present, will still remain in the precipitate, and may be examined under the microscope.

2. To detect *carbonate of lime*, an excess of ammonia must be added to the preceding filtered liquor, in order to precipitate all the phosphates present, and then filter. To this filtered fluid add a small quantity of a solution of oxalate of ammonia. If chloride of lime, derived from the original carbonate, be present, the lime will unite with the oxalic acid, and be precipitated in the form of oxalate of lime, which may be verified under the microscope; and the ammonia will combine with the hydrochloric acid, forming hydrochlorate of ammonia, which remains in solution. If no precipitate occurs on the addition of the oxalate of ammonia, it is an evidence that the calculus contained no carbonate of lime.

3. To detect *phosphate of lime*, take the precipitate left on the filter, in the preceding (2) experiment, and dissolve it in acetic acid; then add solution of oxalate of ammonia, which, as in the last experiment, precipitates crystals of oxalate of lime, leaving phosphoric acid in the solution. Filter the solution to remove the oxalate of lime deposited, which may be verified under the microscope.

4. To the filtered liquid remaining from the preceding (3) experiment, add a little ammonia, which precipitates the ammonio-magnesian phosphates, filter, and examine as named under *ammonio magnesian phosphates*. See *Phosphates.*—5. To find *phosphoric acid* and thus verify that the lime in the calculus existed in the form of a phosphate, a solution of chloride or other salt of magnesia, together with some ammonia is to be added to the last filtered liquor, when, from double decomposition, the double ammonio-magnesian phosphate will be formed and precipitated, which may be examined as stated in experiment 4.—If it be desired to determine approximatively the

relative proportions of each principle found in the calculus, not only must the calculus be weighed, but also its portions, A, B, and C, the filters previous to filtering, and the filters with the precipitates collected on them after having dried them. Thus, if the calculus weighed 40 grains, and the precipitate, from one-third of it, weighs 4 grains (after deducting the weight of the filter previously), the whole calculus will have contained about 12 grains of said precipitate, and so of each one.

Portion C. It is hardly useful to examine this portion, unless from the indications given during the examination of Portion A there is reason to believe in the presence of uric acid or urates. By means of an agate mortar, reduce Portion C to the finest powder possible; then place it in a large test tube, a porcelain, or a sufficiently large platinum, capsule, and add 200 or 300 times its weight of distilled water. Boil it for 15 or 20 minutes, and filter while hot. On cooling, the crystals, which are insoluble in cold water, are precipitated, and may be examined under the microscope. See *Uric Acid* and *Urates.* Urates are decomposed by hydrochloric and concentrated acetic acids, etc., and uric acid always loses its original character by the action of nitric acid; when once decomposed, their original recomposition becomes impossible,—hence the value of water in this experiment, which dissolves without decomposing them, and affords much less chance for error than by any other means, in determining the true character of the urates contained in calculi submitted to analysis, besides allowing a second analysis of each substance separately.

Cystic Oxide, Xanthic or *Uric Oxide,* and *Fibrinous Calculi,* are rarely met with. They require other processes of analysis, which see under their respective heads; also for other principles that have been referred to.

The following table and explanations will be found useful as a guide in the qualitative analysis of urinary calculi. The names in italics must be referred to in other parts of the work for further explanations and details. The figures and letters refer to those of the same character existing in other parts of the table, and, to the article "Observations, etc.," which follow it.

Table VI.

Qualitative Analysis of Urinary Calculi.

Finely pulverize the calculus, and divide it into three portions.	A, one portion, to determine the general composition of the calculus.	*a.* Organic. *b.* Inorganic. *c.* Mixed.
	B, a second portion, to determine the composition when it is..	Organic.
	C, a third portion, to determine its composition when it is..	Inorganic, or mixed.

Portion A.

Put this portion in a crucible, or on a platinum capsule, and heat it to redness.	*a.* There is no residue, or only a trace.	Organic. See *Portion B.*
	b. A part disappears, leaving a large or very considerable residue.	Inorganic or mixed. See *Portion C.*

Remarks. During this operation, observe the following particulars. *Odling.*

Carbonization. All urinary calculi, under heat, undergo a slight carbonization, and become transformed into a black powder (nitrogenized charcoal), due to the organic matters they contain.—In *oxalate of lime* calculi, it is very slight, and the charcoal disappears to give place to an abundant white pulverulent residue.—In *phosphatic* calculi, the carbonization is more complete, and the coal does not burn up so readily.

Decrepitation. This is always very slight; when, at the same time, a white smoke is given off, and considerable agitation is determined in the heated powder, the presence of urate of ammonia may be inferred.

Odor. *Oxalate of lime* calculi rarely emit any odor; the others evolve it particularly those of *cystine* (odor *sui generis*).

Volatilization. If the powder becomes almost wholly volatilized, a search may at once be made for *uric acid*, without any farther preamble. See *Portion B.*

Alkalinity. When the heat from an alcohol lamp has been used, any alkalinity manifested by the moistened residue upon test paper, is probably due to carbonate of soda, proceeding from the decomposition of the urate of soda.

Fusion. The heat from an alcohol lamp, often suffices to fuse *mixed earthy phosphates.*

Effervescence. To the residue previously moistened, add a drop or two of hydrochloric or nitric acid. If effervescence occurs, it indicates the presence of a carbonate, either having existed as a salt in the calculus, or resulting from the decomposition, by the heat, of a *salt with an organic acid,* as, oxalate of lime, fixed alkaline urates. In the latter case, the effervescence is usually feeble. According to Beale, calcined ammonio-magnesian phosphate gives effervescence.

Portion B.

N. B. Obs. I, II, etc., refer to the Observations following these Tables.

The calculus is found to be organic; it may be composed partly, or wholly, of the following elements:
- Uric Acid.
- Urate of Ammonia.
- Cystine.
- Xanthine. Fibrin.
- Urostealith, and protein compounds.

1. A sample of the powder, B, is treated in a capsule by nitric acid; we gently heat it; if effervescence is produced, and especially if the solution diluted with distilled water, gives the *murexide* reaction. See *Uric Acid*. → **Uric Acid, or Urate of Ammonia.**

Heat a sample of the powder in a tube with solution of caustic potassa or soda. I. Obs.
- No Ammonia is disengaged. → **Uric Acid.**
- Ammonia is disengaged. → **Urate of Ammonia.**

2. The result of the treatment by nitric acid is a lemon-yellow residue; it does not become red when sprinkled with ammonia, but passes to orange-red upon the addition of a solution of caustic potassa or soda. → **Xanthine. II. Obs.**

3. The preceding reactions have not been obtained; the calculus contains neither uric acid, urate of ammonia, nor xanthine. A new sample of it is treated by ammonia, then filtered, and a drop of the filtered liquid is allowed to evaporate spontaneously on a glass slide. If, under the microscope, hexagonal tables and other analogous crystalline forms, are observed .... → **Cystine.**

4. The heated calculus evolves a strong odor of burnt horn, the same as all albuminoid matters; a sample of it is to be dissolved in caustic potassa, then precipitated from its solution by acetic acid. The precipitate redissolved by an excess of the acid, is precipitated anew by ferro-cyanide of potassium. *Méhu.* → **Fibrin. III. Obs.**

5 The calculus has fused without decomposing, swells up and evolves a very strong odor, recalling that of a mixture of shell-lac and benzoin. Treated by nitric acid and heated, a feeble disengagment of gas occurs; the residue, evaporated to dryness and treated with ammonia or potassa, becomes dark yellow. → **Urostealith. IV. Obs.**

Portion C.

A sample of Portion C is treated in a capsule by nitric acid with heat, the solution is then diluted with distilled water, and the whole gently evaporated to dryness over a water bath.—Carefully add to it a little ammonia.
- Reaction of murexide. → The calculus contains uric acid, or urates. → See *b*, below, and 1, *Portion* B.
- The reaction does not occur. → The calculus does not contain uric acid, or urates. → See *c*, below, and 2 and 3, *Portion* B.

b. Detection of Uric Acid and Urates.

The Portion C being reduced to an impalpable powder, is boiled for 20 minutes with 300 times its weight of distilled water. Agitate from time to time. Filter the boiling liquor, being careful, however, to leave in the capsule the earthy deposit that has not been dissolved.

> Filtered liquor. See 7.
>
> Deposit on the filter. See 6.
>
> Deposit not dissolved. See c, below.

6. If a deposit remains on the filter, examine it chemically and microscopically to ascertain whether it consists of uric acid.

> See *Uric Acid.*

7. Place a drop of the filtered solution (Obs. V) upon a glass slide, or a piece of platinum foil, and allow it to evaporate; if there remains no residue, pass to c, below. In the contrary case, take another part of the filtered solution, and boil it in a test tube with solution of caustic potassa or soda, at the tenth. If we recognize the ammoniacal odor, the alkaline reaction of moist reddened litmus paper held at the orifice of the tube, and the formation of white vapors of hydrochlorate of ammonia, when a glass rod moistened with hydrochloric acid is held over the orifice,

> *Urate of Ammonia.*

8. Evaporate another part of the filtered solution to a very small amount, add a few drops of nitric acid to it, and gently evaporate to complete dryness. If there is a rose-colored residuum, becoming purple on the addition of ammonia,

> *Uric Acid.*

9. Heat this last residuum to redness so as to reduce it to ashes, which divide into 2 equal parts x and y.

x. This part is to be treated by acetic acid and filtered. Add a drop or two of ammonia to the filtered liquid to neutralize it; then add a drop of solution of hydrochlorate of ammonia, and treat by an equal volume of a solution of *oxalate of ammonia.* If a white precipitate is formed (Obs. VI),

> *Urate of Lime.*

Filter the preceding fluid, or if there is no precipitate treat it as it is by *ammoniacal phosphate of soda,* and a large excess of ammonia; if there is formed a precipitate of ammonia-magnesian phosphate in stellate or feathery crystals, or resembling fern leaves (under the microscope). (Obs. VII),

> *Urate of Magnesia.*

y. Treat the other portion by hydrochloric acid, diluted with a very little distilled water, and allow a drop, placed on a glass slide, to evaporate spontaneously; examine the residue under the microscope, and if cubic crystals of *chloride of sodium* are observed,

> *Urate of Soda.*

c. Detection of Oxalates, Carbonates, and Phosphates.

The earthy deposit left in the capsule or vessel (*b*) is treated by concentrated hydrochloric acid; it is then boiled, observing whether the solution is effected with effervescence, which indicates the presence of a carbonate, if it is complete, etc. The solution obtained, dilute it with distilled water, filter, and divide it into two portions, *e* and *f*.

Portion e. *Detection of the Oxalate.*	Add an excess of ammonia to it.	There is a precipitate.	This precipitate contains one of the following salts:	*Oxalate of lime. Phosphate of lime. Ammonio-magnesian phosphate. Carbonate of lime* changed into chloride.
		There is no precipitate.	Remains in solution. The calculus is formed of........	*Carbonate of lime.*
	Then add an excess of acetic acid to it.	The whole is redissolved.	There is no oxalate of lime in the calculus. The dissolved precipitate indicates the presence of	*Phospate of lime. Ammonio-magnesian phosphate.*
		The whole is not redissolved. (Obs. VIII.)	*Oxalate of lime.*	Separate the precipitate and ascertain its microscopic and chemical characters.
Portion f. *Detection of Carbonate and Phosphates.*	1st. Add an excess of ammonia to it; filter; the filtered liquor contains the carbonate of lime in the state of chloride of lime. Add to this liquor, *oxalate of ammonia*, drop by drop.	No precipitate.	No carbonate in the calculus.	
		A precipitate.	It is formed of oxalate of lime, and indicates..........	*Carbonate of lime.*
	2d. To the precipitate on the filter, add an excess of acetic acid; if oxalate exists, it remains, (counter proof, the phosphates are dissolved; the liquor obtained is neutralized by a drop or two of ammonia 'Obs. IX), then add hydrochlorate of ammonia and oxalate of ammonia.	No precipitate.	The calculus contains no phosphate of lime. There remains in the liquor......	*Ammonio-magnesian phosphates.* See 3d.
		A precipitate.	Oxalate of lime is formed, indicating..................	*Phosphate of lime.*
		Remains in solution.	Indicates.............	*Amm. mag. phos.* See 3d. *Phosphoric acid.* See 4th.
	3d. Filter, or if there is no precipitate treat the liquor as it is by an excess of ammonia, which will precipitate the last phosphate held in the liquor, if any exists.	The precipitate will be composed of stellated feathery-like or fern-like crystals. (×300).	Indicates.	*Ammonio-magnesian phosphate.*
	4th. Acidulate the filtered liquor with nitric acid, and then treat it with an equal volume of *molybdate of ammonia.*	A yellow precipitate.	Indicates (the result of decomposition of the phosphate of lime).	*Phosphoric acid.* (Obs. X.)

Observations Regarding the Preceding Table.

I.—This should be boiled for some time, holding the test tube with a wooden forceps, or engaged in a copper ring. The ammoniacal odor may be perceived, or else a strip of moistened red litmus paper may be held near the orifice of the tube, but without touching it, and which will, more or less rapidly, become blue; on drying it, it resumes its redness.

II.—Calculi of *xanthine* have been very rarely observed; they are clear brown, and quite hard; by friction they acquire a shining, waxy appearance, and are generally formed of amorphous, concentric layers, easy to separate. John Davy states that the urinary concretions of spiders and scorpions are almost wholly formed of xanthine.

III and IV.—Fibrin and urostealith are very rarely met with. In the recent state, calculi of urostealith are soft and elastic, and feel like caoutchouc; on drying they shrink up and become brittle.

V.—Upon allowing a drop of the solution, placed on a glass slide, to evaporate spontaneously under a bell glass, the microscope will sometimes reveal some of the known forms of the *urates*.

VI.—The white precipitate is formed of oxalate of lime, but in the form of small grains agglomerated in little groups; this is due to the rapid precipitation which does not allow the time required for the formation of the regular crystals. In neutralizing with ammonia, too much is sometimes added, and the liquid becomes turbid; the addition of a few drops of acetic acid will dissolve this slight precipitate and clear up the fluid.

VII.—Like the preceding, this precipitate is very slow to form when there are only traces of magnesia. After having well agitated it, a few drops may be placed in a clear sample tube, closed with a cork, and be allowed to rest for 24 hours. If, at the end of this time, a small crystalline deposit occurs, this will show, under the microscope, the regular forms of *ammonio-magnesian phosphate*, and the existence of traces of magnesia may be asserted.—These crystals must not be confounded with groups of rosaceous crystals which sometimes appear, when distilled water, rendered ammoniacal, is treated, by phosphate of soda. (*Phosphate of Soda and Ammonia. Microcosmic Salt.*)

VIII.—This precipitate of oxalate of lime, suddenly obtained, does not present the regular octohedral form, but shows itself under the form of a collection of very black square or roundish points, sometimes with small prisms cut basiled, and uniting in horse-shoe or × form. This black powder is very characteristic, and its chemical characters should be verified.

IX.—Same observation as at VI, for the neutralization with ammonia. When we do not ulteriorly search for soda, we may neutralize with carbonate of soda.—A few drops of hydrochlorate of ammonia is added, because magnesia is not precipitated by oxalate of ammonia in presence of the salts of ammonia.

X.—Examined some time after its formation, this precipitate is observed in the form of yellow spherules, with black outlines, agglomerated by plates.

Calculous Oxide. See *Cystic Oxide.*

Cancerous Fragments in Urine. Matter from cancerous ulceration of any part of the urinary tract generally exists in the urine in the form of more or less blood corpuscles and coagula, and numerous cancer cells, giving to the urinary deposit a thick, dirty, blood-stained appearance. No pus corpuscles, or but very few, are present. The cells are in small masses, and of various shapes, nucleated, caudate, oval, spindle-shaped, or irregular; the nuclei often being large; the cells may also enclose secondary ones. This is almost always from cancerous disease of the bladder, which is ordinarily of the villous kind; and in fragments of it, under the microscope, the capillary vessels forming the villus may be seen, the epithelial coloring being either present or absent.

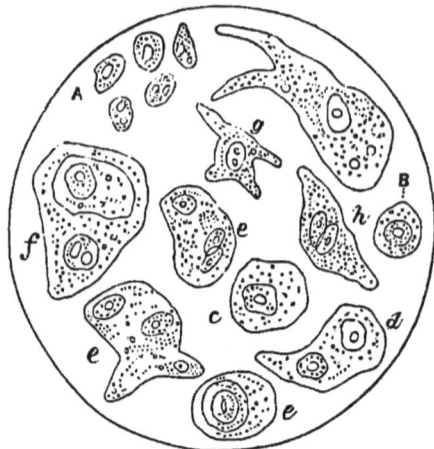

Fig. 12.

A. Five free cancer nuclei.
B. Small cancer cell.
c. Large cancer cell.
d. A cell with two nuclei.
e, e, e. Compound or mother cancer cells, containing two, three, or more nuclei.
f. A mother cell, containing a simple nucleus, and a nucleated cell.
g. Irregular and bifurcated cancer cells, the most usual forms.
h. Cells containing double nuclei; cancer of the bladder invariably contains this variety.

Great care must be taken not to confound these cells of cancerous growth with epithelial cells of the urinary passages, there being considerable resemblance between them. In the diagnosis, not a few cells, but the entire character of the deposit must be taken into consideration, together with the concomitant symptoms, as pain in the bladder, more or less profuse hemorrhages from this viscus, difficult micturition, complexion, constitutional symptoms, etc.—Should the malignant affection be located in the kidney, no dependence can be placed upon these deposits in the urine.—In melanotic cancer, the action of oxygen upon the urine, or the addition of nitric or chromic acid to it, when freshly passed, will turn it black; though all black urine does not always indicate malignancy. Cancer cells, blood corpuscles, renal tube casts, etc., may be examined almost as perfectly as they exist in their natural fluids, by placing them in a solution of 20 grains of white sugar in 4 fluidrachms of distilled water, to which 2 or 3 drops of pure crystallized carbolic acid is added; or, still better, a solution of pure glycerin 27 minims in 4 fluidrachms of distilled water.

Carbolic Acid. *Phenic Acid.* This acid is rarely, if at all, found in human urine, except after the internal or endermic administration of tar, carbolic acid, etc., when the urine assumes an olive-green, a dark, or blackish color, and may also emit the characteristic odor of the acid. This ódor may be observed upon evaporating the urine at a low temperature. It should, however, be remarked that pure phenic acid has been administered internally without any appreciable effect upon the urine, and it has been suggested whether the acid giving rise to the changes in the urine might not be mixed with cresylic acid. Again, free carbolic acid dissolved in alcohol, forms with ammonia a phenate of ammonia, or an artificial aniline; and, very probably, the blue and other dark tints of the urine may be developed by the formation of this ammoniacal phenate in it. Staedeler supposes carbolic acid to be one of the odorous principles of urine. To determine carbolic acid from creosote, A. M. Read states, that a solution of creosote in glycerin becomes cloudy when water is added, and the creosote separates; while a solution of carbolic acid in glycerin presents no such effect upon the addition of water. Among the tests for detecting the presence of carbolic acid, may be named the following:—1. Put a little chlorate of potassa in a test tube, and cover it with hydrochloric acid; after a reaction of about 1 minute, dilute the mixture with $1\frac{1}{2}$ volumes of distilled water. The gas which forms upon the surface must be removed by careful blowing, and then ammonia is gently poured on so as to float upon the surface of the acid liquid. Again blow, to remove the vapors of chloride of ammonium, that are generated, and then allow a few drops of the suspected urine to flow down along the side of the tube. If carbolic acid is present, the upper colorless layer of liquid will assume a color varying from the darkest brown through all shades of red-brown, blood-red, rose-red, according to the amount of carbolic acid present. One part in 12,000 may be detected. *C. Rice.*—2. Bromine water added to the liquid tested, if carbolic acid be present, gives an immediate bulky precipitate of tribromophenol. This will detect 1 part of carbolic acid in 43,700 parts of fluid. *Landolt.*—3. A solution of nitrate of protoxide of mercury, containing *very minute* traces of nitrous acid, when boiled with a solution containing carbolic acid, gives a reduction of the mercurial salt, and, sooner or later, according to its dilution, the liquid assumes an intense red color. One part in 200,000 may be detected. *P. C. Plugge.*—4. Agitate the suspected urine with ether, decant the layer of ether, and allow it to evaporate on a watch glass; an oily residue is obtained. A part of this dissolved in a little distilled water, and then placed in contact with perchloride of iron assumes a purplish color, and, with ammonia and alkaline hypochlorites, a blue tint. That which remains upon the watch glass, submitted to the action of fuming nitric acid, forms picric acid, which dyes silk yellow. *Putrouillard.*

Carbonate of Ammonia. *Ammonium Carbonate.* See *Alkaline Urine.*

Carbonate of Lime. *Calcic,* or *Calcium Carbonate.* When human urine is alkaline from the presence of carbonate of ammonia, it sometimes contains a small portion of carbonate of lime, which is precipitated with the earthy phosphates in an amorphous condition, insoluble in water. It is always present in small quantity, in the form of black or yellowish grains, presenting concentric and radiating striæ like the transverse section of the trunk of a tree. The grains, however, are often so small that these striæ can not be seen. Occasionally it is seen in small globular spheres, discs, and cornucopia-like crystals. In the urine of the horse it is more often observed, giving a jumentous aspect to it, and the spherules of its aggregated acicular crystals are larger than in the human liquid. It rarely forms a sediment, a urinary calculus, or a portion of one; these calculi are very hard and smooth externally, of a grey, yellow, or bronze color, and rarely larger than a hazlenut. Upon the addition of acetic or hydrochloric acid, to the well-washed sediment, the lime is dissolved, and carbonic acid gas is evolved; and when all the acid gas has been driven off, the addition of oxalate of ammonia to the solution gives a precipitate of oxalate of lime. The crystals may be known by their strongly refracting light, by disengaging carbonic acid gas in contact with acids, and by polarizing light. A dark amorphous sediment of *urates* may be determined by the addition of an acid, which is sooner or later followed by the formation of uric acid crystals; this never occurs when carbonates are thus acted on.

Carbonate of Magnesia. *Magnesic or Magnesium Carbonate.* The formula varies according to the relative proportions of the precipitants, the temperature of the solutions, and of the desiccation. This salt is probably sometimes contained in the urine, as it has been observed in urinary calculi. It is stated to be more common in the urine of herbivorous animals. When present in the urine, it may be detected by evaporating this fluid, calcining the residue, and then dissolving it in water. The portion which remains undissolved, is separated by filtration, and dissolved in dilute hydrochloric acid. If it be a carbonate, an evolution of carbonic acid gas will occur. It may be lime or magnesia. Neutralize a portion of the solution with ammonia, then add a few drops of solution of phosphate of soda, when, if magnesia be present, crystals of ammonio-magnesian phosphate will be formed, and may be recognized under the microscope.—To detect carbonate of magnesia in a calculus, dissolve a portion of this in a little dilute hydrochloric acid, and then add ammonia to form hydrochlorate of ammonia. Treat the liquid by solution of phosphate of soda, and from double decomposition the same result occurs as stated in the preceding instance.

Carbonate of Silver. *Argentic, or Silver Carbonate. Test for Uric Acid.* To the suspected urine add a few drops of nitrate of silver solution, to precipitate the chlorides and phosphates, which interfere with the reaction; quickly filter to remove these, and then add solution of carbonate of soda. A precipitate of carbonate of silver ensues, of a *grey color*, if uric acid be present. Tannic acid gives a similar reaction, but it can be determined from uric acid by the ink-black color occasioned by the addition of chloride of iron.—The above test may be made by dissolving a little uric acid in carbonate of soda, put a drop or two on paper, and add a drop of nitrate of silver, which will give the characteristic grey stain.—This will detect the $\frac{1}{17500}$th gramme of uric acid.

Carbonate of Soda. *Sodic*, or *Sodium Carbonate*. It has been estimated that from 4.86 to 11.66 grammes of soda pass in the urine during every 24 hours, the amount varying according to the quantity of soda salts ingested with the food. Carbonate of soda is not generally considered a healthy constituent of urine, being due principally to the kind of food that has been eaten, as fruits containing malates, lactates, acetates, citrates, and tartrates, which during their passage through the organism, become converted into carbonates; a part of the carbonate of soda may escape by urine, when the ingestion of a large quantity of these fruits has occasioned an abundance of this salt in the blood. It is found in alkaline, but not ammoniacal, urine. Carbonate of soda may exist in urine and not be detected in the ash, the carbonic acid having been removed by the heat, or by decomposition; on the other hand, the decomposition of lactates and oxalates by incineration may give rise to carbonate of soda when none existed in the urine. In the analysis of urine for carbonate of soda, it will always be proper to precipitate all the phosphates; then the presence of the soda carbonate in the urine, or in its ash, may be ascertained by the means named under *Carbonic Acid, Carbonates*, and *Potassa*. However, from the combination of this alkali (soda) with other organic acids, etc., it may prove quite difficult to determine the precise quantity of carbonate of soda present in the urine.

Carbonates. In addition to the carbonates above referred to, that of potassa has also been met with in urine. Traces of carbonate of potassa have been observed in urinary calculi, especially in those formed of uric acid, pure or united with ammoniacal or other urates. The presence of these salts appears to be due to the ingestion of vegetable aliments which contain salts of these alkalies [potassa and soda] with vegetable acids, as, potatoes, herbaceous plants, fleshy fruit, cherries, strawberries, apples, grapes, etc., which, during their passage through the organism, become converted into carbonates. The presence of carbonic acid in the urine is readily known by the addition of dilute acetic or hydrochloric acid to the urine, which occasions effervescence. If the carbonate of lime be previously separated from the urine, and then effervescence ensues on the addition of an acid, carbonate of soda (or potassa) is probably present. Another method is to evaporate the urine, calcine the residue, apply the acid to the remaining ash, and if effervescence occurs, a carbonate is present. To determine whether it is lime, magnesia, potash, or soda, see the *Urates* of these bases, also the table under *Calculi*.

Carbonic Acid in Urine. This acid is present in solution in fresh urine. To detect it pass some pure hydrogen through the urine, and then conduct into it pure lime or baryta water, which, if a minute quantity of acid be present, will occasion a turbidity of the urine, or, if the carbonic acid be in considerable amount, there will be a precipitate of carbonate of lime or baryta. If the insoluble carbonate of baryta, thus precipitated, be collected on a filter, washed, dissolved in hydrochloric acid, and precipitated by sulphuric acid, on weighing the precipitate of sulphate of baryta thus obtained, the amount of carbonic acid in the urine may be calculated therefrom.

Carmine. This substance is employed in solution for the purpose of staining or coloring cells, epithelia, casts, etc. It colors epithelia and non-granular leucocytes a rose, the color being especially fixed upon the nuclei. Non organized bodies, as well as vegetable elements—spores, bacteria, etc., are not colored by it. If, after having colored certain elements, a drop or two of acetic acid (1 of acid to 10 of water) be added to the preparation, the color becomes fixed exclusively upon the nuclei and is more brilliant. Various preparations of carmine, are employed, and among them the following: —1. *Ammoniacal Solution of Carmine.* Dissolve a few grains of pure carmine in a little ammonia, and then dilute with distilled water. Filter into a flask, and any undissolved portion of carmine remaining on the filter may be preserved for other uses. Place the flask, uncorked, under a bell glass, allowing it to remain thus for 12 or 18 hours, that the ammonia may be evaporated; or a quicker method of removing this alkali, is to neutralize the solution with a few drops of acetic acid until it has only a very slight ammoniacal odor. In either case, a small amount of carmine may be precipitated, which, when the solution is nearly exhausted by use, may be redissolved with a few drops of ammonia, and prepared as above. See *Picro-carminate of Ammonia.*—2. *Beale's Solution of Carmine.* Place carmine 10 grains in a test tube, and add ammonia 30 minims. Agitate and heat over a spirit lamp until the carmine is dissolved. Boil for a few seconds, and allow it to stand an hour or two; then add pure glycerin, 2 ounces, distilled water 4 ounces. Should the carmine at any time become deposited, add a few drops of ammonia. See *Double Staining.*

Carnine. See *Hypoxanthin.*

Carnivorous Animals. The urine of these animals is clear, transparent, of an acid reaction, containing uric acid, alkaline and earthy phosphates, and urea. Prolonged abstinence causes the uric acid or the urates to disappear, urea alone persisting.

Carrots, used freely as an article of diet may give rise to oxalate of lime in the urine.

Caseine. An organic albuminous substance, coagulable by acids, but not by heat, supposed to have been detected in chylous and kiesteinic urine, but the existence of which has only been satisfactorily demonstrated in milk.

Casts of Bright's Disease. See *Renal Casts.*

Cauliflower. When eaten, this vegetable gives a very unpleasant odor to the urine, and also a few crystals of oxalate of lime.

Cell, or *Utricle,* that peculiar formation upon which, either in an isolated or aggregated condition, the existence of animals or vegetables depends— "the ultimate organized unity of animal life." The cell is an imperforate vesicle or membranous sac, enclosing a liquid cell-sap, and having a thin, delicate, albuminous wall. The vegetable cell has, in addition, an external or thick, strong, cellulose wall or layer. When cells develop other cells, they are termed *parent* or *mother cells*, and the newly formed cells, *daughter cells*. Cells are also termed *corpuscles,* and very minute cells are frequently called *cellules*.

Cellular, Having reference to, or constituted of, cells.

Ceramuria. A name applied to urine giving deposits of earthy, and earthy alkaline, phosphates.

Chalk. This substance may be placed in urine accidentally or designedly; its tests are those named for *carbonate of lime;* examined under the microscope some of the microscopic organisms entering into the formation of chalk may be detected, as, foraminiferous shells, etc.

Chemical Reaction of Urine. This has reference to the acid, neutral, or alkaline condition of the urine, and is generally determined by reddened litmus paper for the alkaline, and blue for the acid. Neutral urine has no effect upon either of these test papers. See *Acidity of Urine; Alkalinity of Urine.*

Chemical Reagents. The various chemical reagents used in the investigation of urine, referred to in this work, will be found respectively under the heads of the articles themselves: thus, for "Compound Solution of Iodine" refer to *Iodine;* for "Ammoniacal Tincture of Carmine," refer to *Carmine;* for "Solution of Acetate of Lead," refer to *Lead*, etc.—In using chemical reagents and test fluids, the practitioner should be careful to have them accurate, and to know with exactness the strength of every test fluid. His graduated vessels, pipettes, etc., should be correctly marked, and every vessel, pipette, etc., should be kept clean when not in use; these necessary attentions will enable him to obtain positive and satisfactory results, which can not be had by an opposite course.

Chloral. Liquor Potassa added to fluids containing chloral, renders them turbid, and evolves a chloroformic odor. See *Urochloralic Acid.*

Chlorate of Potassa. *Potassic or Potassium Chlorate.* This salt is detected in the urine, and in the saliva, by means of sulphate of indigo, a solution of which is decolorized by a liquid containing $\frac{1}{500}$th of chlorate of potassa. Dissolve indigo in sulphuric acid, and add this solution, drop by drop, to a small amount of the urine to be examined. The solution is decolorized as long as any non-decomposed chlorate of potassa remains, but as soon as the chlorate is completely destroyed, the urine assumes the blue color of the indigo solution. By always operating upon the same volume of urine, its richness in chlorate may be comparatively judged of, by the number of drops required.—The tincture of litmus, in presence of sulphuric acid, is equally decolorized by chlorate of potassa; but it is less sensitive, not acting upon a fluid containing $\frac{1}{70}$th of chlorate.—Chlorate of Potassa promptly and markedly reduces the quantity of urea, and is useful in boils when an excess of urea is present in the urine.

Chloride of Ammonium. *See Hydrochlorate of Ammonia.*

Chloride of Sodium. *Sodic, or Sodium Chloride.* Common salt is ordinarily present in healthy urine mixed with traces of chloride of potassium.

Being very soluble, it is never found in urinary deposits. Its presence is due to the fact that nearly all food and water contain more or less of it. It not only proceeds from the food, but, likewise, from the destructive assimilation of the tissues, nearly all of which contain some of it. Suppress the food, and the salt will rapidly diminish in the urine, and fall to a regular discharge of two or three grammes daily. When diminished in amount, or wholly absent, the health of the person deteriorates. Urine containing salt, in health, averages sp. gr. 1,015; and from 100 to 300 grains of salt passes per twenty-four hours, according to the quantity ingested. The normal quantity passed is, according to Robin, and Beale, 3 to 8 parts in 1,000 of urine; to Vogel, 10 to 13 grammes per day, or, 6 to 8 grammes of chlorine. The urine of persons dying is almost wholly deprived of salt. In testing the urine for salt, this fluid should be perfectly clear and transparent, or be rendered so by decantation or filtration. Should albumen be present, it must be coagulated by heat or nitric acid, and then removed by filtration. For exact estimation the method referred to under quantitative analysis may be pursued; but for the detection or approximate estimation of the salt, *Nitrate of Silver Solution No. 1* will be sufficient.

‡ Detection.

1.	2.	3.
Acidulate 10 c. c. of the urine with 2 or 3 drops of *Nitric Acid.*	Now add the solution of Nitrate of Silver. No. 1, drop by drop, until it no longer produces any turbidity.	A white curdy precipitate shows the presence of *chlorine* (in the form of chloride of silver), insoluble in acids; soluble in ammonia.

Remarks.—1. The acidulation with nitric acid is to prevent the precipitation of phosphates, phosphate of silver being very soluble in acids. Care must be had not to add an excess of nitric acid, as it will precipitate uric acid. It is generally the better plan to boil the urine and filter it, previous to acidulating it with nitric acid.—2. The argentic solution must be added until a precipitate is no longer produced, which is very readily ascertained by observing that a drop of the solution fails to occasion any farther turbidity of the urine; then the precipation of the chlorine is complete.—3. The precipitate should be entirely soluble in ammonia. If an insoluble portion remains, an excess of the argentic solution has been added, and phosphate of silver has been precipitated; a little nitric acid added will dissolve this phosphate.

Quantitative Analysis.

‡ *Approximative.* If the preceding process for detection is conducted in a graduated test tube, by allowing the precipitate to rest for twelve or twenty-four hours, its height in the tube may be read off. By repeating this daily,

with fresh samples of urine, the variations in quantity of the salt may be ascertained for each day. Thus, if 10 c. c. of urine give a precipitate of 2 grammes per day, and if the whole amount of urine passed during the day be 1,050 c. c., we will have in this urine about 210 grammes of chloride of sodium. By recording these amounts, there will be no difficulty in determinating the approximate quantity of this salt passed daily. It has been ascertained that, on an average, 1 c. c. of a tested urine of twenty-four hours' standing, corresponds to 62 milligrammes of chloride of sodium.

Exact. The reagents employed in this more tedious and troublesome analysis, are, *Nitrate of Silver Solution, No.* 2, and *Solution of Neutral Chromate of Potassa.* The estimation of chloride of sodium by nitrate of silver is based upon the fact that, in a *neutral* solution, containing chloride, phosphate, and chromate, the salt of silver precipitates the acids of these salts in the following order: 1st, hydrochloric acid; 2d, chromic acid; 3d, phosphoric acid. A urine containing chlorides and phosphates, must have a chromate added to it, which will give, with the argentic salt, a characteristic reaction, apprising the operator that the precipitation of the hydrochloric acid is terminated.

Preliminary Steps.

1. We must be certain that the urine is not albuminous; if this be the case, coagulate the albumen by heat, and filter.—2. On no account must there be any free acid in the urine, because of the ready solubility of the chromate of silver.—3. The *urine must be neutral* to litmus, or slightly alkaline. If it be acid, carefully neutralize it by a drop or so of ammonia, conveyed on a glass rod. Avoid adding too much, because the precipitate is dissolved in it. If the urine be very alkaline, it can be exactly neutralized with acetic acid; to effect which, a piece of blue litmus paper is placed in the urine, and the acid added, drop by drop, on a glass stirring rod, until the paper commences to become pale. This neutralization is delicate; the first time one attempts it, he may be obliged to repeat it anew.

These preliminary steps observed, the operation is very simply effected, as follows:—To 5 c. c. of the urine, add a few drops (3 to 5) of the *Solution of Neutral Chromate of Potassa* (see *Potassa*), and agitate the liquor to thoroughly mix the fluids. Then, by means of a pipette graduated into cubic centimetres and tenths, drop into the mixture *Nitrate of Silver Solution, No.* 2, constantly stirring, until a persistent reddish tint is produced. The operation is now terminated. Every c. c. of the argentic solution that has been used in this process, indicates the presence of 1 centigramme of chloride of sodium, or 6.06 milligrammes of chlorine.—This is a very beautiful reaction. The liquid, which is at first of a light canary-yellow color, exhibits, at those points on which the silver solution falls, reddish spots, which disappear. when the fluid is stirred, so long as any chloride of sodium remains in it. But, as soon as this chloride is wholly decomposed by the silver solution, the very next drop that is added, produces a permanent reddish color of chromate of silver, which is soluble in nitric acid.

Corrections.—This method, thus simplified, gives too high a figure for the chloride of sodium, because there are always other substances beside chlorine that have been precipitated by the silver solution. *Hardy.*—Our investigations made on titrated solutions of chloride of sodium, and estimated by this method, then upon urine previously tested and treated with known weights of sea salt, have led us to allow as constant, an error of an excess of 1 gramme per litre. There is an advantage in operating upon only 5 c. c. of urine; the estimation is more exact, because the change of color is manifested more rapidly. When the quantity of chloride of sodium in 5 c. c. of urine is ascertained, the entire quantity for 1 litre can be calculated, from which subtract 1 gramme.—It should, also, be remarked, that the red color will not appear at first; if the silver solution be carefully added, drop by drop, constantly stirring the fluid, the tints of coffee and milk, chocolate, brown-red, and brick-red, will be observed to appear in succession. If the operation be stopped at the brown-red coloring, too high a figure will be obtained, having dropped in the urine nearly a cubic centimetre too much. The milk and coffee tint is that at which the operation should be terminated, when it does not pass unperceived; but with a little attention it will not be permitted to escape observation. In fact, it announces that a trace of chromate of silver is produced, which, spread throughout the whole mass of the fluid, gives this particular tint very different from the initial color.—The tubes must be cleansed with ammoniacal water. (*Marais.*)

Other Methods.—1. This is based upon the fact that *chlorine* gives a soluble, and *urea* an insoluble, compound with peroxide of mercury, while that of chlorine has a greater affinity for mercury than urea has; therefore, if pernitrate of mercury be added to a solution containing chlorine and urea, the chlorine will first combine with the mercury, and no precipitate of the compound of urea and mercury will take place until all the chlorine has been saturated; and if we observe how much of the mercurial solution has been used before a precipitate commences to form, we can at once ascertain the quantity of chloride present. It is necessary, in this operation, to first remove the phosphates from the urine, which is effected by means of the *Baryta Solution.* The reagent is the standard *Solution of Pernitrate of Mercury.* The process is as follows: To 40 c. c. of urine add 20 c. c. of Baryta Solution; when the phosphates have all become precipitated, filter through a filtering paper rendered feebly acid by diluted nitric acid, and then allowed to dry. Place 15 c. c. of the filtrate into a beaker glass, and carefully add a drop or two of nitric acid to just produce a faintly acid reaction. Now, keeping the urine constantly in motion by means of a glass rod, allow the pernitrate of mercury solution to fall into it, from a graduated burette, drop by drop, until a slight persistent turbidity (precipitate) ensues. Then from the scale of the burette read off the quantity of the mercurial solution required to effect this precipitate, every 1 c. c. of which corresponds to 10 milligrammes of chloride of sodium. By calculation, it is easy to determine the whole amount of this salt in the urine of 24 hours.—Enough of the baryta

5

solution must always be added to the urine, to precipitate the whole of the phosphates.—2. For greater accuracy, 10 c. c. of the urine are placed in a platinum crucible, to which 1 or 2 grammes of pure nitrate of potassa, free from chlorine, are added. After the potassic salt is dissolved, the whole is gradually evaporated to dryness, and then exposed to a low red heat, until the carbon is completely oxidized, and the contents of the crucible are white. After cooling, these are to be dissolved in distilled water, a drop or so of nitric acid added to the solution to faintly acidulate it, a small quantity of carbonate of lime solution being then introduced to again restore neutrality to the acidulated solution; and then the chloride may be estimated by one of the preceding processes.—As by far the greater part of the chlorine in the urine is in combination with sodium, the above processes may answer to determine the amount of chlorides present.

‡ *Microscopic Examination.* Crystals of chloride of sodium are cubical, similar to those of iodide and bromide of potassium, but assuming several varieties. The presence of urea modifies their crystallization from urine; they are then found in large crystals in the form of daggers or crosslets, either isolated or grouped perpendicularly one above the other. When in octohedral forms, these crystals may be distinguished from those of *oxalate of lime*, by their solubility in water. The crystals of chloride of sodium disappear under the polarizing apparatus at the same period when the field of the microscope becomes dark, which will aid us in distinguishing it from other crystals of similar form, which give colors, or may be seen under the polarizer, when the microscopic field becomes obscure. This salt in the urine may be examined by placing a drop of fresh clear urine upon a

Fig. 13.

A. Chloride of sodium, in combination with urea, and evaporated quickly from urine.

B. Chloride of sodium, crystallized from distilled water, and resembling oxalate of lime; never exists in urine, and is soluble in water, while the oxalate is not.

C. Chloride of sodium crystallized slowly from urine, also resembles oxalate of lime, but differs in being soluble in water.

D. Chloride of sodium resembling crystals of cystine from slow evaporation of urine.

glass slide, and then allowing it to evaporate under a bell glass, to one-fourth or one-sixth of its volume; to serve as a comparison, a drop of a solution of salt may, at the same time, be placed by the side of it. See *Fig.* 13.

Clinical Import. The amount of chloride of sodium in the urine, varies according to circumstances. The maximum amount is found during the day, and the minimum during the night; influenced, however, by diet, and mental or physical labor. Chlorides increase in the urine according to the amount of chlorine compounds introduced into the organism, whether in health or disease. The ingestion of a large quantity of water, and all causes which increase or diminish the secretory activity of the kidneys, temporarily increase the chlorides. The presence of the crosslets and dagger-like crystals of chloride of sodium in a partially or wholly evaporated urine, afford a safe indication of the presence of urea.—In *acute disease,* there is diminution and sometimes disappearance of chloride of sodium in the urine; the diminution of the chlorides, being in proportion to the intensity of the attack, and, to a great extent undoubtedly, to the suspension of food ; and their disappearance will announce the production of serous effusions, or of inflammatory exudations. This fact is not special to pneumonia. Beale intimates that chloride of sodium accumulates at any point in which inflammatory changes are occurring. In *intermittent fever,* Vogel states that there is an increase of the chlorides during the access.—In acute articular rheumatism, when the chlorides suddenly disappear from the urine, and albumen can be detected in this fluid, pericarditis may rapidly ensue. In cholera, the chloride of sodium diminishes greatly, its subsequent increase is a highly favorable indication.—In *chronic disease,* the results are variable. A diminution of the chlorides indicates debility of the digestive power, if there does not exist another way of elimination, as, serous diarrhea, or hydropic effusions. A considerable increase will indicate diabetes insipidus.—It must be remembered, however, that these indications are not wholly reliable.—Chloride of sodium is the antidote of bromide of potassium, promptly antagonizing its action.

Chlorides. These never form spontaneous deposits in the urine, the chlorine being combined with soda, potassa, lime, or magnesia, forming salts that are very soluble. The presence of which must be ascertained by chemical processes referred to elsewhere. See *Chloride of Sodium ; Chlorate of Potassa; Chloride of Ammonium; Chlorine.*

Chlorine. The amount of chlorine discharged in healthy urine per 24 hours, varies, according to circumstances, from 5 to 8 grammes. An approximate estimation of its amount may be made as follows : Prepare a test fluid by dissolving pure fused nitrate of silver, 1 grms. 861 (grammes), in distilled water 87 c. c. Of this solution 30 c. c. will correspond to 129 milligrammes of chlorine. To perform the operation, take some of the urine, and remove albumen, if present, by the addition of a little acetic acid, and boiling. Separate the coagulated albumen by filtration. Acidulate the clear filtered urine with nitric acid, and carefully measure into a beaker 7.4 c. c. of this acidulated urine. The silver solution being placed in a burette or pipette graduated into tenths of a cubic centimetre, is allowed to fall, drop by drop, into the urine, which must be constantly stirred, as long as a white precipitate occurs. The moment the silver test solution ceases to occasion a precipitate, the process is terminated. The amount of test solution used may now be read off upon the burette, and as 30 c. c. of it represents 129 milligrammes of chlorine, we ascertain by calculation the amount existing in 7.4 c. c. of the urine; and by knowing the whole amount of this fluid discharged per day, we can, therefrom, determine the entire amount of chlorine contained in it.

Chloroform. This agent is sometimes employed to remove fatty matters, to dissolve and separate cholepyrrhine from biliverdin when these are present in the urine, and for other purposes in the investigation of urine. Chloroform may be found in the urine of persons who have taken it (or chloral) internally, as well as of those who have inhaled it for the production of anæsthesia,—by the following process: Pass a current of air bubblingly through the urine, so that it becomes charged with the vapor of chloroform disengaged from this fluid; the current thus charged, is directed through a porcelain tube heated to redness. The chloroform vapor is decomposed, and the resulting chlorine, being passed through a Liebig's carbonic acid apparatus, with bulbs, containing a solution of nitrate of silver, of determined strength, forms therein a precipitate of chloride of silver. From the weight of this body, the quantity of chlorine, and, lastly, that of the chloroform are deduced.—According to Baudrimont the cupro-potassic liquids employed to detect the presence of sugar in urine, likewise form precipitates, when this fluid contains a little chloroform, chloral, or other substances that may engender chloroform. M. Limousin relates a case in which an error was made in the analysis of the urine of a person recently chloroformized. This urine was declared glucosuric, because of the precipitate given when treated by a cupro-potassic solution. An error of this character might lead to serious consequences.

Chlorophyll. The green coloring matter of plants, somewhat analogous to that of the urine ; it is a substance of a waxy nature, soluble in ether or alcohol, which discharges its green tint; is not affected by water; becomes a yellowish-brown by tincture of iodine, and more or less deep blue by sulphuric acid. Preparations containing chlorophyll lose their green color when mounted in chloride of calcium.

Chlorosis. In this affection the solids of the urine are more or less diminished, and, sometimes, even when the urine is nearly colorless, an abundance of *urohematin* may be detected by adding one-fourth its volume of strong nitric acid to the fluid, and then boiling; it becomes red, if urohematin be present.

Cholalic Acid is a product resulting from the decomposition of both cholic and cholinic acids. It has a bitterish sweet taste like gall, is insoluble in water, soluble in 27 parts of ether, and readily soluble in boiling alcohol, and in alkalies. It crystallizes from alcohol in octohedra and tetraheda, belonging to the quadratic system, and which are of a vitreous brilliancy.

Cholepyrrhin. See *Bile Pigments.*

Cholesterin. This is a peculiar fatty substance forming a normal element of brain and nerve tissue ; it is also met with in the bile, the blood, and other animal fluids. Occasionally it has been detected in the urine, combined with other fats, proceeding from renal fatty degeneration. When it exists in the urine, the sediment must be collected and dried over a water bath, and then digested in a mixture of alcohol and ether. Filter this solution ; concentrate it by evaporation, and upon cooling crystals of cholesterin will form in more or less profusion. These are of variable size, rhomboidal or rectangular, extremely thin, frequently showing a break on one of the borders, and ordinarily imbricated one over the other. Many of the crystals have their obtuse angles truncated. They are readily determined under the microscope, with a power of 300 or 400 diameters. If the crystals are colored, they should be redissolved in boiling alcohol, filtered through animal charcoal, and upon the cooling of the liquid, they will form in colorless plates. Boiled a long time with nitric acid, a resinous mass is formed, which, by prolonging the boiling, becomes converted into *cholesteric acid,* $C_8 H_4 O_4$ under the formation of acetic, butyric, and capronic acids, etc. Cholesterin is a neutral body, white, pearly, insipid, inodorous, insoluble in water, slightly so in cold alcohol, very soluble in boiling alcohol, and does not saponify with alkaline liquors. It is heavier than water. Sulphuric acid in contact with dry cholesterin, on a white porcelain plate, gives a beautiful play of colors, as, different shades of orange, red, purple, and green. When pus has stood for several days, crystals of cholesterin may be detected in it. See *Fats.*

Choletelin.—A substance, existing in the urine of persons laboring under hepatic disease, being derived from bile pigment by oxidation.

Cholic Acid. *Glycocholic Acid.* Various bodies have been described under this name. Cholic or Glycocholic Acid crystallizes in extremely fine needles, which are at first voluminous, but shrink together on drying. It is a bile acid, scarcely soluble in cold water or ether, more soluble in boiling water and alcohol. It has a sweet taste, a feebly acid reaction, and, at an elevated temperature, melts, and is decomposed, evolving a peculiar odor. Caustic alkalis, and alkaline earths, decompose it. Solution of sugar mixed with its solution, and concentrated sulphuric acid then added to it, a purple-violet color occurs at a heat of 122° to 158° F. Its salts have a neutral reaction. Boiled with a strong solution of hydrate of baryta it is decomposed, and yields *cholalic acid* and *glycocoll.*

Cholinic Acid. *Taurocholic Acid.* Remains in solution, after the deposition of cholic acid from bile, under the action of dilute sulphuric acid. It has a bitter-sweetish taste, an acid reaction, and under the action of caustic alkalies is decomposed, yielding *cholalic acid* and *Taurine,* differing, in this respect, from cholic acid.

Cholochlorine. See *Bile Pigments.*

Choloidic Acid. *Choloidinic Acid.* This acid is obtained by the decomposition of cholic acid when boiled with dilute hydrochloric or sulphuric acid. It is solid, amorphous, white, inodorous, insoluble in water, scarcely soluble in ether, and very soluble in alcohol. It forms salts with bases, which, with the exception of the alkaline, are insoluble. When boiled for a long time in dilute hydrochloric or sulphuric acid, it becomes changed into *glycocoll* and *dyslisine*, $C_{24} H_{36} O_3$.

Cholopheine. See *Bile Pigments*.

Chromaturia. Discharge of abnormally colored urine.

Chromogen. A peculiar colorless substance met with in indigoferous plants, which, under the action of acids and ferments, yields indigo blue, a peculiar kind of sugar, and a small amount of other matters. The term has also been applied to *indican* found in urine.

Chylo-serous Urine. See *Chylous Urine*.

Chylous Urine. Urine having a milky, turbid, and opaque appearance, due to the presence of fatty matter in a molecular form (but rarely any oil globules), fibrinous particles enclosing red-blood corpuscles, perhaps, also, albumen or albuminoid substances. When much blood is present a rose tint is imparted to the color. Upon standing for some time, a tremulous coagulum occurs, succeeded by flakes, the fatty matter floating, like cream, upon the surface of the liquid. Sometimes the urine coagulates within the bladder and is voided with difficulty and more or less pain. Heat, and nitric acid, coagulate chylous urine. If the urine be mixed with ether, the fatty matter is dissolved, and the urine assumes its natural color and transparency. If the ether be evaporated, it yields a yellowish uncrystallizable fat, solid or fluid. Under the microscope, the fatty particles in chylous urine require, to be seen distinctly, the highest powers, at least a power of 450 diameters. The appearance of chylous urine is very irregular, sometimes coming on gradually, at other times breaking out suddenly. It may last a short or long time, then disappear, and return again after a certain interval. Its cause is unknown, though several hypotheses have been advanced. Rest, vegetable diet, and fasting lessen the milky character of the urine, while meats appear to increase it; it also varies in amount during the day, the urine being more natural in the morning, and chylous after meals, etc. The urine is neutral, or faintly acid, soon becoming alkaline. The presence of chylous urine is not necessarily an unfavorable indication, nor does it denote disease of the kidneys unless accompanied with coagulated fibrin, renal casts. The disease is more common in tropical climates, and is apt to be attended with pains in the loins and epigastrium, lassitude, emaciation, etc.; occasionally no unpleasant symptoms, nor impairment of health are present; and again it has been observed during the presence of certain maladies, as, epilepsy, erysipelas, diabetes, tuberculosis, etc. No satisfactory method of treatment is known. When death occurs in instances of chylous urine, it is generally the result of some accompanying, but independent, malady.

Chyluria. A discharge of chylous urine, without any apparent indications of renal disease.

Cinnamic Acid. This acid, formerly mistaken for benzoic acid, is a constituent of Tolu, Peru, and some other balsams, of storax, sweet gum, Botany bay resin, etc. When taken internally it becomes transformed into hippuric acid, which is then found in the urine.

Coddington Lens. Fig. 14. A simple microscope, consisting of a glass sphere, around the center of which a triangular groove is cut and then filled with some opaque matter. When properly constructed, it possesses a large field of view, with slight aberration, and a good magnifying power, and is so arranged that it may be safely carried in the vest pocket for bed-side examinations. It is superior, for this purpose, to the Stanhope lens. However, as there are excellent pocket compound microscopes manufactured at the present time, for clinical and other examinations, to be made away from the physician's office, and which are furnished at moderate prices, it will be much better to procure one of these, when it can be done, as, from their magnifying powers, they are adapted for nearly all urinary investigations.

Fig. 14.

Coddington Lens.

Cod-liver Oil. Taken internally, lessens the free acidity of the urine, and likewise diminishes the amount of urea and uric acid. *Beneke.*

Coffee. Taken internally, lessens urea and uric acid; also the amount of chlorine in the urine. The pulmonary carbonic acid is increased by it. Caffein has the same influence in a minor degree. According to Dr. Hammond, coffee is more of a stimulus to the brain faculties than tea.

Colchicum. The internal use of this agent, lessens the quantity of urea and uric acid in the urine, and sometimes diminishes the amount of water, as well as of the earthy phosphates. It increases the solids in the urine.

Colica Nephritis. *Nephritic Colic.* Intense pains in the renal region and along the course of the ureters, due to acute inflammation of the kidneys, or to the descent of a calculus along the ureter into the bladder.

Collection of Urine. In ordinary investigations the first urine of the morning will be sufficient, especially for the mere determination of the presence or absence of certain normal or abnormal elements. But where as much accuracy as possible is required, the whole amount of urine passed every 24 hours should be collected and kept by itself, examining it as soon as possible after its discharge, and, likewise, several hours afterwards. Certain kinds of urine speedily decompose, while others do not change for a long time, and may be properly examined in 24 or 48 hours after being voided. It does not make so much difference as to the vessel in which the urine is contained, so that it is clean, and will admit of being covered to prevent the entrance of extraneous substances. A half gallon magnesia jar, or similar bottle, will answer the purpose very well; and this may be graduated by a diamond writing point, or by means of fluoric acid, into fluid ounces or cubic centimetres, so that the practitioner can at once read off the amount at a glance. Glass jars are made especially for this purpose, holding 2,000 c. c., graduated into 5, 10, or 50 c. c. The greatest difficulty in collecting all the urine of every 24 hours will be found with children, and with patients prostrated from severe illness, or who are paralyzed, or delirious. For women and children, special forms of apparatus are employed (urinals); for the others, watchfulness is required, and, after as near a determination as possible of the amount of urine not collected or lost, 20 per cent. may be allowed for any error in the estimate. Yet, in these cases, it can not be expected that the same certain or satisfactory conclusions can be arrived at, as in those instances in which the entire amount of the urine of 24 hours is collected,—though useful and important information may always be obtained.—In public institutions for the sick, and not unfrequently in private practice, it becomes necessary or desirable to examine each specimen of urine passed at different times during the night and day. Several small vessels may be kept for this purpose. All vessels should be carefully and thoroughly cleansed immediately after their contents have been removed, or when they have been satisfactorily investigated.

Fig. 15.

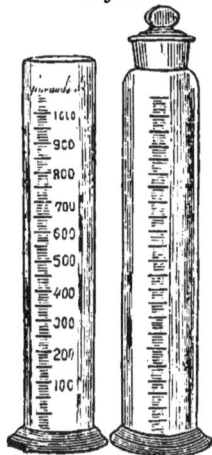

Graduated Glass Jar.

When the urine has stood the required time, it should be decanted from its sediment, and this be then placed in a smaller vessel for subsequent examination; or the sediment may be collected on a filter. For small amounts of urine with deposits, 4 fluid ounce conical glasses with lip and foot, will be found useful; the bottom of these vessels, inside, instead of terminating in a small convexity should end in a point. The height and diameter of these glasses should be large enough to allow the use of an urinometer, when it is desired to ascertain the sp. gr. of the urine. See *Preliminary Remarks*, page 13.

Color of Urine. Healthy urine is of an amber color, due to certain coloring substances present in it. When disease is present, this fluid will present various tints according to the malady and the existing circumstances. The urine may be pale, greenish, straw-yellow, brown, red, or dark. These tints may be due to an excess or diminution of the normal coloring matters, to an increase or decrease of the quantity of water, to the presence of blood, coloring matters of bile, and other abnormal coloring agents, and, lastly, to certain articles of diet or medicine which impart coloration to the urine.

According to Harley, who has carefully investigated the coloring matters of the urine, the normal coloring agent is a substance derived from the blood, and held in solution by the urine, its quantity fluctuating according to the state of the health, and to the circumstances influencing the organism. And as he has been able to detect a common character in the composition of this agent, *biliverdin, chlorophyll, draconin, hematin, indigo, and melanin,* viz., the presence of iron, he concludes that, with a few exceptions, the various colors that have been observed in urine, yellow, red, green, blue, brown, and black, are derived from a common source, a white radical (proceeding from the blood), of which they are simply different grades of oxidation, and may, therefore, be considered under one head. Even the green biliverdin of the bile may result from oxidation of the red hematin of the blood. He has named the normal coloring matter of the urine, *urohematin;* and, according to its degree of oxidation, the quantity present in the urine, and the action of certain substances contained in, or added to, the urine, are its various tints developed. Thus, in one urine, nitric, hydrochloric, and sulphuric acids, may occasion the same color, while in another urine, each one of these acids may color the urine a tint differing from that caused by the others. He states that the quantity of urohematin passed every 24 hours, may be considered "a tolerably exact measure of the destruction of blood corpuscles," being "an index to the tear and wear of the tissues," and "the best measure we at present possess of the rapidity with which burns life's lamp." As the rule, the darker the urine, the more serious the condition of the patient.

Vogel, and many other authors, Scherer, Polli, Virchow, etc., consider the urinary pigments (and biliary) as products of the decomposition of hematin, basing their views upon the following reasons:—The coloring matter of the blood is destroyed with difficulty; the blood extravasated in the teguments (ecchymoses) like that which is left to various influences external to the body, tenaciously holds its color, with more or less modification. Hence, it is not probable that the hematin, which has been used and rendered unfit for the purposes of the organism, is eliminated from the body as a colorless substance. Moreover, the only colored excretions of the body are the urine and the feces, and, consequently, the urinary pigment, or the bile pigment (as modified in the feces), or even both of them, may be considered as products of decomposition of hematin. From this, it follows that, the quantity excreted of these pigments, may give the measure of the intensity of the destruction of the blood globules. Thus, in all acute febrile diseases, the quantity of the coloring matter of the urine is greatly increased. This increase is still greater in septic fevers. Now, a diminution of the blood globules and an anemic condition are the consequences of all these diseases. On the contrary, in those cases where the formation of the blood globules is less active, as, in chlorosis, anemia, neuroses, etc., the quantity of urinary pigment is much below the normal standard, the urine being very pale.—By establishing a scale of colors, Vogel has proposed a process of proximative estimation of the urinary pigments.

In health, the urine fluctuates between a pale straw and a brownish-yellow tint, the depth of color being due to the amount of water holding the coloring matter in solution; thus, after large draughts of water, or beer, the urine becomes very pale and abundant; during periods of profuse perspiration, this fluid is darker and in diminished quantity; in lessened perspiration, as, during winter, the urine is nearer its average normal color, and in increased amount, etc. However, a urine, pale, or of a normal color, is not invariably an indication of health, as, it may contain an abnormal quantity of urohematin, or, of one or more of its derivatives, determined only by proper investigation. High or dark-colored urine almost invariably indicates the existence of a more or less grave pathological condition of the system.

The exceptions to the preceding statements, are the *accidental colorations* of the urine after the ingestion of certain vegetable and other matters, as, aloes, arseniureted hydrogen, blackberries, cactus opuntia, campeachy wood, carbolic acid, creosote, gallic acid, gamboge, indigo, logwood, madder, picrotoxin, raspberries, resin, rhubarb, santonin, senna, tar, turpentine, etc.

A dark-colored urine, when diluted with water, will, according to the amount of dilution, give all the colors more commonly observed in this fluid, thus strengthening the hypothesis that these various tints are only dilutions of one and the same character of coloring agent. From this, Vogel has furnished, as a standard for the approximative determination of the amount of urohematin existing in a given urine, a table of nine different shades of color. See *Vogel's Table of Colors.*

Coloring Matters. The coloring agents of the urine, may be considered as normal, abnormal, and accidental. The *normal,* are *urohematin* and *uroxanthin;* the *abnormal,* are not so readily named, as some of them have been found in healthy urine, being changed conditions of the two normal colors,— however, they may at present be considered as follows: *Bile acids, bile pigments, blood corpuscles, hematin, hemoglobin, melanin, methœmoglobin, pus, uroerythrin.* See *Urobiline.*

Composition of Urine. Urine is a complex and variable liquid, being influenced, as regards its composition, by health, diet, exposure, mental or physical labor, etc. It consists, on an average, of 21.11 parts of water and 1 part of solids. The solids are in solution when the urine is voided, but may be deposited on standing, or by the use of chemical agents. The various organic or nitrogenized principles held in solution, are, the debris of the organic changes, as, *urea, uric acid, creatin, creatinine, ammonia, hippuric acid, xanthine, hypoxanthine, sarcine, normal pigment, mucus, unoxidized sulphur* and *phosphorus,* etc.,—the eleven, last mentioned, together with minute quantities of oxalic and lactic acids, being frequently classed under the head of *extractive matters;* also, *chlorides, phosphates, alkaline* and *terreous sulphates,* proceeding either from the interchanges of the materials of the organism, or derived from the external world with the food and drink. These materials circulating in the blood,—the remains of the nutritive processes, either extraneous to it, or useful in part, must be expelled through the renal filter. The urine also contains carbonic acid gas and nitrogen.

The above refers to the composition of healthy urine; a morbid condition of this fluid varies considerably in its composition,—there may be an absence, a diminution, or an excess of one or more of its physiological elements, or, some new and unnatural principle may be mixed with the urine, as, *albumen, sugar, fat, cystine, blood, pus, fibrine, epithelial cells, spermatozoids, biliary matters,* etc.

Confervoid Growths in Urine. See *Algæ; Fungi.*

Consistence of Urine. Urine in health is perfectly fluid, like water, readily flowing through, and dropping from, a tube of exceedingly small calibre. Disease may render it thick, viscid, and so tenacious that it will not drop from a tube or vessel, but will form ropy, stringy masses on attempting to drop or pour it; this is the case with urine, alkaline from ammoniacal decomposition, and containing pus. Chylous urine is often of such increased consistency as to form a thick, firm, jelly-like mass. I have met with several cases in which the urine was passed of a white color and of the consistence of cream, forming a thick jelly-like mass shortly after being voided; it contained no fat of any kind, but an incredible amount of mucus and phosphates.

Copaiba. Or rather its copaivic acid, when taken internally, is eliminated by the urine in large amount, imparting a peculiar odor, *sui generis*, to this fluid. Nitric acid added to such urine occasions a copious precipitate of an oily or gelatinous character, and which has been mistaken for albumen. However, it consists chiefly of copaivic acid, separated from its combination with soda or potash by the nitric acid, and which is deposited together with urates or uric acid. The urine, containing copaivic acid, is very apt to have its color heightened, its quantity increased, its quality altered, its appearance somewhat turbid, and its taste bitter.

Copper. *Cuprum.* This metal, when it or one of its salts, is taken internally, passes off in the urine, as well as in the feces, and may be detected in cases of poisoning by it. It may be detected as follows: 1. Place the urine in a vessel upon a water bath, and treat it with chlorate of potassa and fuming hydrochloric acid, until both the chlorate and the organic matters are wholly destroyed. To the pale straw-colored liquid thus obtained, add an excess of ammonia to render it alkaline, which imparts a smoky-brown hue to it. Filter to remove any precipitate. Evaporate, on the water bath, until the filtrate is perfectly dry, and the residue that remains is then to be moistened with pure nitric acid (sp. gr. 1.5), placed in a china capsule, and heated to redness, until no charcoal remains. Dissolve the ashes in hydrochloric acid, and boil, adding a little nitric acid, to keep up the highest possible oxidation of the iron and copper present. Add an excess of ammonia to this acid solution, and remove the resulting precipitate of hydrated oxide of iron by filtration. If, upon adding acetic acid to the filtrate, it becomes bluish, or if it furnishes a reddish turbidity, or reddish-brown deposit with ferrocyanide of potassium, copper is present. A clean piece of iron, as a knife blade, will, when held in the solution, acidulated, become covered with a deposit of metallic copper. Sulphureted hydrogen, or sulphide of ammonia, will occasion a brown or black precipitate of sulphide of copper, provided the alkaline liquid is acidulated with hydrochloric acid.

Copper is likewise employed in the investigation of urine for the detection of sugar, being one of the reduction tests for this purpose. Grape sugar, in presence of a solution containing copper, precipitates this in the form of a brick-red suboxide; but it should be likewise remembered that the presence of glycerin, cellulose, uric acid, chloroform, and leucine, produce in different degrees, a reduction of the oxide of copper, while albumen, interferes with the reaction. The several copper tests that have been proposed are based upon the same principle (viz., the property that grape sugar has, in the presence of free alkali, and at an elevated temperature, of depriving oxide of copper of one-half its oxygen, and thus converting it into a red suboxide), and are only modifications of each other; it is unnecessary to name all of them, only those more commonly employed. Among these the best are :—1. *Fehling's Solution.* a. Take of pure, air-dried sulphate of copper 34.64 grammes, distilled water 200 grammes; mix. b. Take of neutral tartrate of potassa 173 grammes, solution of caustic soda, sp. gr. 1.12, 600 grammes; mix.—Add the copper solution, a, to the alkaline liquid, b, in small quantities at a time, and when this is done, add enough distilled water to make the whole measure exactly 1 litre. It requires exactly 10 c. c. of this clear, violet-blue liquid to be reduced by 0.05 gramme (or 50 milligrammes) of sugar. This solution may be used for both the qualitative and quantitative examinations; it is subject to precipitation when kept any length of time, hence it is better to put it into smaller bottles holding 60 or 80 grammes, seal them, and keep in a dark place in the cellar. Label "*Fehling's Cupro-potassic Solution.*" *Detection of Sugar.* ‡ E. Boivin and D. Loiseau, have observed that when 1 c. c. of this test liquid is added to 50 c. c. of distilled water, and boiled for a few minutes, it becomes wholly decolorized, which result may be prevented by the presence of a small amount of a calcareous salt; this is important to bear in mind, when using Fehling's or Pavy's solution for minute quantities of glucose. In consequence of the changes effected in this test-liquid, which give rise to error, the following modification has been advised, which, it is stated, will keep perfectly, may be boiled for a long time with or without pure cane sugar, and may be exposed to diffused daylight, without depositing any suboxide of copper: 1 gramme neutral tartrate of copper, 40 grammes pure hydrate of sodium, and 50 grammes distilled water. The neutral tartrate of copper is procured by treating a solution of sulphate of copper, with neutral tartrate of potassium, washing the precipitate with pure distilled water by decantation, and then carefully drying it at 212° F.—W. L. Classen, states that the determination of glucose by the cupro-potassic tartrate, is a much more certain and reliable method than by the saccharimeter. E. Pollaci states that the urine should always be treated by a solution of acetate of lead, to precipitate its coloring matters and any tannin that may be present, as these substances interfere with the exactness of the result with the cupro-potassic test of Fehling (and others), by exerting a reducing action upon this test the same as the sugar itself.—E. Baudrimont states that chloroform in the urine gives a precipitate of chloral, etc., which may lead to an error in supposing sugar present when this is not the case.—2. See *Pavy's Test or Solution.*—3. Harley employs two solutions; one, a solution of potassa, sp. gr. 1.060; and the other, a solution of ten grains of pure sulphate of copper to the fluid ounce of distilled water. — *Cupric*, or *Copper Sulphate*, are recent terms for sulphate of copper. See *Ammonio-oxide of Copper.*

Corpora Amylacea. Microscopic roundish bodies with concentric layers, giving the peculiar violet color with iodine and sulphuric acid, have been found in the urine; their indications are not known.

Cotton. Very careless observers, and occasionally very careful ones, and especially in hospitals where it is often almost impossible to prevent, will find fibres of cotton in the urine, which may be mistaken for renal casts. Patients who are careless in collecting their urine, may likewise have these fibres in this liquid. Under the microscope these fibres have a flat limp appearance, with a dark medullary part, often looking like narrow glassy cylinders, varying in diameter from $\frac{1}{3000}$th to $\frac{1}{1000}$th of an inch. They are not dissolved by potassa, are dissolved by *ammonical solution of copper*, and give the blue color of iodide of cellulose with iodine and sulphuric acid; first carefully applying the acid (which causes them to swell), and then the iodine water. Nitric acid does not cause cotton to swell.

Creatine. This substance exists in the blood and in the fleshy tissues in small amount, and has been detected in the urine to the quantity of from 194. to 420. milligrammes per 24 hours. It may be considered an excretion, holding an intermediate position in the retrogressive tissue-metamorphoses between the most complex protein bodies and the simpler forms, urea, etc., being more closely allied with the latter. In the grave forms of uremia, instead of finding ammonia or urea in the blood, creatine, creatinine, and other extractives are present, and which may in a later stage of histolysis be converted into urea and uric acid. At present this substance is of no practical value in affording any clinical indications.—Creatine may be obtained, by neutralizing 200 or 300 c. c. of fresh urine with a little lime water, and then adding solution of chloride of calcium to precipitate all the phosphoric acid. Filter, and concentrate quickly in a water bath so as to eliminate the greater part of the inorganic salts by crystallization; separate the liquid from these salts, and add to it $\frac{1}{24}$th its weight of a concentrated syrupy solution of chloride of zinc, free from acid, and set it aside for several days. Crystals of creatine are formed, also creatinine mixed with chloride of zinc. Filter the liquid to separate the crystals and wash them with warm water; then dissolve them in boiling water, and treat with freshly precipitated and well washed hydrated oxide of lead, until the reaction is alkaline. The oxide of zinc, and insoluble basic chloride of lead are removed by filtration, the filtrate decolorized by animal charcoal, and then evaporated to dryness. The residue is then treated with boiling alcohol, which dissolves the creatinine, and leaves the creatine which may be recrystallized from its solution in boiling water.

Simple evaporation of urine on a water bath may give these crystals in plates or prisms according to the degree of evaporation; they are pearly or micaceous, transparent, brilliant, and splendidly color polarized light, which will distinguish them from salt crystals.

Creatine crystallizes in transparent, glistening, rectangular prisms belonging to the clinorhombic system, and which are usually connected with each other in tufts or groups. They have a pungent taste, are inodorous, permanent in the air, soluble in 74.4 parts of cold water, very soluble in boiling water, dissolve in 9,410 parts of alcohol, and not at all in ether. At 212° F., they lose their water of crystallization, and at a higher temperature are decomposed. They have no reaction on colored reagents, do not form salts with acids, and dissolve in baryta water without change, crystallizing again from the solution. Boiled for a long time in baryta water they are converted into urea, and sarcosine; boiled in concentrated acids, they are converted into creatinine, parting with 2 atoms of water.

Creatinine. Although having the same composition as creatine, less 4 equivalents of water, its properties are very dissimilar. It is found in the muscles in small quantities, and in a greater amount in urine, from 5 to 10 grains per 24 hours, and is the product of the natural or artificial decomposition of creatine. Like creatine and urea, it may be considered as an excrementitious substance, and is of no practical value in a clinical point of view, as far as known. To obtain it, take the alcoholic solution from which the creatine was directly procured, by filtration, and carefully evaporate it,—crystals of creatinine are formed. The evaporation should not be too slow, for under long exposure to heat, and in dilute solution, creatinine, by taking up 4 equivalents of water, becomes reconverted into creatine. Should the urine employed in the preceding process for obtaining these two extractives, contain albumen, this must first be coagulated, and then separated by filtration. Creatinine crystallizes in the form of colorless, brilliant, right rectangular prisms of the clinorhombic system. They have a pungent, ammoniacal taste, a strong alkaline reaction, form crystallizable salts with acids, and are soluble, in 11 parts of water at 68° F., in much less boiling water, in 100 parts of cold alcohol, and in a very small quantity of ether. Boiling alcohol dissolves the greater part of them, most of which become recrystallized as the solution cools. If urine containing creatine be allowed to stand exposed to the air for 2 or 3 weeks, the creatine will be converted into creatinine. Aqueous solution of creatinine gives reddish crystals with bichloride of platinum, and a white flocculent precipitate with corrosive sublimate. R. Maly gives a process for procuring large, brilliant, and hard crystals of hydrochlorate of creatinine by precipitation with bichloride of mercury, which it is not necessary to repeat.

Creosote. Creosote, as well as phenic acid, is stated to occasion a black deposit in the urine; and, although there is a close alliance between the creosotic compounds and indigo, the matter is but imperfectly understood. Mineral creosote gives a blue color with a weak and very slightly ammoniacal chloride of iron, while vegetable creosote gives a green color succeeded by a brown. In observing these cases of coloration in the urine, from this agent, it is important to know the character of the creosote employed, the presence or absence of cresylic acid, and, likewise, of phenic acid. See *Carbolic Acid.*

Crimnodes Urina. Urine depositing a sediment resembling bran.

Cryptophanic Acid. This acid, detected in urine by Thudicum, was considered by him to be the cause of the acid reaction of this fluid. The process for obtaining it is, to add milk of lime to the urine until it is alkaline, filter, acidify with acetic acid, and evaporate to a syrupy consistence. Allow a crystalline deposit to form, and when terminated, decant, and mix the clear liquor with strong alcohol; after the resulting precipitation has ceased, purify the dark, impure cryptophanate of lime by repeated solutions in water, filtration, and reprecipitations with strong alcohol. Finally, add acetate of lead to a solution of the salt, then add alcohol to precipitate the cryptophanate of lead from the liquid, and decompose this salt by the addition of an equivalent amount of sulphuric acid. The acid forms a transparent, amorphous, gummy, and nearly colorless mass, readily soluble in water, less so in alcohol, and decomposes the carbonates with effervescence. Aqueous solutions of its earthy salts yield an abundant white precipitate with nitrate of mercury. Thudicum believes that this last reaction shows a fault, or liability to error, in the determination of urea by Liebig's method, probably, to the amount of 5 or 10 per cent. in excess, and which renders a correction of this process (allowing for cryptophanic acid) necessary. It is a tetrabasic acid.—Pircher in applying the method of fractional precipitations to cryptophanic acid, is disposed to view it as only a mixture of principles.

Cubebs. When taken internally, imparts its odor to the urine, and is stated to occasion precipitates in this fluid, resembling albumen.

Cuminic Acid. Is voided by the urine unchanged. So are cumarinic acid $C_9 H_8 O_3$, camphoric acid $C_{10} H_{16} O_4$, and anisic acid $C_{18} H_8 O_3$.

Cyanate of Ammonia. *Ammonic,* or *Ammonium Cyanate.* When urea is decomposed by an elevated heat among the products of this decomposition are cyanate and carbonate of ammonia. If a solution of the hydrated cyanate of ammonia be evaporated to dryness, both the cyanic acid and ammonia become lost, and urea remains. By digesting this in alcohol, and evaporating the filtered solution, pure urea is obtained.

Cyanourine. *Urocyanine.* This is a term applied by Braconnot to a blue or violet coloring matter observed in urine.—An indigo-color now termed *uroglaucin,* though at one time applied to uroglaucin and urrhodine in combination.

Cylinders. See *Renal Tube Casts.*

Cystic Oxide. See *Cystine.*

Cystine. *Cystic Oxide. Vesical Oxide.* This is an abnormal ingredient of urine, but rarely found. It may occur as a sediment, and occasionally it forms small calculi, of which it is the chief ingredient. Urine containing it is generally paler than healthy urine, frequently of an oily appearance, with occasionally a greenish tint, and usually of low sp. gr. It is generally neutral, rarely acid, but soon becomes alkaline. Urea and uric acid are apt to be deficient; and the odor of the urine is similar to that of sweetbriar, but on decomposing, it exhales sulphureted hydrogen, and some ammonia. The sediments more commonly observed in urine containing cystine are, ammonio-magnesian phosphates, mucus, epithelia, and sometimes oxalate of lime; but it may be present when these are absent. Cystine is not ordinarily observed among old persons,—generally, during middle age and any period previous thereto; and, it has been stated, that its presence in the urine may be found in certain families, the disposition to its formation passing from one genera-tion to another.

When cystine is held in solution in a neutral or alkaline urine, a deposit may not occur, until acetic acid has been added, which precipitates it either in an amorphous form, or in imperfectly formed six-sided crystals. When occurring spontaneously in acid urine the crystals are in hexagonal plates, occasionally in quadrangular, and sometimes in multangular plates or rosettes, with sharply crenate margins, darker in the center than at the cir-cumference. The crystalline plates are thin, and often lie one over the other, and, probably, the rosettes are formed in this manner. (See Fig. 16.)

Fig. 16.

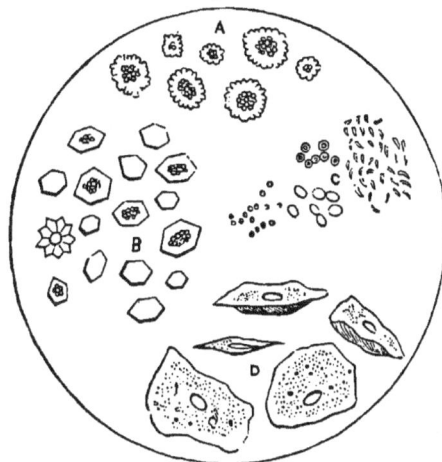

Cystine forms white, or pale fawn-colored, amorphous deposits; its crystals pre-sent a pale-yellow, or amber color, opaque in bulk, transparent in thin plates, and free from taste, odor, or reaction. Exposed to light and air it changes its color, becoming of a sea-green hue. It is soluble in ammonia, the caustic alka-lies and their carbonates, and in nitric, hydrochloric, sulphuric, oxalic, or phos-phoric acids; insoluble in acetic, citric, or tartaric acids, water, alcohol, and in carbonate of ammonia. It forms salts with the min-eral acids. Carbonate of ammonia precipitates it from its acid solutions, and

A. Cystine, as an urinary deposit.
B. Cystine, crystallized from an ammoniacal solution.
C. Vibriones, found in urine of debilitated persons.
D. Scaly epithelium from the vagina.

acetic acid from its alkaline. Its solution in ammonia evaporated sponta-neously, gives rise to magnificent crystals, which, like the others, color polar-ized light. The crystals may be determined from those of uric acid by their solubility in ammonia, and from those of chloride of sodium by their insolubility in water, and in not disappearing when the urine is heated.

Cystine more commonly exists in form of calculi having a white or yellowish, crystalline, wax-like substance, tasteless, neutral, and gritty between the teeth. Sometimes other matters mixed with it give a green or bluish color. Subjected to dry distillation, it yields hydrocyanic acid and carbonate of ammonia, a thick, disagreeably smelling oil, and leaves a spongy charcoal. It burns with a greenish-blue flame, emitting at the same time a characteristic, acid, disagreeable odor. Its other properties are named above. In the examination of a cystic calculus, if a small portion be placed upon platinum foil and burned, it may be recognized by the greenish-blue flame, the thick, white fumes, the peculiar sickening acid garlicky odor, and by its staining the platinum surface of a dark greenish-blue color, which disappears under further heating. Cystine, similar to taurine or choleic acid, is a sulphurous substance, containing 25 per cent. of sulphur, and appears to form under the influence of active emotions, fatigue, prolonged watchings, etc. Its clinical importance is not known, further than the danger of the formation of gravel and calculus.

Cystirrhea. Catarrh of the bladder.

Cystitis. Inflammation of the bladder.

Cystolithic. Uric vesical calculi, or pertaining to such calculi.

Cystorrhagia. Hemorrhage proceeding from the blood vessels of the bladder.

Cysts of Echinococci. Cysts containing hydatids (*Echinococcus hominis*) are occasionally formed in the kidneys, varying in size from 2 to 30 millimetres; they are filled with a fluid and some granular matter, in the latter of which fragments of the echinococcus may be found. The cysts have an external wall, varying in thickness from 0.22 to 2.5 millimetres, according to the size of the cyst, and which, under the microscope, appears to be formed of numerous thin layers of structureless membrane. The innermost membrane, termed *germinal*, is thin, transparent, tough, and homogeneous, and in which the echinococci are developed, being very small, ovoid, animated beings, barely visible to the naked eye. These beings are stated to constitute the "encysted phase" in the development of a minute tapeworm frequently found in the dog; they consist of a caudal vesicle furnished with a pedicle for attachment, and a tenia-like head furnished with four suckers and a double crown of hooklets. Each cyst may contain six, eight, or ten of these beings, or it may be barren. It is extremely rare that these cysts are found entire in the urine discharged, being ordinarily partially or wholly disintegrated, so that the hooklets alone may, or may not be detected, and, perhaps, flakes of membrane, and fragments of cysts. In the passage of these from the kidneys to the bladder, various unpleasant and painful symptoms are sometimes experienced, as well as during their discharge through the urethra. The detection of these hyatids or their fragments in the urine,

6

would indicate their development in the kidneys, or in the vicinity of the urinary apparatus.

Cytoid. Like a cell or corpuscle. A term applied to cells of similar aspect and nature to the white corpuscles or *leucocytes* observed in blood, and which are found in lymph, mucus, chyle, pus, etc.

D.

Damaluric Acid. An acid discovered in the urine of man, the cow, and the horse, by Staedeler. It is an oily fluid, heavier than water in which it is slightly soluble, has a valerianic acid odor, reddens litmus, and forms salts with bases, and is supposed to be one of the odorous matters of urine.

Damolic Acid. An acid discovered in urine by Staedeler, and, like the previous one, supposed to be one of its odorous principles.

Detritus. The residuum of worn-out or disorganized tissues, considerable of which, in the animal system, is eliminated through the urine.

Diabetes. *Glucosuria. Glycosuria.* A disease characterized by excessive thirst, progressive emaciation, and abundant discharges of urine containing grape sugar, and termed "diabetes mellitus." When the saccharine matter does not exist in the abundant urine, it is termed "diabetes insipidus." Sulphide of calcium ⅛th grain, and sugar of milk 3 grains, mixed and given for a dose, and repeated three times daily, has been found serviceable in diabetes.

Diuresis. A profuse discharge of urine.

Divulsio Urinæ. Urine that is slightly turbid or cloudy.

Double Staining is a process by means of which blood corpuscles, cells, casts, and the various tissues of the human body, are brilliantly and permanently colored for microscopic investigation; some tissues assuming a green, and others a blue, red, or pink color, under the process. It was introduced by F. Merkel, of Germany, and improved upon by Drs. W. F. Norris and E. O. Shakespeare, of Philadelphia. The fluids employed are:—1. Take of pure carmine 10 grains, pulverized borax 40 grains, distilled water 10 fluidrachms and 40 minims. Place the articles in a glass or Wedgewood mortar, and rub them well together; let the mixture stand for 24 hours, carefully pour off the clear liquid, and keep it in a well closed vial. *Stains red.*—2. Take of indigo-carmine 40 grains, pulverized borax 40 grains, distilled water 10 fluidrachms and 40 minims. Prepare similar to 1. *Stains blue.* The sections to be stained, having been first hardened, if necessary, and well washed with water, are soaked for 2 or 3 minutes in alcohol, and are then placed in a liquid composed of equal quantities of the above two fluids. After remaining in this liquid about a quarter of an hour, they are removed and at once placed in a saturated solution of oxalic acid, from which they are also removed in about 10 or 12 minutes, and washed with water until the washings are neutral. The stained sections, cells, or corpuscles may subsequently be mounted in Canada balsam. Mr. W. H. Walmsley of Philadelphia prepares very beautifully double-stained vegetable, and other, tissues.

Dropsy, Renal. Dropsy due to renal disease.

Dumb Bells. See *Oxalate of Lime.*

Dysuria. A painful, difficult emission of urine, usually accompanied with a sensation of more or less heat or scalding.

E.

Enæorema. The pendulous cloud or nubecula observed in urine while cooling.

Enuresis. Involuntary micturition; incontinence of urine.

Epinephelos. Urine that presents a cloudy appearance.

Epistasis. Substances that float upon the surface of urine. See *Hypostasis.*

Epithelial Cells. See *Epithelium.*

Epithelium. The thin, delicate covering of mucous surfaces, and which consists of minute cells of different forms The flattened form of epithelium is termed *scaly;* that from the vagina and external part of the female urethra, consists of large, irregular, and frequently uneven, ragged-edged cells, each containing a distinct nucleus, and if acetic acid be added to them, the granules within the cells become more indistinct. The cells are frequently folded over at the sides. See *Fig. 16.* The epithelium from the urethra is termed *columnar, prismatic,* or *cylindrical,* and is generally mixed with that in the vicinity of the meatus which is chiefly of the scaly form. See *Fig. 22,* page 129. Around the prostate they are fusiform, caudate, and irregular. The epithelium from the bladder varies ac-

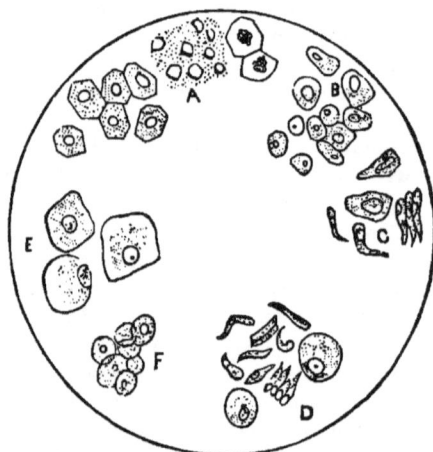

Fig. 17.

A. Glandular epithelium from the kidneys.
B. Tesselated epithelium from the pelvis of the kidney.
C. Epithelium with large and distinct nuclei, from the ureters.
D. Columnar epithelium, from the fundus of the bladder.
E. Large flattened cells, with a very distinct nucleus and nucleolus, from the trigone of the bladder.
F. Epithelium from the bladder.

cording to the part from which it is derived; from the fundus, it is *columnar*, mixed with large oval cells (*Fig.* 17); from the trigone, the cells are large, flattened, and have a distinct nucleus with nucleoli. (*Fig.* 17.) The epithelium from the ureters, is of the columnar variety, with a large and distinct nucleus; some are fusiform. (See *Fig.* 17.) The epithelium of the convoluted portion of the tubes, is termed *glandular* or *secreting;* the spherical cell is nearly double the size of a blood globule, contains a large, distinct nucleus surrounded with an apparently granular substance,—the outline of the cell, or cell membrane, is very indistinct. In the straight part of the uriniferous tube, the epithelium is flatter, more resembling the scaly form, and with a more distinct outline. The epithelium from the pelvis of the kidney is of the *tesselated* or *pavement* variety, consisting of small, round, and oval, thin flat scales, or cells, united at their edges without overlapping each other. (*Fig.* 17.)

These epithelial cells, from the mucous membrane lining the urinary passages, are found in larger or smaller quantity in normal urine, and in urinary sediments; sometimes these epithelia are isolated, at other times they are contained in mucus blood, or pus. They may be found in the enæorema or mucous filaments of normal urine, mixed with some rare mucous globules, with granulations of urates, or, with ammonio-magnesian-phosphatic crystals among persons who drink alkaline waters, the whole being held in a small quantity of mucus.—When found in the sediments, they are lost in the midst of organized or other elements that form the mass of the deposit. By treating the deposit with acetic acid, the phosphates and the urates are dissolved (precipitated uric acid appearing only after a short time), and the deposit becomes clear. The epithelial cells can now be readily distinguished, by their form and size, from other organized elements. Yellowish epithelial cells present in urine would lead to testing for biliary coloring matters.

Diagnosis. Epithelial cells must not be confounded with leucocytes or blood globules, which are much smaller; with urinary casts, which have a special form and are much larger; nor with the spores of fungi, and certain voluminous infusoria (volvox). In some rare cases, cancer cells, or tuberculous masses may be taken for epithelia, and reciprocally. But other symptoms enable us to diagnose the cancerous or tuberculous affections of the urinary passages, long before their elements appear in the urine. Sometimes the strangeness and the variety of the cellular forms found in the urine are such that a mistake is not possible; for instance, in cancer of the uterus with propagation to the bladder.—It is not always possible to satisfactorily determine the precise origin of each epithelial variety met with in the urine, because of the diversity of their forms, and the very variable dimensions of the epithelial cells of the urinary mucous membrane. It will be attempted in the following table to classify their distinctive characters, so that when these type forms are found, one may be enabled to trace them to their origin.

VII. Table of the Typical Characters of Epithelial Cells met with in the Urine.

1. Round or oval epithelial cells, a little swollen.	Very large (0 mm .016 to 0 mm .0.33) with a single large nucleus, ordinarily 0 mm .011).	Urethra.
		Bladder (bas fond).
	With 2 nuclei, or one nucleus and nucleiform granulations.	Pelvis of the kidney. Ureters (superficial layer).
	Much smaller (about 0 mm .011 to 0 mm .015).	Kidney (rarely in an isolated state). Bladder (near the neck).
2. Epithelial cells lamellar, very thin polygonal.	A large nucleus, and often two nuclei and nucleiform granules.	Pelvis of the kidney and ureters (superficial layer).
	Very large lamellæ (0 mm .022 to 0 mm .045) but with very small nucleus (0 mm .006).	Vagina, and external genital parts. Urethra (near the external orifice).
3. Cylindrical, fusiform, or battledore epithelial cells, with tail longer or shorter and more or less bent.	Bladder. Pelvis of the kidney. Ureters.	

Of course, this table can not be an absolute guide for the diagnosis of epithelia; intermediary forms and dimensions exist that escape any classification; we have had especially in view the indication of some starting points, so that, in the midst of these capricious elements, one may be enabled to bethink himself. No dismay should be felt at the measurements, as the micrometer of the microscope will readily determine them; beside, by taking as a measure of comparison, with the same magnifying power, the dimensions of a leucocyte or of a blood globule, it becomes very easy to appreciate the differences in size of the epithelial cells, with sufficient precision.

Examination of Epithelia. Be it understood that, we suppose the general characters of epithelia and of the structure of the urinary mucous membrane, are known, and will, therefore, merely indicate a very simple method by which epithelia may be procured from the urethra during life. With individuals who have just undergone a sitting for dilatation of the urethra by bougies, upon compressing the canal a milky fluid escapes, consisting of oil globules, leucocytes, and a mass of epithelial cells detached from the urethral mucous membrane by the repeated frictions. But it is much better to introduce a bougie, coated with glycerin, to a certain depth in the urethra, allow it to remain a few minutes, and then after having withdrawn it, compress the canal; it is useless to pass the bougie as far as the bladder. Sufficient epithelia will thus be procured for examination, and without the non-desirable oil globules. And upon these epithelial cells may be practised their mensuration.—To obtain vesical epithelia not mixed with that from the urethra, practise cathcterism when there is but a small amount of urine in the bladder, and in the urine thus collected will be found more or less cells.

‡ *Action of Reagents.* While acetic acid develops two, and more commonly three, nuclei, in leucocytes, it simply causes the nuclei of epithelial cells to become more conspicuous by rendering the protoplasm surrounding them paler.—Carmine colors epithelia quite well when they are not too granular. A very excellent reagent is the *picrocarminate of ammonia*, prepared as directed under *Ammonia.* If a drop of this coloring reagent be added to a sample of epithelia, on a glass slide, their nuclei will assume a delicate rose tint, while the rest of the cell will be pale yellow.

Clinical Import.—An increased amount of epithelium from the urethra indicates catarrhal or specific inflammation of the walls of this canal; from the bladder, more or less catarrhal inflammation of the lining membrane of this viscus; from the vagina, leucorrhea, or specific inflammation. When epithelia are found in the urine, coming from the ureters or kidneys, a more or less serious affection of these organs is indicated according to the amount of cells present, in connection with albumen, blood, renal casts, pus, and local or constitutional symptoms. For numerous plates showing these various epithelia, with important information regarding their indications, treatment, etc., the reader is referred to "Basham on Dropsy," and "Johnson on Diseases of the Kidneys."

Esbach's Method. This is a quick and accurate process for the quantitative analysis of urea, devised by M. Esbach, of the Necker Hospital, Paris, who claims that the results obtained do not vary more than 1.6 per cent. The apparatus employed is as follows:—

1. A *generator*, consisting of a cylindrical glass tube, having a bulb blown upon its upper third, to the capacious neck above which a ground glass stopper is very accurately fitted. From the upper part of the bulb a narrow tube passes, permitting communication between this part of the generator and the top of the gasometer. At the bottom, the generator is closed, and is attached to one side of the gasometer, at about its middle, by means of a solid glass stem. A small test tube, capable of holding 1.5, or 2, c. c. of urine, accompanies the instrument, and which is sufficiently narrow to be readily introduced within the neck above the bulb.—2. A *gasometer*, likewise a cylindrical glass tube, about twice the length of the preceding one, and somewhat larger in diameter, graduated, open below, and communicating, at its top, with the generator, by means of the narrow tube above referred to.—3. A *baroscope*, consisting of a barometer U tube, with its open extremity drawn out to a fine point, and its closed end dilated into an elongated bulb in which is contained a chemically inert gas; this gas is separated from the atmosphere by a column of mercury, above which is a drop of water that keeps it always in a state of saturation. Here are realized the conditions under which a gas is collected in the presence of water. The graduation of this instrument expresses the resultant of the atmospheric pressure, temperature, and aqueous tension.

The reagent (sodium hypobromite) is prepared by carefully measuring and adding together water 60 c. c. and bromine 2 c. c., and then, sodic hydrate (36°) 40 c. c. As this is not a permanent mixture, it is better to prepare it as needed. Some care should be had to avoid the irritating vapor of bromine, which may be effected by covering the bromine, in its flask, with water, and removing therefrom the quantity required by means of a graduated pipette.

The manner of using is as follows: A glass jar, sufficiently large to contain the instrument (generator and gasometer combined) in a vertical position, and with an arrangement near its bottom into which the gasometer may be fitted, rendering it immovable in any direction except upward and downward, is filled with sufficient water; the instrument is immersed in this water to a depth that will bring the level of the water, measured by the lower line of the meniscus, to a mark made on the narrow tube forming a communication between the generator and gasometer. This having been done, by means of a small glass funnel, pour the reagent into the generator until it rises as high as to the commencement of the lower part of its bulb. Now quickly introduce, with the thumb and index finger of the left hand, the small, narrow test tube into which 1 c. c. of the urine has been previously added, and immediately close the generator air tight by its glass stopper, at the same time dexterously pushing down the narrow test tube by means of the knob-like end of the stopper. The evolution of nitrogen commences at once, with considerable violence; to relieve the pressure the instrument is raised upwards, but not out of the water. The water in the gasometer sinks, as it is lifted up, and causes an aspiration, thus facilitating the disengagement of the gas. In about a minute the reaction is completed. The gasometer is now pushed downwards until the level of the water, inside and outside of it, coincides, and the value of the gas is then read off on the graduations of this tube. Now correct for temperature, barometric pressure, and tension of aqueous vapor, and it will be found that 1 centigramme of urea gives 3.4 c. c. (or 34 divisions) of nitrogen at 0° and 760 millimetres. To avoid all these necessary calculations, the baroscope is consulted, and the true percentage of urea present is ascertained by referring to the baroscopic tables accompanying the instrument.

Should albumen be present in the urine, as it is also decomposed by the reagent it should be separated, by boiling the urine in a test tube, and then filtering; if, however, the urine be alkaline, it must first be acidified by the careful addition of. a drop or two of acetic acid. Uric acid, also creatinine, are decomposed by the reagent, but more slowly and not so instantaneously as urea, hence, any error from this cause would be inappreciable.—W. Brewer, No. 43 Rue St. André-des-Arts, Paris, furnishes these instruments and the baroscopic tables.

Ethereal Solution of Peroxide of Hydrogen. *Ozonic Ether ?* Saturate rectified ether, to which a small amount of alcohol has been added, with a strong solution of peroxide of hydrogen; agitate gently, allow it to rest for a time, decant the ethereal solution, and keep it in well stoppered vials. This solution will keep for many months without deterioration, and, as it is highly inflammable, it must not be brought near a light, or fire. This is used in medicine as a disinfectant, stimulant, etc., and is frequently employed as a test. *Richardson.*

Solution of peroxide of hydrogen may be made as follows: 1. Dissolve finely powdered peroxide of barium in dilute hydrochloric acid to saturation. To the solution, when filtered and cooled, add baryta water until the silica and foreign oxides have been thrown down, and a faint precipitate of hydrated peroxide of barium appears. Again filter to remove the precipitates, and to the clear solution add a quantity of strong baryta water, which precipitates crystalline *hydrated peroxide of barium.* Collect this precipitate on a filter, pour on water until the washings are free from hydrochloric acid, and preserve it in a moist state in well stoppered glass vessels.

2. Treat the moist hydrated peroxide of barium with dilute sulphuric acid, constantly stirring, and adding the acid gradually, and continue this process until only a trace of free acid remains. Allow the sulphate to subside, filter the liquid, and then cautiously remove the slight excess of acid by the careful addition of dilute baryta water. Keep this *solution of peroxide of hydrogen* in well closed vessels. *Jul. Thomsen.*

The presence of peroxide of hydrogen in the fluid may be determined by the following methods:—*a.* Add a few drops of the solution of ammonio-nitrate of silver, wholly free from excess of ammonia, to some solution of peroxide of hydrogen, and then expose the mixture to a boiling heat; instant turbidity will appear, from the silver being reduced to its metallic state. *Boettger.*—*b.* Dissolve calcined titanic acid in boiling sulphuric acid, and then pour the solution into a large quantity of pure water; filter, to remove the precipitated hydrated titanic acid, which must at once be redissolved in dilute sulphuric acid. If a small amount of peroxide of hydrogen in solution be added to a little of this titanic acid mixture, it will develop an orange or yellow color. *Schœn.*—3. Fresenius states that a "solution of peroxide of hydrogen may be easily prepared by triturating a fragment of peroxide of barium (about the size of a pea), with some water, and adding it, with stirring, to a mixture of about 30 cubic centimetres of hydrochloric acid and 120 cubic centimetres of water. The solution keeps a long time without suffering decomposition."

Extractive Matters. Imperfectly defined animal matters, pre-existing in the blood, and met with in the urine when certain pathological conditions are present. Dr. Owen Rees recommends tincture of galls to be added to the urine (from which albumen, if present, should be removed by coagulation and filtration), which immediately causes a precipitate of these matters; after five or ten minutes, another precipitate ensues which is due to the action of the alcohol in the tincture upon the potassa and terreous salts. These extractives do not exist in normal urine. Their presence is indicative of some active lesion in the urinary apparatus, which permits these matters to transude.

In albuminaria, the extractive substances exist in the urine for some time previous to the presence of albumen, and also subsequent to its disappearance; thus enabling us to detect the malady in its incipient period, as well as to determine the necessity for a further continuance of treatment; as long as they exist in the urine, the prognosis is unfavorable. They may be found in urine at a very early period, when renal congestion or irritation exists, but are absent in nephritic calculus.

It may be proper to state that two orders of elements have been designated by authors as extractive matters: One, crystalloids, capable of detection and possessing well known chemical properties, as, creatinine, hippuric, succinic, phenic, taurylic, damaluric, and damolic, acids, etc., existing in normal urine; the other, observed only in pathological urine, as lactic, and butyric acids, allantoin, leucin, tyrosin, xanthin, hypoxanthin, etc. The source of these extractives is found in the incessant molecular destruction of the tissues; this origin is common to them together with urea and uric acid. They are the products of disassimilation having attained different degrees of oxidation, and may be said to represent, as it were, the ashes of the animal focus. Their physiological part is finished, they can no longer serve either in nutrition or in the formation of new tissues, and they are expelled as foreign bodies of no utility. The condition in which they are found indicates a weakness of the organism powerless to convert them into urea, the last degree in the scale of oxidations. Their presence in the urine, even in its physiological state, fully settles this view.

One is reluctant to admit, and it is contrary to the sound ideas of physiology to suppose, that, in an organism normally performing its functions, the histological elements do not attain their complete evolution. If, in the normal state, we encounter in the excrementitial fluids bodies of various degrees of oxidation, it is because they are qualified to appear in them, and they there occupy their natural and legitimate place. The difference of their origin (fibrous, muscular, nervous, parenchymatous, tissue) explains the multiplicity of their forms. In fevers, in the urine of which these extractives are always present in a plus or minus degree, a constant and most prominent character of commencing and progressive convalescence is a diminution of their quantity [as well as of that of uric acid], and an increase of chloride of sodium; this diminution of the extractive matters would appear to prove that the oxidation of the nitrogenous substances in the organism was more perfect. The increase of the chloride of sodium is probably due to the stronger and more extensive diet of the convalescent.

Hepp comprises under the term "extractive," the group of organic matters found in the urine, with the exception of urea and uric acid, in cases where we may determine these last separately. Chalvet designates as extractive matters, both of the blood and the urine, all the elements, with the exception of urea, that are soluble in absolute alcohol.

Extraneous Matters in the Urine. These are of two classes: 1. Those bodies that accidentally enter the urine after it has been discharged from the bladder, as, *hair* of various kinds, fragments of *linen, cotton, silk,* or *wool,* of *tea leaf,* of *starch,* of *feathers,* of *wood splinters,* of *oil globules,* of *chalk,* and of *sand.* —2. Those substances that enter the urine previous to its elimination, as, coloring and odorous matters from food, or medicinal agents, as, *copaiba, rhubarb, iodine, lead, mercury,* etc. The physician should especially make himself familiar with the character and appearance of those belonging to the first class, by microscopical examinations of them outside of the urine, that he may not be led into error from their presence in this fluid. Caustic potassa, or soda, 5 or 10 parts to 100 of water, dissolves wool and silk, but not flax, hemp, or cotton. Neutral chloride of zinc at 140° F., dissolves silk. Ammonio-oxide of copper dissolves the cellulose of cotton, hemp, and flax, but does affect silk or wool.

F.

Fatty Matters. These are occasionally found in the urine in different states, as, in a molecular state, noticed under *chylous urine;* in the form of concretions (*urostealith*) ; and in the form of globules, as when milk or fat is added to the discharged urine, or when derived from the organism, existing in the urine in a free state, entangled in renal casts, or, in the form of fat cells. Sometimes, being dissolved by other constituents in the urine, a chemical examination alone will be able to detect it. When the fat is observed to float upon the surface of the urine, it may be derived from within the organism, but should remind the practitioner that its presence may also be the result of catheterism, use of unclean urinary vessels, or, of design.—When the fat is held in the urine in the form of drops, in suspended granules, enclosed in epithelial cells, or in the products of exudation, renal casts, it indicates a probable fatty degeneration of the kidneys, or of the epithelial covering of the urinary apparatus, or, according to C. Bernard, an excess of fat in the blood.

Fat or oil globules when once seen and studied under the microscope, will never be confounded with any other substance. They present the form of smooth, roundish, flattened disks; sometimes, when compressed, polyhedral; they strongly refract light which gives them a sharp, dark outline with transmitted light, and, with reflected light, a whitish center, and a shining silvery outline. See *Fig.* 22, page 129, and large organic globules, *Fig.* 30, page 163. A good method of accustoming one's self to recognize these globules in this fluid, is to strongly agitate some freshly voided urine, with a little milk, and then examine a drop or so under the microscope. If a urine containing fat be evaporated to dryness over a water bath, and some ether be then added to the residue, it will dissolve the fat, and on being allowed to evaporate spontaneously, the ether will disappear, and the fat will remain. A few drops of this on fine paper will leave a permanent grease spot; it may also be tested chemically.

Remarks on Fats. The immediate principles constituting the suets and the fats, present no crystalline form in the organism, and can not be separated by the scalpel, nor by washings with water. These principles possess properties so similar, that great efforts are necessary to enable one to separate them from each other, that is to say, to analyze the tissues which they form by their combination. Suets, indeed, present organic characters which, as far as known, have their analogue nowhere else. Thus, it is recognized that the suet of beef, for instance, is composed of three substances,—stearin, margarin, and olein. Two of these substances are solid and melt only at an elevated temperature; the third, olein, is a fluid substance, a limpid oil, solidifying only at a very low temperature. It would seem that nothing could be more simple than to extract this oil by pressure, or by prolonged contact with some porous body, such as bibulous paper; but it is not so, whatever care may be taken, it will be impossible to remove from the suet a trace of olein. If bibulous paper be greased by pressure at a gentle heat, the portion of the grease retained by the paper will be equally composed of the mixture of the three substances, and in the same proportions as they existed in the suet before the experiment.

In the *analysis* of a fatty tissue, the first thing is to extract its suet; for this purpose, the fatty matter must be macerated in boiling water, and be broken up to remove the suet contained in the cells. The fat, set free, floats upon the surface of the water, forming a cake on cooling, which is to be removed. Suets always contain olein, stearin, and margarin; the latter two vary in quantity according to the nature of the tissue, the part of the body from which it is taken, and according to the animal to which it belongs.—In order, therefore, to be enabled to distinguish these principles, a small portion of the cake of suet must be dissolved in boiling alcohol. As the alcohol cools, but before it is completely cold, crystals of stearin will be observed to form first; a short time after, the margarin crystallizes. These two substances may be readily distinguished under the microscope. The olein remains in solution in the alcohol, and some of it is likewise attached to the crystals of stearin and margarin, from which it may be removed by pressing these between sheets of bibulous paper which absorb the olein, and from which it may be extracted by ether.

In certain cases this mode of analysis may suffice, but it is generally preferable to transform the suet into a soap, by boiling it with a dilute solution of soda or potassa ; the olein, margarin, and stearin, then become converted into oleic, margaric, and stearic acid, combined with the base employed. This solution of soap is then decomposed by hydrochloric acid, and now the characters of oleic, stearic, and margaric acids, which it is greatly more easy to separate from each other, are much more clearly marked, than the olein, stearin, and margarin, from which they have been derived.—To separate these three acids, dissolve the mixture in boiling absolute alcohol. While cooling, stearic acid crystals form, and may be removed; when cold, those of margaric acid form, and the oleic acid remains in the solution. The crystalline acids may be thoroughly separated from each other by successive crystallizations, and be examined microscopically.—Stearic acid melts at 167° F., and margaric acid, at 132° 8′ F., and the acid that will melt at just one of these points of fusion, may be considered perfectly pure.—To obtain the oleic acid pure, saturate it with a little litharge to form a soap of lead, dissolve this in ether, and by means of sulphureted hydrogen decompose the oleate of lead, and a perfectly pure oleic acid will be the result. Stearate and margarate of lead are not soluble in ether.

Feathers. Fragments of feathers are frequently observed in urine, probably derived from the bed or pillows. They may be recognized by the shaft of the feather, and the branched character of the barbs proceeding from it.

Fehling's Solution. See *Copper.*

Fermentation. See *Alkaline Urine.*

Fibrin. Fibrin is sometimes met with in urine. It occurs in bloody urine, either in the form of more or less voluminous clots, including some blood corpuscles and leucocytes, or, in the state of colorless coagula, sometimes solid, at others, gelatinous. The fibrinous coagulum is sometimes so large that it can not possibly be attributed to the blood in the urine as its sole origin. Vogel relates the case of a woman affected with Bright's disease, whose urine, several hours after its emission, presented a fibrinous coagulum, of a very pale-red color, at the bottom of the vessel. This coagulum contained too few globules to attribute its presence to anything else but an abnormal exudation of the fibrinous fluid (blood plasma).—In certain cases of cantharidal cystitis consecutive to the application of large blisters, shreds of pseudo-membranes are found in the urine, or greyish fibrinous pellicles, often accompanied with vesical mucus, leucocytes, and some blood corpuscles.

Under the *microscope,* coagulated fibrin possesses somewhat different characters, according to its age. Shortly after its coagulation it presents a very distinct fibrillary aspect, more or less like felted network, and slightly granular. After a short time, the fibrillary disposition disappears, and then the fibrin is seen in the form of a granular, amorphous matter, with or without a lamellar and stratified tendency, or else, it separates into small granular, split fragments, and more frequently polyhedric, irregular, with blunt angles. —Certain filaments of mucus floating in the urine, and presenting a fibrillar or striated appearance may be mistaken for fibrin. All doubt may be removed, by adding a little acetic acid to them, which does not change the aspect of the mucus (it rather increases the striation), while it causes the fibrin to swell, rendering it gelatiniform and transparent, thus allowing an examination of the anatomical elements imprisoned by it during its coagulation, as, epithelia, leucocytes, blood corpuscles. The presence of fibrin in the urine indicates an exudation of fibrinous fluid from some part of the surface of the urinary apparatus, more generally of the kidneys.

Fibrinous Calculus. Calculi have been called fibrinous, from having given the chemical reactions of fibrin. They are sometimes hard and brittle, or, like wax, of a yellowish-brown, or dark-reddish-brown color, rough, uneven surface, soluble in potassa, from which the fibrinous substance is precipitated by an excess of acid; soluble in acetic acid under heat, which solution gives a precipitate with ferrocyanide of potassium. A red heat leaves very little residue, but when blood is also present, iron may be detected in the ash. These calculi are very rare.

Filamentous Urine. Urine in which are observed filamentous or thread-like matters.

Filter. In all cases of filtration the best chemical filtering paper should be employed, and as it is frequently the case that the filter has to be ignited with the precipitate remaining on it, the best method is to cut a number of small filters, of the same size exactly, and from the same sample of filtering paper, and keep them in a covered box where no particles of dust can collect upon them. To determine the weight of the ash of such filters with sufficient accuracy for analytical purposes, ten of these filters should be ignited until every trace of carbon is consumed; then weigh the resulting ash, and divide the weight found by ten, which will give, with sufficient precision, the average quantity of ash left by each filter upon incineration, and which amount is to be deducted from the weight of the residue remaining after ignition of a precipitate and filter.—When there is no danger of a reduction of the precipitate by the action of the carbon of the filter, the precipitate with the filter are burned together. When there is danger that such reduction may occur, the precipitate and the filter are separately ignited, having first removed as much of the former from the latter as possible. The two ashes are then united and weighed, deducting the weight of the filter, as stated above.

Flax Fibres. Fragments of linen in the urine present the striated aspect of the woody fibrous portion of the flax entering into their composition. The fibres are jointed at intervals, have a round, solid appearance, and their broken ends look like a brush composed of minute fibrillæ. They are slowly soluble in *ammoniacal solution of copper*, but are not soluble in liquor potassa or soda.

Foreign Bodies in Urine. See *Extraneous Matters in the Urine.*

Fuchsin. *Anilin Red.* Take of crystallized fuchsin 1 centigramme, absolute alcohol 20 to 25 drops, distilled water 15 cubic centimetres; mix. This colors almost instantly, and without altering the tissues. It is especially adapted for the study of epithelia and pale delicate cells, rendering them more distinct. If a drop of this red solution be dropped upon a preparation of elastic fibres, all the elements become colored, but as soon as the preparation is washed with distilled water acidulated with acetic acid, the color disappears, and remains only upon the elastic fibres. This reaction is useful in determining and characterizing the elastic fibres in certain specimens of expectoration. Alcohol, however, soon extracts this coloring matter from the specimen.—Half a grain of fuchsin dissolved in four fluidrachms of distilled water, is preferred by many microscopists to the mixture with alcohol, or with acetic acid.

Fungi. These are vegetable growths found in urine, after it has stood for some time, none of which, with the exception of sarcinæ, are in the urine when voided; the most common are the *penicilium glaucum* or mould fungus, the *torula cerevisiæ*, yeast or sugar fungus, and the *sarcinæ ventriculi*. See *Vegetable Organisms*. The spores or reproductive organs of these are what we find in urinary deposits. According to Van Tieghem (*Recherches sur la fermentation de l'urèe et de l'acide hippuric*, Paris, 1864), the alkaline fermentation of urine is due to the development of a torulace consisting of globular cells united in the form of beads; these cells being very small (0 mm .001), are not granular, and no difference is found between their envelop and their contents. This ferment multiplies by gemmation, and it is never developed upon the surface of the liquid, but in its interior, or upon the walls of the vessel containing it. It is found mixed with the white deposit formed by the precipitated salts.—The organs of vegetation (mycelium) composed of more or less partitioned and ramified tubes, are only met with in urine left for a long time to putrefaction.

G.

Gallic Acid. ‡ Add a little cane sugar to the urine, and when dissolved dip a strip of filtering paper into the fluid and allow it to dry. If now, by means of a glass rod, a drop of concentrated sulphuric acid be placed upon the paper, an intense violet color will be produced, if gallic acid exists in the urine.

Gamboge. This substance taken internally will impart a yellow color to urine.

Garlic. The ingestion of garlic, as well of onions, communicates a peculiar odor to the urine, with an increased amount, in some instances, of crystals of oxalate of lime.

Gas in Urine. According to M. Morin 100 volumes of urine contain on an average 2.44 volumes of gas, consisting of carbonic acid 65.40, oxygen 2.74, and nitrogen 31.86. The first named, being a product of combustion, has its percentage increased after violent exercise.

Globules. The blood globules are red, and among them are found pale or white globules called *leucocytes*. in the proportion, in a normal state, of 2 or 3 to 1,000 of the red globules. Bernard considers the red globules to be the respiratory element of the blood, and the white its plastic element.

Globulin. The colorless, albuminous fluid, or coagulable matter of the blood corpuscles.

Glucogene. *Hepatic Dextrine.* An animal starch, strongly resembling that from the vegetable kingdom; all albuminous substances must pass through the transitional stage of glucogene, before they can be converted into sugar.

Glucose. *Fruit Sugar. Fructose.* By some the term is applied to grape sugar. It is the uncrystallizable sugar of honey, grapes, and fruits, but not found in diabetic urine, although frequently used to express the saccharine matter in that fluid. Glucose may, however, be converted into grape sugar by molecular changes. See *Grape Sugar; Sugar.*

Glycerin. This agent renders almost all histological elements more transparent, and has a somewhat analogous action in this respect to acetic acid. It renders nuclei and elastic fibres very evident, in the midst of elements it has caused to swell and become more transparent. It is very useful in the investigation of hairs which it clears so as to allow the detection of vegetable parasites which at times infiltrate their roots, and even their medullary canal. Mixed with equal parts of acetic acid, glycerin constitutes a valuable agent in the study of all parasites, especially the Acarian. It is also used in combination with gelatin, and other agents, for preserving or coloring various specimens.

Glycocholic Acid. See *Bile Acids. Cholic Acid.*

Glycocoll. *Glycocine. Glycin. Sugar of Gelatin.* May be artificially obtained from gelatin by the action of mineral acids; from hippuric acid by boiling with hydrochloric acid; and from *cholic acid,* which see. It forms large, hard, colorless crystals, of the oblique rhombic system, containing nitrogen but no sulphur, which are sweet to the taste, dissolve in 3 or 4 parts of cold, and less of boiling, water, are unchangeable in the air, and fuse and decompose at a high temperature. Glycocoll unites with acids and bases.

Glycose. Another name for sugar in diabetic urine. See *Glucose.*

Grape Sugar. The crystallizable sugar of grapes, fruit, and honey, and, likewise, forms the sugar met with in diabetic urine. It may be separated from evaporated diabetic urine by boiling the resulting extract in alcohol. Grape sugar is white, inodorous, less sweet than cane sugar, gritty between the teeth, crystallizes in rhomboidal prisms, soluble in $1\frac{1}{3}$ parts of cold water, freely in boiling, insoluble in absolute alcohol, but soluble in diluted, and its solution turns a ray of polarized light to the right. At 212° F., the crystals melt and lose their two equivalents of water, and at 284° F., they are converted into *caramel,* $C_{12}H_{18}O_9$. See *Sugar.*

Gravel. A minute form of *calculi.* Small concretions which form in the kidneys, pass through the ureters, and are discharged with the urine. They more commonly consist of urates and animal matter; occasionally of oxalate of lime, phosphates, etc.

Gravidin. A urinary deposit with pregnant women, the decomposition of which, as stated by Stark, results in the formation of the pellicle upon the surface of the urine, known as *Kiesteine,* which see.

Guanine. A substance occurring in small amount in guano, and also met with in human urine. It is one of the intermediate series occurring in the regressive metamorphosis of nitrogenized tissues, which terminates in the formation of urea, uric acid, and carbonic acid. It is a yellowish-white, crystalline powder, without taste or odor, insoluble in water, ether, and alcohol, feebly soluble in hydrochloric acid, and in caustic soda, without action on vegetable colors, and can bear a temperature up to 392° F. without decomposition or loss of weight. Its reactions are similar to those of *xanthine*.

H.

Hair. This is occasionally met with in urine, rarely originating from its formation in the urinary passages, but more frequently in cysts, discharging into the bladder; and much more commonly, from its introduction through vesico-vaginal fistula, or designedly. Hair from blankets, from cats, etc., is sometimes found in the urine; the best method for promptly determining them, when present, is to examine preparations of various kinds of hair under the microscope, that they may not be mistaken for renal casts, etc., when in the urine.

Hæmaphæin. A name given, by Simon, to the amber-yellow coloring matter of healthy urine, and supposed by some to be modified hematin. See *Urohematin.*

Hæmatin. *Hæmatosin.* Hematin is a coloring matter of the blood, wholly distinct from *hemoglobin*, and is a product of the decomposition of the latter blood-coloring substance. It is of a blueish-black color, with a metallic luster; the black-chocolate colored, or, coffee-like matters vomited in certain affections, owe these dark colors to its presence, as do likewise certain fluids in a pathological condition. An old blood stain does not give the spectrum of hemoglobin, but that of hematin. Hematin may appear in the urine with blood corpuscles (see *Blood in Urine*), or, the corpuscles may be disintegrated and dissolved in the urine, giving more or less color to this fluid, according to circumstances (see *Urohematin*). Alkaline solutions of hematin are of red-brown color in mass, and green in a thin layer; acid solutions, present a red-brown color whatever may be the thickness of a layer or mass.—Recent investigations by Paquelin and Jolly, appear to have shown that blood corpuscles contain iron in the form of tribasic phosphate of the protoxide, and that their coloring matter, hematosin, contains no iron whatever.

Hæmatinuria. Urine containing not blood corpuscles, but only the coloring matters of blood (hematin); it is met with in several affections, as purpura, malignant scarlet fever, pyæmià, scorbutus, etc. The urine is usually of a chocolate color. The blood is supposed to be in a decomposed or dissolved condition.

Hæmatocrystallin. Blood (or hemogloblin) crystals, prepared by placing a drop of blood on a glass slide, and after an exposure to the air for 10 or 15 minutes, a drop of water is added, the whole breathed upon 4 or 5 times, then covered with thin glass. and allowed to evaporate slowly in sunlight. *Funcke.* It is preferable to use defibrinated blood. The crystals are of the prismatic form, belonging to the rhomboidal system, and require from an

hour to several days for their formation, according to the different specimens
of blood. A spark of electricity passed through blood, also gives rise, after
a longer or shorter time, to the formation of these crystals; the blood of the
Guinea pig crystallizes the most readily.

Hæmatoglobin. *Hemoglobin, Hemoglobulin, Hematoglobulin, Cruorin.*
Terms applied to the red coloring matter of the blood, which is the only
element of this fluid that contains iron; it may be separated into *globulin*
and *hematin.* Hematoglobin exists in the blood of all vertebrated animals, being enclosed in the red corpuscles of this fluid; it likewise exists
in the blood of some of the invertebrata; but, with a few exceptions, not in
the form of red globules, but in solution in the plasma, as, with the blood
of the earth worm. It is found in the blood of some annelids, the larvæ of
certain insects, in branchipus stagnalis, planorbis corneus, etc.

Disintegrated blood consists of the globulin (the colorless albuminous fluid
of blood corpuscles), and the red coloring matter of blood, which may, by
certain manipulations, be obtained in crystals—*hæmatocrystallin.* Hematoglobulin, or disintegrated blood, coagulates at 204.4° F., while albumen
requires a temperature of about 145° F.; the former, exposed to heat, becomes
deoxygenated and converted into hematin and globulin. It is not positively
determined whether the disintegration of the blood corpuscles takes place in
the circulation, or at some other place (the kidneys); but the presence of
hematoglobin in the urine augurs more or less unfavorably, according to its
persistency and amount. In typhus, scorbutus, purpura, etc., its presence in
large amount is an indication of danger, more especially should urinary
suppression and discoloration of the skin ensue.

As to the presence of the coloring matters of blood in the urine, without
blood globules, Vogel gives his opinion as follows: "The passage of the
hematoglobulin into the urine may be explained thus,—In the organism some
blood corpuscles are being continually decomposed during the metamorphosis of the tissues, and consequently hematoglobulin is set free. When this
metamorphosis proceeds in a normal manner, the small quantity of hematoglobulin set free in the blood, is, in its turn, transformed; the globulin,
serving for the nutrition of the muscles and other protein tissues, is finally
eliminated from the body in the form of urea and uric acid. The hematin
becomes equally changed and oxidized, and, probably, terminates by being
separated from the organism under the form of urinary and biliary pigments,
so that when the metamorphosis of the tissues follows its normal course,
hematoglobulin never passes into the urine. But when, under pathological
influences, large quantities of blood globules are suddenly decomposed, the
amount of hematoglobulin then existing in the blood is so great that the
whole of it can not undergo the normal changes referred to above; and,
under these circumstance, a part of the hematoglobulin may pass unchanged
into the urine, just as is observed to be the case with other substances not
ordinarily found in this fluid, as sugar, bile matters, and probably albumen,
which, when in excess in the blood, may appear in the urine."

7

Clinical Import. The presence of hematoglobulin in the urine, indicates an excessive decomposition of blood globules, due to one of two causes:—1st. The cause of the blood decomposition is *temporary,* the destruction being limited to a greater or smaller quantity of blood corpuscles ; the prognosis is favorable.—2. The cause of the decompositon is *permanent;* there is then produced a real dissolution of the blood, endangering life, and the prognosis is unfavorable or doubtful.—From the observations of Meckel, Heschl, Frerichs, and especially from the excellent investigations of J. Planer, we learn that in certain cases, and very probably when a great quantity of hematoglobulin is set free, a granular pigment may accumulate in the blood, and, by obstructing the capillary blood vessels, especially those of the brain, occasion serious consequences. It appears advisable, therefore, before giving a prognosis in these cases, to examine the blood under the microscope, to ascertain whether it contains any such granular pigment. M. Gubler has based his hypothesis of *hemapheic icterus,* upon the rapid and abnormal destruction of the corpuscles, with accumulation of coloring matter in the blood. See *Blood in Urine. Microspectroscopy. Urohematin.*

Hæmatoidin. Small rhomboidal prismatic crystals found in the decomposed blood of ligatured blood vessels, and in that of extravasations; they are the last product of decomposed blood, are red or yellow, insoluble in water, alcohol, ether, acetic acid, glycerin, dilute mineral acids, and dilute alkalies. They are derived from hematin. By some erroneously considered the same as hematocrystallin.

Hæmaturia. The discharge of blood by urine ; the red corpuscles being present in their natural form. See Table, p. 250.

Hæmin. Crystals that may be obtained from fresh, putrid, or dried blood, and even from the oldest blood stains, and hence of value only as a test for blood. To procure them, separate the stain (from a supposed blood stain), place it in a test tube, add a drop or two (or more, according to amount of stain) of glacial acetic acid to it, boil it for a few seconds, filter a drop upon a glass slide, add another drop of acid, and set aside in a warm place to evaporate. They crystallize in rhomboidal tables of a blackish-brown, or, rarely light-brown color, and dissolve in caustic potassa.—Erdman macerates the suspected stain in water, and slowly evaporates the solution on a glass slide; a minute crystal of common salt and a drop of glacial acetic acid are then added to it, and again gently evaporated to dryness over a spirit lamp; when cool, a drop of the acid is again added, and under the microscope, if the stains were of blood, will be seen hemin crystals, in the drop of acid. They vary from yellow to red according to their thickness, and are soluble in caustic potassa.

Hæmochromogene. A provisory name given by Hoppe-Seyler to a substance intermediary between hematoglobulin and hematin, and found during the decomposition of the former. Its color is red purple. The existence of this substance has not been satisfactorily investigated; probably it is identical with the coloring matter in urine, the *urochrome,* of Thudicum. Hemo-chromogene is formed by the action of alkalis or acids on hematoglobin, when protected from the air; under the action of the air, hematin, and not hemo-chromogene is the result of the decomposition.

Heat. Heat applied to urine in a test tube, causes a precipitate of *phosphates*, or of *albumen;* if the *former*, the addition of a little nitric dissolves it; if the *latter*, it is not dissolved. If a precipitate is already formed in the urine, and is dissolved when this fluid is heated, it is formed of *urates;* if the precipitate does not become dissolved, and the addition of acetic acid causes it to disappear, it consists of *phosphates;* or, if the acid fails to dissolve it, *cystine* or *oxalate of lime* may be present. The addition of nitric acid, instead of acetic, will cause a disappearance of the precipitate, if it consists of *phosphates*, or *oxalate of lime* crystals, while *cystine* remains intact. If caustic potassa be added to the urine containing a precipitate, and on heating it, ammonia is disengaged, the precipitate may consist of *ammonio-magnesian phosphates.*

Hipparia. Urine containing an excess of hippuric acid.

Hippuric Acid. Traces of this acid exist in normal urine, in the form of alkaline hippurates. But it is sometimes met with uncombined in the deposit of a urine, and can be detected and recognized under the microscope. Certain articles of vegetable diet, as apples, prunes, mulberries, cloudberries, cowberries, etc., as well as Peru and Tolu balsams, quinic, benzoic, and cinnamic acids, give rise to it in human urine. It may also be found during a milk diet, and in the urine of diabetic patients, according to the following process of M. Icery:—Precipitate the uric acid by hydrochloric acid, and examine the precipitate obtained under the microscope; if hippuric acid be present, long colorless prisms with four surfaces will be seen, which may be tested by the reactions hereafter named. If ten grains of benzoic acid be administered to a person on going to bed, hippuric acid will be found in the next morning's urine. This acid exists in large amount in the urine of herbivorous animals. Fresh urine must be taken, because, under the influence of putrefaction it becomes converted into benzoic acid. It may readily be procured for examination in the urine of the horse or cow. There is nothing positively known regarding the clinical importance of this acid in the urine. It proceeds, in part at least, from metamorphosis of the nitrogenized substances of the body.

Urine containing hippuric acid is neutral, feebly acid, or alkaline, of sp. gr. 1.006 to 1.008, having an odor like whey, and often containing ammonio-magnesian phosphate. This acid can be obtained by evaporating the urine containing it to a few drops, then adding about half the bulk of hydrochloric acid, boiling the deposit that occurs in a few hours, in alcohol, then filtering, and evaporating. Hippuric acid is yellowish when mixed with animal substance, but when perfectly pure, is colorless. It is soluble in ether, from which it is precipitated in needle-like crystals or prisms by evaporation; more soluble in alcohol, and less soluble in cold than in hot water. Its solutions redden litmus paper. It is soluble in warm nitric acid, and in hot sulphuric and hydrochloric acids, from which it crystallizes on cooling; but if these acid solutions be boiled for a time, on cooling benzoic acid is formed in crystals, and glycocoll remains in the solution. A strong heat fuses it into a red oily substance having the odor of Tonka beans, at the same time decomposing it into benzoic acid and benzoate of ammonia; at a higher heat it evolves an intense hydrocyanic acid odor, and leaves a porous, combustible, coaly mass.—It may be distinguished from ammonio-magnesian phosphates by adding a drop of acetic acid to the drop of sediment on a glass slide, which dissolves the phosphates, and leaves the hippuric. acid crystals intact. —It may be determined from uric acid (sometimes the needles of hippuric acid are fixed like spears on the larger crystals of uric acid), by collecting

Fig. 18.

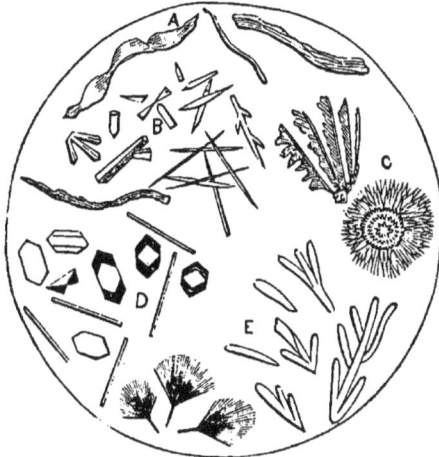

the sediment on a filter, boiling a small portion of it in alcohol, which dissolves only the hippuric acid. Upon allowing a drop of this alcoholic solution to evaporate on a glass slide, crystals of hippuric acid will be obtained, and which may be further distinguished from uric acid by not giving the murexide reaction.

Microscopical Characters.— Hippuric acid crystallizes in the form of rhomboidal prisms, which are long, glistening, transparent, four-sided, parallel to the longest axis, and the ends presenting 2 or 4 bevelled surfaces; sometimes they are in fine needles and scales. Their elementary form is always a right rhombic prism. They may sometimes be mistaken for crystals of uric acid, or of the triple phosphates. See *Fig.* 18.

A. Ordinary forms of hippuric acid, when benzoic acid is administered.
B. Different forms of hippuric acid in healthy urine.
C. Crystals of hippuric acid, by evaporation of an alcoholic solution.
D. Crystals do., from an aqueous solution.
E. Crystals do., after the action of hydrochloric acid on urine holding hippuric acid in abnormal proportion.

Husemann's Test. To a sample of the suspected urine add an equal volume of concentrated sulphuric acid, and allow it to stand for 12 or 15 hours at a temperature of 62° to 75° F.—or, still better, expose it for half an hour to a temperature of 212° F. On cooling, a faint violet-red color will be observed, if *morphia* or *narcotina* be present; a light red, if *brucia;* and a blue purple, if *codeia.* Now add a drop of nitric acid to the fluid (or one of the following, chlorine water, chloride of iron, solution of chlorinated soda, or a small piece of nitrate or chlorate of potassa). A blue to violet red is soon produced, changing into a dark red if *morphia* be present; the same reaction if *codeia*. A bright pink red or carmine indicates *narcotina*. If the liquid contain *brucia*, it must be warmed, and chloride of tin be added to it, when a yellow color will gradually appear, which changes to an intense purple on the addition of nitric acid. This test has been proposed chiefly for the detection of morphia, of which it will indicate the presence of one-hundredth of a milligramme.

Hydrated Deutoxide of Albumen. A modified albumen first obtained and described by Dr. Bence Jones. The urine gave no precipitate on boiling, nor with nitric acid; but when the urine became cool a precipitate occurred, which was immediately redissolved by heat. A similar substance occurs in the buffy coat of inflamed blood, in the secretion from the seminal vesicles, and in the albuminous fluid of pus. As this was found in a case of mollities ossium, Dr. Jones suggests examining for it again in similar cases; it might be detected by the addition of nitric acid to the urine, causing it to assume a red color.

Hydrochlorate of Ammonia. *Chloride of Ammonium. Ammonic*, or *Ammonium Chloride.* This salt when taken internally passes out unchanged in the urine, though some observers have not been able to find it in this fluid. It is, like other chlorides, always in solution. It dissolves mucus, but does not coagulate albumen, and, according to Beale, its presence in saccharine urine prevents the precipitation of the suboxide of copper, when the copper tests are employed for the detection of sugar; in these cases, a solution of potash caused an evolution of ammoniacal fumes and a precipitate of the suboxide. This salt can be detected in urine containing only one of the two urinary salts, viz., urate, or ammonio-magnesian phosphate, and whichever of these two salts is present must first be separated, that the hydrochlorate only may remain. This salt crystallizes in octohedrons, but in presence of urea it crystallizes in the regular system of the cubic type, and sometimes in elegant arborescent forms, with swollen extremities; the swollen extremity of each arborization distinguishing it from those of chloride of sodium,—though there may, probably, be a mixture of the two chlorides. But little is known concerning the part performed by this salt in the urine. To detect chloride of ammonium, Simon states that the following appears to be the most appropriate method:—Evaporate the alcohol extract of urine, dissolve a portion of it in water, and add a solution of caustic baryta. If ammoniacal salts are present, a strong odor of ammonia will be developed. Neither pure urea, nor the nitrate, on being similarly treated, gives off this ammoniacal odor. But the detection of ammonia does not prove its existence in the urine as a chloride. See *Ammonia.*

Hydrochloric Acid. This acid destroys calcareous substances, renders the margin of blood globules clear and distinct, also fibrin, which it causes at first to swell, and then dissolves it; acid 1 part to 4 parts water. It also dissolves cystine, phosphates, hippuric acid, and oxalate of lime, and precipitates albumen, and uric acid from its salts; it also renders evident the contents of naviculæ and their disposition, causing them to become blue or green, and likewise arrests the movements of vegetable cells. A little water added to concentrated hydrochloric acid, sufficient to render it less fuming, dissolves the intercellular substance of the kidneys, and thus enables to readily separate the tubes; this usually requires from 12 to 24 hours.—The same as *nitric acid*, hydrochloric acid vapors render ammoniacal vapors white. Hydrochloric acid should be used, under the ordinary microscope, as seldom as possible, because it very rapidly injures the metallic mounting of the objectives; and when it has been thus used, the examiner must not forget to immediately cleanse the objective with a piece of fine old linen, or soft chamois skin. With a chemical microscope, in which the stage and glass slide or vessel is above the objective, the danger from corrosion is considerably less. These remarks equally apply to all mineral and corrosive acids and vapors. It must be recollected that hydrochloric acid favors the acid fermentation of urine, and the development of certain kinds of microscopic vegetable organisms which act as yeast cells on the urates, rapidly decomposing them. Nitric acid entirely prevents this action.

Standard Solution of Hydrochloric Acid. For the Determination of Lime. Take 1 gramme of carbonate of soda, heat it to redness, allow it to cool, weigh it, and dissolve it in distilled water. Add a few drops of tincture of litmus to color the solution blue, heat the fluid to boiling to drive off free carbonic acid, and, still keeping at boiling heat, add from a graduated pipette or burette, dilute hydrochloric acid, until the blue color has passed into a pale red, that does not disappear with continued boiling. Suppose that, by calculation, we find that 1 litre of the dilute hydrochloric acid employed, corresponds with 41.6 grammes of carbonate of soda; then 457 c. c. of the acid will saturate exactly 18.9 grammes of carbonate of soda. Therefore, we measure off 457 c. c. of the dilute hydrochloric acid, and mix it with enough distilled water to make the whole measure exactly one litre, or 1,000 c. c. of standard solution. Of this solution 1 c. c. exactly neutralizes 10 milligrammes of lime. It will always be proper to ascertain the accuracy of the solution by one or two confirmatory testings with carbonate of soda.

Hydrothion. *Sulphureted Hydrogen.*

Hydruria. Urine deficient in solid matters. Diabetes.

Hyperchlorosodie. An excess of chloride of sodium in the system.

Hyperuresis. Incontinence of urine.

Hyperurorrhea. Increased flow of urine. Diabetes Insipidus.

Hypochlorosodie. An absence or great diminution of chloride of sodium in the system.

Hypostasis. The spontaneous precipitation of sediment in urine.

Hypoxanthin. *Sarkine. Carnine.* Hypoxanthin invariably accompanies xanthin in the organs and in urine, and is a lower oxidized product than xanthin or uric acid, though it never forms a spontaneous urinary deposit.

Like xanthin, creatine, etc., it forms one of the intermediate steps in the regressive metamorphosis of azotized tissues. Scherer found it in the spleen, in the urine of leucocythemia, and in the heart, and other organs. It was named hypoxanthin from the fact of its containing one equivalent less of oxygen than xanthin, to which it bears a strong resemblance. Between the hypoxanthin derived from the organs, and that from the urine, there exists, according to Thudiaum, some differences in a few of their reactions, although alike in all other respects. As nothing is known of its indication when existing in urine, a further reference to it is unnecessary.

I.

Incontinence of Urine. *Enuresis.* Incapability of retaining the urine, which flows out involuntarily.

Indican. See *Uroxanthin.*

Indigo. See *Color of Urine. Uroglaucin. Urrhodin.*

Indigo-glucin. A saccharine substance separated from uroxanthin or indican when this is decomposed by mineral acids, in order to procure its blue and red pigments. This indigo-glucin is capable of reduction by oxide of copper, but has no action under the fermentation test.

Indigose. See *Uroxanthin.*

Infusoria. See *Monad. Bodo.*

Inosite. *Muscle Sugar.* A hydrocarbon or unfermentable sugar found in muscle and flesh, corresponding with grape sugar, but containing 2 atoms more water when crystallized. It has not been detected in healthy urine, but has been in diabetic, and albuminuric. A similar substance was detected by Vohl, in the unripe beans of *Phaseolus vulgaris*, and which he named "phaseomannite." It may be obtained as follows:—To 30 or 60 grammes of diabetic urine add a saturated solution of the tribasic acetate of lead as long as a precipitate ensues. Filter the mixture, and wash the precipitate on the filter so long as any sugar passes, and to the clear fluid, which contains the inosite, add solution of basic acetate of lead which precipitates an insoluble lead compound of inosite. Let the fluid stand for 24 hours, filter, and wash with distilled water so long as any soluble matter is extracted. Mix the precipitate with 30 or 60 grammes of distilled water, and pass a current of sulphureted hydrogen through the mixture so long as an insoluble precipitate of sulphide of lead occurs, and filter. Evaporate the clear filtrate to nearly dryness, add a drop of nitric oxide of mercury, and apply heat; if inosite be present the fluid assumes a fine rose color,—if ordinary urine be employed, in which inosite is rarely found, uric acid interferes with the test, and consequently the first precipitate must be made with saturated solution of neutral acetate of lead; as well as in albuminous urine, which, however, must first be deprived of its albumen by acetic acid and heat. *Gallois.* 2.

Inosite may readily be detected by evaporating a few grammes of urine to a syrupy consistence, and then adding a few drops of nitrate of mercury; if inosite be present a yellowish precipitate ensues which, on being gently heated, assumes a pink color. On cooling this color disappears, but reappears under heat. The nitrate of mercury is made by dissolving one part of mercury in two parts of nitric acid, and then adding one part of distilled water.

Inosite forms in colorless, transparent, cauliflower-like crystals massed together in groups, which belong to the clinorhombic system; they appear as groups of fine needles, as stars, or fan-shaped; though sometimes they are found single, and 3 or 4 lines in length. Exposed to the air they lose their water of crystallization, and become dull and opaque on drying. Inosite has a sweet taste, is readily soluble in water, but insoluble in alcohol, or ether. Boiling alcohol dissolves it, but on cooling it crystallizes in small glistening particles. It fuses at 410° F., forming a clear liquid mass, which, if rapidly cooled, crystallizes in needles, but if cooled slowly hardens into a horny amorphous mass. It does not undergo vinous fermentation with yeast; with putrid cheese it yields butyric and lactic acids It does not brown with potash, nor reduce the oxide of copper, like grape sugar. If its solution is evaporated on a platinum spoon, with nitric acid, nearly to dryness, and the residue be treated with ammonia and a little chloride of calcium, and then cautiously evaporated to dryness, if even ½ milligramme of the inosite be present, a vivid rose-red color is produced. This reaction is not produced with cane, grape, or milk, sugar.—Inosite has been found in the urine of albuminuric, syphilitic, and diabetic patients, but its clinical importance has not been ascertained. Perfectly healthy urine does not contain it. Gallois, who has bestowed much attention upon this subject, is of opinion that inosite in the urine may be the result of an incomplete performance of the gluco-genic function of the liver.

Inosuria. The voiding of urine containing inosite.

Iodine. Tincture (as well as solution) of iodine, in contact with starch gives an intense blue color. It indistinctly colors all organized elements, as, epithelia, leucocytes, urinary cylinders, etc., both cells and nuclei, a greenish-yellow, rendering the vibratile cilia of spermatozoids, epithelial cells, and infusoria more distinct under the microscope, at the same time arresting their movements; the vegetable elements, as, spores, and filaments of algæ and fungi (leptothrix, etc.) have their contents colored brown by it, while the envelop is not affected. The tincture gives a yellow-brown color to prostatic nitrogenized concretions, as well as to most of the nitrogenized substances termed, "albuminoids."—*Solution of Iodine; Iodine Water*, used in urinary investigations is of two strengths, viz.:—1. Take of iodine 5 parts, iodide of potassium 15 parts, distilled water 3,000 parts (all by weight). Mix the last two and then add the first. Label, "*Solution of Iodine for Coloring.*"—2 Take of iodine 1 part, iodide of potassium 1 part, distilled water 50 parts; mix. This iodine water enables us to detect a peculiar form of degeneration, amyloid infiltration, at the same time it allows us to verify the amyloid bodies of the brain, and to distinctly characterize starch grains derived from the vegetable kingdom. It forms a beautiful blue color with pure starch, guaiac resin, and the pollen of plants; the color changes from violet to purplish when the starch is mixed, or altered by fermentation or heat. The presence of a carbonate, or of an alkali in a solution containing starch, entirely prevents the reaction of iodine, hence, we should be careful to acidulate the solution before submitting it to the reagent.

When iodine is present in urine it may be detected by the following processes:—1. To 5 grammes of the suspected urine, add from 2 to 5 drops of a mixture of pure nitric acid, of 36° (sp. gr. 1.32), and concentrated solution of hypochlorite of lime an equal part. If the urine is colorless, and contains iodine in some quantity, it will instantly assume a yellowish color, due to the iodine set free. If the amount be very minute, add 2 or 3 grammes of bisulphuret of carbon, and strongly shake the mixture, the fluid becomes turbid, red droplets collect on the walls of the tube, and finally fall and form a more or less thick layer on the bottom. The intensity of the color is proportioned to the quantity of iodine dissolved in the bisulphuret. *Degauquier.*—2. If the urine be alkaline, first neutralize it with sulphuric acid, and then add a solution of starch (or starch paper may be dipped into it). This effected, add chlorine water, drop by drop, when the blue color of the iodide of amylum formed, will become manifest. There is no necessity for adding the sulphuric acid unless the urine be alkaline; yet, as urine is apt to possess certain organic elements that prevent the reaction existing between iodine and starch, and that even decompose the amylum iodide when once obtained from this reaction, it will generally be found better to disorganize these elements by the addition of nitric, or sulphuric acid, previous to the application of the test. If a solution of nitrate of silver be added, instead of the chlorine water, a yellowish-white precipitate of iodide of silver at once occurs.—Other tests might be named, but it is not necessary.—3. Prof. Rienzi uses the following reagent: 1 gramme of bichloride of mercury is dissolved in 30 grammes of distilled water; to this a solution of iodide of potassium is added in sufficient quantity to redissolve the precipitate formed. Five drops of this mixture will detect infinitessimal traces of quinia in half a cubic centimetre of urine two hours after 50 centigrammes have been taken internally. The addition of 5 drops of sulphuric acid to the urine will reveal the presence of quinia an hour and a half after 15 centigrammes of the alkaloid have been taken. We can thus determine whether the patient is still under the influence of this medicine. See *Bromine and Iodine in Urine.*

Iron. Iron is a constituent of urohematin, one of the coloring matters of urine, and, if a large amount of healthy urine be acted on, more or less iron (traces) may be detected. Thus, take one or two litres of normal urine, and evaporate to dryness; continue the heat until the whole is reduced to an ash. Dissolve the ash in a little very pure hydrochloric acid, triturate it in a glass or porcelain mortar, and after 15 or 20 minutes, add half its volume of distilled water, boil for a few seconds, filter, and divide the filtered solution into two portions. To one portion add a drop or two of nitric acid, boil, and add, drop by drop, sulphocyanide of potassium (1 part dissolved in 10 parts of distilled water), according to the quantity of iron present, the fluid will assume a scarcely observable reddish tint to a deep red.—To the second portion, add a drop of nitric acid, boil, dilute with a little water, and add, drop by drop, ferrocyanide of potassium (1 part to 10 of distilled water), according to the quantity of iron present, an immediate precipitation of Prussian blue will occur, or blue flakes or light clouds of this blue will become visible in from one to several hours. This last part of the process requires neutralization of the fluid, previous to adding the ferrocyanide of potassium, by solution of carbonate of potassa, because acid liquors are very apt to color the ferrocyanide blue.—Unless the hydrochloric acid used in this operation be very pure, no satisfactory result will be obtained; it is difficult to procure this acid free from iron, and the presence of this metal in it would render the operation useless. See *Urohematin.*

Preparations of iron taken internally, greatly increase the amount of this metal in the urine, and may be detected by the ordinary tests for iron with the usual reagents; or, 10 c. c. of the urine may be evaporated, in a small porcelain capsule on a water bath, to dryness, and then gradually heated to redness. The matter swells up, disengages odorate products, and when all disengagement has ceased, there remains only a black coal, which should be allowed to cool in the capsule. With a small wooden or platinum spatula scrape the coal from the capsule, pour some chemically *pure* hydrochloric acid upon it, triturate with a small glass or porcelain pestle, and then allow it to rest for 10 or 12 minutes. Now add one-half its volume of distilled water, boil for a few seconds, constantly stirring, and then filter. Divide the filtered liquid into two portions, and proceed as in the method just named above. Iron, in considerable quantity, can be found in the urine, after the tartrate of iron, and other of its salts, have been taken for several successive days. We may, consequently, study comparatively the facility of absorption of the martial preparations. Many of them simply pass through the digestive tube and are passed with the feces; in such case, they will not be found in the urine, and it is very presumable that they do not enter the circulation. From thence may be comprehended the uselessness of prolonging a treatment which is often exceedingly inconvenient. On the contrary, if iron, in a greater amount than exists in the coloring matter, be detected in the urine, it renders it absolutely certain that the metal has been absorbed; and when it ceases to manifest itself under the ordinary reactions, it is because it is no longer absorbed.

Quantitative Analysis.—Evaporate in a platinum vessel, 100 c. c. of urine, and heat to perfect carbonization; burn this until the whole of the carbon is driven off and a perfectly white residue remains. When cool, dissolve the mass in very pure hydrochloric acid, and heat it; add water to it, and a little, q. s. sulphite of soda, and boil until the fluid becomes colorless, and no trace of sulphurous acid can be detected in it. Carefully pour into a clean flask, and add enough distilled water to bring the solution to about 60 c. c. in quantity. When thoroughly cool, place the flask upon a sheet of white paper, and, from a graduated pipette, drop into it the standard solution of permanganate of potash, until the fluid assumes a pale rose-red color, when the process is terminated. Now, if 1 c. c. of the permanganate solution corresponds with 0.0005 gramme of iron, and, in the preceding operation, but 3 c. c. of it were employed to occasion the test color, the 100 c. c. of urine, from which the ash and its solution were obtained and prepared, will contain 3×0.0005 gramme, equal to 0.0015 gramme of iron. This, multiplied by 1.43, will give the corresponding quantity of peroxide, and, by 1.286, of protoxide. Should the red color occasioned by the last drop of permanganate solution disappear after a short time, no attention should be given to it, as it is due to changes not connected with the analysis. The sulphite of soda added to the solution, above described, is for the purpose of reducing the peroxide of iron to the condition of suboxide or protoxide.

Ischuria. Suppression of urine; also, an incapability of voiding it when contained in the bladder.

J.

Jumentous. A term applied to colored, turbid, sedimentary urine, of a strong, rank odor, like that from a beast.

K.

Kidney, Bright's Disease of. See *Bright's Disease.*

Kidneys. These organs, in health, as well as disease, remove from the blood certain effete matters, in the formation of which they have not the slightest influence, as, urea, urates, uric acid, oxalates, phosphates, sugar, etc., the presence of which elements in the urine, according to their character, quantity, and quality, indicate a normal or abnormal condition, not of the kidneys, but of the system generally,—a healthy or unhealthy state of the nutritive or disassimilative process. On the other hand, an actual disease of the kidneys may affect the general system in proportion to its capability of removing the effete matters, referred to, from the blood; of which matters, urea is considered the most important. Albumen in the urine, especially when renal tube casts are also present, indicates, in most instances, a more or less serious structural lesion of the kidneys themselves, and which may affect the character and quantity of the urine in various ways. In the examination of this fluid, the modifications it may undergo from physiological influences alone, the morbid conditions of the organism that may occasion an excess or diminution of its normal elements, or develop new, abnormal matters in it, and the influences exerted upon the nature and amount of its healthy and unhealthy constituents by renal lesions, should be constantly kept in mind.

Kiestein. This is a name given by Nauche to a whitish, iridescent pellicle, often observed upon the surface of urine of pregnant women several hours, or even a day or two, after its discharge. The etymology of the word indicates the relation that was supposed to exist between pregnancy and the appearance of this albuminoid pellicle; a relation that is by no means constant. Indeed, this pellicle manifests itself upon the urine of non-pregnant women, as well as upon that of man, whenever the urine commences to putrefy and become ammoniacal. An entirely analogous phenomenon is produced upon fluids in which exist decomposing animal or vegetable substances. Numerous observations have been made on this subject, with the following result:—In water in which bones are macerated, a peculiar turbidity will be observed, which is determined by the development in the liquid of various infusoria, and minute vegetable cells, cells of Protococcus, and which are endowed with extremely rapid movements. After a day or two, the fluid becomes covered with greenish patches, consisting of these same cells, agglomerated, immovable, and increased in size; if a drop of the liquid be taken, a little deeply, with a pipette, and be examined under the microscope, the green cells will be found still in motion, and will, at a later period, ascend and join those at the surface, thus enlarging the green pellicle, which will soon cover the entire surface of the fluid. At the same time, some of these cells will implant themselves upon the bones, and there develop themselves.—A similar order of action occurs in vegetable infusions. If water, in which a bouquet, for instance, has been steeped, be exposed to the air, this water soon evolves a disagreeable odor, and, if a drop of it be examined under the microscope, thousands of bacteriform infusoria, and volvoces in quantity, will be found swarming and moving about in the fluid drop. Several hours or days afterwards, according to the temperature and various other circumstances, a whitish pellicle will be formed, in which the previously moving bodies will be observed, immovable, and more or less deformed; they are the cadavers of our infusoria of the previous examination.

Now, this is exactly what occurs in urine, which likewise contains nitrogenous matters, and which, sooner or later, undergoes putrid fermentation. Solely because the urine contains more crystallizable salts, will the pellicle be incrusted with them, and so thoroughly that after a short time it will break and settle to the bottom of the vessel, being carried down by the crystals. Then it will be replaced by another pellicle, which will be destroyed in the same manner. The fragments of the pellicle become mixed with the somewhat old sediments, thus rendering it somewhat difficult to draw them up with a pipette. Any one can ascertain the correctness of these statements by allowing urine to pass into putrefaction, and carefully examining the changes that occur.—Kiestein consists of vibrios, spores, epithelial debris, crystals of urates, amorphous phosphate of lime, crystallized ammonio-magnesian phosphates, fatty particles, etc., and is simply an epiphenomenon of the putrefaction of urine, supervening more or less rapidly. Prof. Regnault believes that the production of kiestein, with pregnant women, is due to an excess of nitrogenous materials in the urine. This will explain the more rapid fermentation of the pellicle during the state of pregnancy. But this pellicle has no semeiological signification; it has been observed among anemic, scrofulous, and other patients, shortly after the emission of urine. Of course, the condition of the atmosphere and its temperature have a controling influence in the production of kiestein.

Kletinsky's Cupro-potassic Test. *For Sugar.* Take of pure solid caustic potassa 4 parts, glycerin 3 parts, saturated solution of sulphate of copper 2 parts; mix. Boil in a test tube with the urine, if glucose be present, a reddish-fawn color will be produced. The sugar removes one equivalent of oxygen from the deutoxide of copper, and precipitates the red-fawn colored protoxide. Although this is a very delicate test, from the fact that crystallization of sulphate of potassa from the solution is unavoidable, thus preventing its being at once prepared for quantitative purposes, it presents disadvantages exceeding the advantages.

Knapp's Test. *For Sugar.* This test is an application of the fact first observed by Liebig, that an alkaline solution of cyanide of mercury is completely reduced by glucose to metallic mercury. This test is proposed as superior to Fehling's, as the test liquid is easily prepared and is absolutely permanent, the application is quicker, the results are accurate, and less time is required. Take of pure dry cyanide of mercury 10 grammes, distilled water a sufficient quantity to dissolve it, then add to the liquid a solution of, caustic soda, sp. gr. 1.45, 100 c. c., and dilute with distilled water to measure one litre. 400 milligrammes of cyanide of mercury, in a boiling alkaline solution, is completely reduced by 100 milligrammes of desiccated glucose. 40 c. c. of the alkaline solution of cyanide of mercury (equal to 400 milligrammes of cyanide) are placed in a porcelain capsule and boiled; while boiling, the urine is dropped in until all the mercury is reduced. The quantity of urine required for this reduction will correspond to 100 millgrammes of glucose. In order to know when the mercury is completely reduced, as the urine is carefully added, a drop of the boiling liquid should from time to time be placed upon a clean piece of Swedish filtering paper, and be exposed to the action of a drop of diluted sulphuret of ammonia, or to a vapor from a concentrated solution of this sulphide; when this procedure no longer develops a brown color the reduction is completed.

L.

Lactic Acid has been found in some specimens of healthy urine, in small amount. This has been denied by Liebig, but other investigators have confirmed the statement concerning its presence, under certain circumstances, advanced by Berzelius. To determine the presence of lactic acid in the urine its zinc salt affords the most ready method: Evaporate fresh urine on a water bath to the consistence of thick syrup, and treat this with an alcoholic solution of oxalic acid. Oxalate of lime, soda, potassa, and urea are precipitated, while there remains in the solution, phosphoric, hydrochloric, oxalic, and, if any be present, lactic, acids. The fluid is now digested with an excess of hydrated oxide of lead, and the precipitate of chloride, phosphate, oxalate, and the excess of the oxide of lead, separated by filtration. The filtrate containing lactate of lead, is treated with sulphureted hydrogen, and, after filtration, is boiled with oxide of zinc; again filtered, and allowed to stand until crystals of lactate of zinc form. The presence of lactic acid is readily determined by the peculiar bellied, barreled, or club-shaped form of these crystals; and from them other salts may be prepared. Other processes have been given, but it is unnecessary to name them; Lehmann, in his " Physiological Chemistry," Vol. I, has given a method by which lactate of lime is formed, being recognized by its double brush-like crystalline forms. The clinical signification of lactic acid in urine has not been satisfactorily determined; it may indicate a great supply of lactates to the blood, or, an acid fermentation occurring in the urinary passages. Lactic acid is said to be more or less constantly present in the urine of those who labor under pulmonary catarrhal affections, and in most febrile diseases, or where oxidation is effected imperfectly. Lehmann found it always present in urine containing much oxalate of lime.

Lateritious. A term applied to the brick-dust like sediment of urine, as well as to the urine itself.

Lead. *Plumbum.* In cases of lead poisoning, and the therapeutical use of lead preparations, this metal may pass into the urine; though its detection is not always successful. Parke states that a peculiar yellowish color of the urine is sometimes observed. Ollivier advises the following process for its detection :—500 or 600 grammes of the suspected urine is treated by nitric acid, then heated, and after addition of the same acid, is calcined. The residue is mixed with distilled water, left at rest for several hours, and is then thrown upon a double filter. Pour several drops of hydrosulphide of ammonium into the filtered liquid, until a precipitate is no longer formed; collect it upon a filter, wash, and dry it.—Again treat the dried precipitate with nitric acid and apply heat, which redissolves it   The solution is then diluted with distilled water, filtered, the filtered liquid concentrated, and a little solution of iodide of potassium added to it. If a yellow precipitate falls we may rest assured that it is the iodide of lead, and this proof confirms the first determination by the hydrosulphide of ammonium.—For Reagents, see *Acetate of Lead.*

Leeks. Impart a garlicky odor to urine, when eaten in considerable amount, and also increase the amount of oxalate of lime crystals.

Leptothrix Buccalis. A microscopic, filamentous, chain-like, slowly moving, if at all, alga, observed in putrid or decomposing urine, and in the fluids of the mouth, and of which, bacteria and vibriones are supposed to be various phases of development. Other allied species are also sometimes observed, as, leptomitus. See *Algæ*. *Vegetable Organisms*.

Leucin. *Aposepidin*. This substance is a normal product of some of the organs, as of the spleen, liver, pancreas, lungs, brain, etc.; that is, it has been met with in these organs after death, but whether as a morbid product of decomposition, is not satisfactorily determined. It is likewise encountered in diseased conditions of these organs, in pus, thickened toe nails, perspiration of the feet and axillæ, etc., being the result of decomposition; with exception of that in the spleen. It appears, therefore, to be both a physiological and pathological element of various fluids and solids of the body, being frequently found in association with *tyrosin*. Leucin, as well as tyrosin, belongs to the transition products in the metamorphosis of albuminous or nitrogenized substances. Dr. Harley considers them to be likewise the products either of the arrested, or of the retrograde, metamorphosis of glycocholic (crystallizable) and taurocholic (non-crystallizable) acids, and standing in the same relation to each other in disease as these two bile acids do in health. In urine containing leucin and tyrosin, there is always a decrease in the amount of urea. When leucin is present in urine, it may be detected by evaporating and concentrating the urine, and then examining a drop of it, when cool, under the microscope. Or, when present in minute quantity, it may be obtained by treating a large amount of fresh urine with solution of subacetate of lead, filtering to remove the precipitate formed, and then removing all traces of lead from the filtrate by sulphureted hydrogen. Again filter, evaporate the clear filtrate over a water bath to dryness; extract the residue with boiling alcohol, filter, and evaporate the filtrate over a water bath to the consistence of syrup. On standing for a day or so, tyrosin will crystallize in needles; at a later period, leucin will appear in yellow spheres. In this operation the urine must be used as soon as voided, because decomposing animal matter converts leucin into valerianic acid; and if the urine be albuminous, before proceeding to the operation, it should be heated, and the coagulated albumen be removed by filtration.—It is unnecessary here to describe the methods for obtaining pure leucin, as it is almost invariably impure when obtained from urine.

Microscopical Characters. Impure leucin, as first obtained from animal fluids, appears in yellow spheres or disks, with striæ radiating from their centers, strongly reflecting light, and having a tendency to aggregate and form warty-like masses, the margins of the spheres in contact with each other appearing as if fused together. Sometimes they resemble oil globules, but these are more highly refracting, present a broader marginal outline, and are soluble in ether. Neubauer observes that the spheres of impure leucin are in part concentrically striped, and some of them finely pointed. These spheres may be determined from microscopic crystals of lime, by floating in water—the lime crystals sink.

Chemical Characters. Impure leucin is partially soluble in water, less so in alcohol, and insoluble in ether. It is readily soluble in concentrated acids and alkalis. Acetic acid renders it more soluble in water and alcohol.—Pure leucin is white (the yellow coloring of that found in urine being due to bile pigment), in crystalline scales, inodorous, tasteless, has a fatty feel to the fingers, and possesses the above named characters in great perfection. When cautiously heated to 338° F., in a glass tube open at both ends, it sublimes in thick flocculent fumes, without previous fusion, and which, like oxide of zinc, float along the tube and escape into the air. A solution of nitrate of mercury does not cause a precipitate in an absolutely pure solution of leucin,—any precipitate thereby formed indicates the presence of *tyrosin*, that is, if the supernatant fluid has a reddish or a rosy-red color. Scherer gives the following test for leucin, pure, or even not quite pure:—Place a small quantity on a platinum spatula, add a little nitric acid, and carefully evaporate. A colorless and almost imperceptible residue remains. Treat this residue with a few drops of caustic soda solution, and apply heat, to dissolve the leucin, which. according to its purity, will form a transparent or yellow fluid. Concentrate this fluid over a spirit lamp. and in a short time it becomes converted into an oily-looking drop, which rolls about on the spatula, neither moistening nor adhering to it. However, enough leucin can rarely be had from a urine under investigation, to admit the application of this very characteristic test ; in such an investigation, we have to rely entirely upon its microscopic appearances.

Clinical Import.—Leucin has been occasionally met with in the urine of patients suffering from typhus, as well from small-pox. Its most important signification, however, is, in poisoning by phosphorous, and in that fatal hepatic malady termed "atrophy of the liver,"—chiefly in its acute form, and frequently in the chronic, also in ramollissement of the liver. In the acute form of atrophy, leucin is usually in abundance in the urine. See *Tyrosin.*

Leucocytes. *White Blood Corpuscles.* Colorless nucleated blood corpuscles observed in the blood, an increased development of which is present in that abnormal condition of the system termed "leucocythemia." It has been supposed that these white corpuscles were formed of chyle and lymph (the plastic element) previous to their being converted into red corpuscles (the respiratory element of the blood). The term *leucocyte* has also been proposed as a synonym for "sarcophyte" and "*cytoid*;" thus, leucocytes in mucus, mucus globules, or, in pus, pus globules, etc. The diameter of leucocytes varies from the $\frac{1}{3000}$th to the $\frac{1}{1400}$th of an inch in diameter. See *Blood in Urine.*—Since the employment of Hayem and Nachet's Hematimetre, in counting the red and white blood corpuscles, it has been observed by Malassez, Fouassier, Bonne, Brouardel, and other investigators, that the formation of pus in a focus always coincides with a great increase of leucocytes in the blood, which rapidly disappears after the opening, and evacuation of the contents, of this focus,—as, with all suppurative diseases.

Clinical Import.—When leucocytes are observed in albuminous or purulent urine, no other element being present, they would lead to a suspicion of inflammation of some part of the urinary apparatus, as, cystitis, nephritis, pyelitis, etc. If they exist in non-albuminous urine, with epithelium and mucous casts, there is probably some irritation or diseased condition of that part of the urinary apparatus from which the epithelia are derived, with a morbid discharge of mucus. If they are but few in number, with crystals of ammonio-magnesian phosphate, the irritation or morbid condition is very probably confined to the bladder alone. See *Mucus; Pus.*

Lime. Ca O. *Calcium.* Ca. Lime (calx) exists in urine in the form of *carbonate, oxalate, phosphate, sulphate,* and *urate;* more commonly as a phosphate. And in sections of country where limestone water, or water impregnated with lime, is used as a common drinking water, it is by no means rare to encounter oxalate of lime and phosphatic calculi. In cases so disposed to such calculi, it would be much better and safer to drink soft or rain water, or else distilled water flavored with toast, a little infusion of tea or coffee, some orange juice, or even a little wine, to render it more palatable. The use of water impregnated with lime, frequently interferes with the action of many therapeutical agents used in the treatment of diseases of the urinary apparatus, and often those of the digestive. I am aware that these views are in non-accordance with those of several writers, I therefore give them simply as the results of my own investigations and experience in practice.

In health about 120 to 200 milligrammes of lime pass with the urine voided in every 24 hours. To determine the presence of lime in urine, filter 200 c. c. of the urine to be investigated into a beaker glass; add ammonia until there is no longer any precipitate, and then carefully add acetic acid to dissolve this precipitate of earthy phosphates, adding a few drops of the acid in excess. To this solution add solution of oxalate of ammonia, and allow the glass, covered, to stand in a warm place until all the precipitate has subsided. Filter to separate the oxalate of lime precipitated, washing it well with distilled water, and preserve the filtrate and the washings for the determination of the *magnesia.*

The precipitate, together with the moist filter, is now placed upon a small platinum crucible, dried, and exposed to a strong red heat until all the carbon is consumed. The lime residue, partly caustic, is introduced into a small flask (using a very little water, if necessary), to which 10 c. c. of the *standard solution of hydrochloric acid* are then added, and heat applied until it is all dissolved, and the carbonic acid be driven off. A few drops of tincture of litmus are then gradually added to color the solution light red; when the *standard solution of soda* is to be carefully added until the blue color returns. —Now, subtract from the 10 c. c. of hydrochloric acid which have been added, the number of c. c. of soda solution that have been employed, and we obtain the number of cubic centimetres neutralized by the lime, each c. c. of which corresponds with 10 mgrms. of lime. If 100 c. c. of urine have been used in the operation, by multiplying the number of c. c. of standard acid neutralized by the lime by 10, we directly obtain the percentage of lime contained in the urine.

8

Bouchardat considers lime a good reagent for the detection of sugar in diabetic urine. He places 50 grammes of the urine in a beaker or matrass, adds to it 5 grammes of lime, and boils the mixture. The fluid assumes a caramel color, being darker in proportion to the quantity of sugar present. By this method he states that 5 grammes of sugar can readily be detected in a litre of urine. The reagent is made by slacking quicklime with water and then putting it into a bottle, corking it tightly.—When the above named boiling does not color the urine, 5 grammes more of the lime is to be added, and the whole again boiled; if it does not then become colored, it only remains to be certain that the lime solution is good. For this purpose, throw into the urine in the matrass, a dessert spoonful of honey, or starch syrup, and boil. The urine must then become decidedly colored, which is evidence that the lime was well calcined in sufficient amount, and that before the last boiling, the urine contained no saccharine substance.

Milk of Lime. This name is applied to a solution of lime in water, the lime being in excess so as to give a milky appearance to the fluid. It should always be thoroughly agitated previous to using it.—See *Carbonate, Oxalate Phosphate, Sulphate,* and *Urate, of Lime.*

Linen. Linen fragments may fall into urine, and be mistaken for casts, etc. See *Flax Fibres.*

Lithate. Another term for urate.

Lithia. *Lithon.* This is the oxide of lithium, Li, some of the salts of which have been used in medicine, as, the carbonate of lithia, which, from its soluble power upon uric acid and urates, has been found useful in calculi, in which these substances enter, as well as in cases where they form a greater or less amount of the deposit in urine. Bromide of lithia, useful in irritability of the bladder, and in the neuroses.

Lithic Acid. Uric Acid.

Lithodialysis. *Litholysis.* The dissolving of a bladder calculus.

Lithonephritis. Nephritic calculus, in which the urine contains urate of ammonia, or uric acid, in form of small calculi or gravel.

Lithos. A stone or calculus.

Lithoxiduria. Urine containing uric oxide.

Lithuria. *Uric Acid Diathesis.* That tendency of the system occasioning the deposition of uric acid and urates in the urine.

Lot. *Lotium.* Synonyms of urine.

M.

Magnesia. *Magnesic or Magnesium Oxide.* This is the oxide of magnesium, Mg. It is met with in the form of carbonate in the urine of certain animals and in human calculi; and almost invariably in the urine of man in combination with phosphoric acid, occasionally, with uric acid. For the qualitative determination of magnesia, see *Calculi, Analysis of; Carbonate of Magnesia; Ammonio-magnesian Phosphate*, the presence of the crystals of which is evidence of the existence of magnesia in the urine. In health, from 120 to 300 milligrammes of magnesia are passed by the urine in every 24 hours. The quantitative determination of magnesia by the solution of perchloride of iron, is not a very accurate one; the determination of it as a pyrophosphate is the better process: Take the filtered liquor, that was preserved during the operation for estimating the lime in urine after having separated the oxalate of lime (see *Lime*), and add ammonia to it until it has an alkaline reaction, and, if magnesia be present, there will be a precipitate of ammonio-phosphate of magnesia. After precipitation of these crystals has ceased, filter, and thoroughly wash the precipitate on the filter with distilled water to which a little ammonia has been added. Collect the precipitate in a platinum capsule, and expose it to a red heat; if any remains on the filter, burn this by itself and add the ash to that of the precipitate. (See *Filter.*) Now expose the whole to a red and white heat, which converts it into pyrophosphate of magnesia. The amount of this pyrophosphate, less the weight of the filter ash, multiplied by 0.3687, will give the amount of magnesia. See *Carbonate, Phosphate, Urate,* and *Biurate Hydrate, of Magnesia; Ammonio-magnesian Phosphate.*

Matracium. A urinal; a vessel or flask for the reception of urine as it flows from the urethra, and of which there are several varieties.

Matula. Same as the preceding.

‡ **Maumene's Reagent,** Cut woolen fabric, as merino, containing no cotton nor linen, into small strips, and soak them for 5 or 10 minutes in a solution of perchloride of tin 1 part, distilled water 2 parts. Dry the strips over the water bath, or in a drying oven. This is used for the detection of rather a strong proportion of sugar in the urine, by letting a drop of the suspected urine fall upon the wool thus prepared, drying it, and then exposing it to the dull red heat of an alcohol lamp, or in front of a hot fire. If sugar be present, a black spot is produced. This will detect 1-20.000th of sugar.

Melanin. *Melanogene. Melanose. Black Pigmentary Matter.* The name given to a solid organic substance characterized by its black or russet-brown color, and which exists as a pigmentary substance in many parts of the body, either in cells, or in the state of free granulations. It has also been met with in the urine in cases of melanotic cancer. When this substance is present in urine this fluid, from slow oxidation, gradually becomes of a dark color; but if an active oxidizing agent be added to it, as, chromate of potassa, sulphuric, nitric, or chromic acid, the freshly passed urine becomes instantly black. Upon being extracted, this substance forms a dark, coherent powder, soluble in concentrated alkalies, insoluble in water, concentrated acetic acid, and in nitric and hydrochloric acids, when these are sufficiently diluted not to decompose it. Ammonia is its best solvent. Upon long standing, or on boiling, it communicates a dark color to water, and many other fluids in which it does not dissolve. It possesses neither taste nor odor, though when heated it exhales a disagreeable odor. It is composed of carbon, hydrogen, nitrogen, oxygen, and less iron than exists in hematin. Melanin in the urine may aid in the diagnosis of obscure cases of melanotic cancer, when accompanied by other corroborative circumstances and symptoms. However, we must be careful not to consider every case of black urine as due to melanotic cancer, as it frequently accompanies other diseases, as, purpura, scorbutus, palustral poisoning, etc., and almost invariably is of an unfavorable signification, indicating a local disorganization of blood corpuscles in the kidney alone, or in the kidney and liver. Indeed, melanin, like the other coloring pigments of the urine, is derived from the coloring material of the blood, and, hence, may be present in urine, as one of the colors of urohematin. Albumen is frequently present in brown and black urine resulting from various forms of disease, as, in jaundice, cardiac disease, etc. See *Color of Urine.*

Melanourin. A name given by Braconnot to dark brown coloring matter in urine from which its cyanourine had been extracted. This urine was of a brownish yellow, but the color disappeared upon heating it, and a deep black sediment was deposited. In this urine, no uric acid was found. It is only one of the coloring matters derived from the blood, or from uroglaucin. See *Color of Urine. Uroglaucin.*

Melanuria. Urine containing black coloring matter. See *Color of Urine. Melanin.*

Melituria. *Melithyperuria.* Diabetes. Diabetic Urine.

Mercury. *Hydrargyrum.* Mercury passes off largely by the kidneys, and may be detected in the urine for a long time after it has been taken. It does not appear to augment the amount of the urinary constituents in health. Thudicum states that when the urine contains mercury, a peculiar albuminous substance is present, as well as a substance giving the reaction of sugar; but, so far as I know, this condition has not been confirmed by other observers. Mercury may be detected in the urine as follows:—1. Strongly acidulate about 350 grammes of the urine with hydrochloric acid (about ⅛th its volume), concentrate by evaporation to a small volume, filter, and then boil the filtered liquid with a small slip of bright copper wire or foil.—Even in very dilute solutions the copper becomes covered with a layer of mercury resembling silver; if the copper be washed, dried, and then heated in a small reduction tube, the mercury volatilizes, yielding a sublimate of metallic globules, while the copper returns to its original red color.—M. Ludwig adopts a somewhat similar method; to half a litre of the urine he adds 2 or 3 cubic centimetres of hydrochloric acid warmed to 140° F., and then adds 5 grammes of granulated zinc, stirring briskly with a glass rod for a minute or so. The mercury amalgamates with the zinc, is precipitated, removed by filtration, washed with hot water, and dried in a water bath. Place the dried substance in a combustion tube drawn out at one end, heat it while a strong current of air is passed through the tube; the mercury volatilizes, is collected in the capillary portion of the tube, and, if allowed to come in contact with iodine vapor, gives fine ruby crystals of iodide of mercury.—2. Since the researches of Merget upon the constant vaporization of mercury at the ordinary temperature,—researches carried on by means of extremely sensitive paper reagents,—we possess a means for readily determining the presence of traces of mercury. He impregnated paper with a solution of nitrate of silver in ammonia, which must be constantly kept in the dark. In contact with mercurial vapors, in the dark, this paper gradually gets darker and darker, until it becomes black from the reduction of the silver salt to the metallic state.— 3. Attach a piece of platinum wire to a nail, immerse it in the suspected urine, and add pure sulphuric acid, until there becomes a gradual evolution of hydrogen gas. If mercury be present, it then becomes deposited upon the platinum wire in the metallic state. After a time remove the wire from the fluid, wash it with distilled water, and then expose it to the action of chlorine vapor, which converts the metallic mercury into the bichloride. Now, having ready prepared a strip of filter paper moistened with a 1 per cent. solution of iodide of potassium, rub this gently with the wire on which the bichloride remains, when a red color will be produced, the biniodide of mercury, and which color may be removed at once by applying an excess of iodide of potassium. *Mayençon* and *Bergeret.*

4. M. Byasson, has proposed the following reagent, for the detection of mercury in urine: Take of bichloride of platinum 2, chloride of gold and sodium 3, distilled water 500; mix, and preserve in dark glass vials, as the salts in solution are reduced by direct light, diffused light, ammoniacal vapors and even by organic matters in the shape of dust. It would probably be as well to completely cover the vial with a black, varnished paper. To obtain the paper reagent, several lines are to be traced upon some white Berzelius paper (filtering paper), or other, by means of a goose-quill pen moistened with the solution, being careful to operate in a dark room, and to avoid touching, the lines made, with the fingers. Allow the lines to dry in a dark place,. as in a tightly closing drawer or box. The lines thus made, will, under the influence of traces of mercurial vapors, rapidly appear, and in from 20 to 30 minutes, assume a deep black tint.—The operation is as follows: The bright copper foil having remained in the acidulated urine in which the presence of mercury is sought for (as referred to heretofore in 1), is taken out and washed with distilled water. It is then dried between folds of bibulous paper and introduced into the bottom of a new, or perfectly cleansed, test tube of copper. At the superior part of this tube, and without being in contact with the copper foil, are placed 2 or 3 pieces of the above-mentioned paper reagent; the tube is then incompletely closed with a cork, and *very gentle heat* applied to its lower part. If mercury was present in the urine, in a few minutes the lines on the paper will appear of a yellow-brown color, soon becoming black.—M. Byasson recommends, as more sensitive than copper foil, the use of a Smithson's pile; gold leaf rolled around a rod of tin. But, the gold leaf will require to be removed, or at least calcined, for each

experiment. So that although the copper foil is less sensitive, it answers the purpose, and has the great advantage of being less expensive. And there is no reason why the copper test tube may not be replaced by one of colored glass, such as are now made, covered, if necessary, by copper foil, which, not undergoing directly the action of the mercurial vapors, may serve indefinitely.—In using this test, we must be careful not to operate in a place where mercury exists in a free state, as, mercurial ointment, mercury jars, receivers, etc.

1. *Standard Solution of Pernitrate of Mercury, Used in the Determination of Chloride of Sodium.* Take of perfectly pure mercury 17.06 grammes, put it into a beaker glass with pure nitric acid, and

Fig. 19.

dissolve it in a sand bath, having the beaker covered with a large watch glass When fumes of nitrous acid cease to be evolved, and a drop of that fluid tested with chloride of sodium ceases to show any cloudiness from formation of subchloride of mercury, the solution of oxide of mercury, thus obtained, may be evaporated in the beaker containing it, on a water bath, until it has a syrupy consistence; then add distilled water sufficient to make the whole measure exactly a litre. Should the separation of basic salt render the solution turbid, it may at once be cleared up by the addition of a few drops of nitric acid. The next step is to graduate this solution—render it volumetrical. Having previously dissolved 20 grammes of pure fused chloride of sodium in one litre of distilled water, and 4 grammes of pure urea in 100 c. c. of distilled water, we

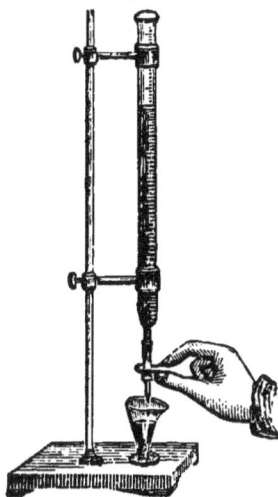

Burette and Stand for Volumetric Analysis.

proceed as follows: Of the previously prepared chloride of sodium, measure 10 c. c. into a small beaker glass and then add to it 3 c. c. of the solution of urea, and 5 c. c. of a cold saturated solution of pure sulphate of soda. Next, fill a graduated pipette or burette with the above-mentioned solution of mercury, and let it fall into the mixture in the beaker glass, drop by drop, constantly agitating it with a glass rod, until a distinct and permanent precipitate or opalescence occurs, when the test is complete.—If 7.8 c. c. of the mercury solution only have been used to effect this precipitate, it is too concentrated to give accurate results, and consequently should be diluted with an equal volume of distilled water. If 15 c. c. of this diluted mercurial solution are required to produce the desired cloudiness in the above mixture in the beaker glass, we must add to every 155 c. c. of the solution of mercury, 45 c. c. of distilled water. By this means, a solution of pernitrate of mercury is obtained, of which 20 c. c. indicate 200 milligrammes of chloride of sodium, or, 1 c. c. of 10 milligrammes. In this process every drop let fall causes a white precipitate in the soda and urea solution, and which is redissolved by stirring the mixture, but, as soon as all the chloride of sodium is decomposed, and the insoluble precipitate of nitrate of mercury commences to form, the opalescence is not removed by stirring but remains permanent, when the process is finished.

2. *Standard Solution of Proto-nitrate of Mercury, Used in the Determination of Urea, by Liebig's process.*—Take of dry, chemically pure oxide of mercury [sufficiently pure to leave no visible residue when volatilized on a platinum capsule], 77.2 grammes, place it in a porcelain basin, and by means of a gentle heat dissolve it in the smallest possible quantity of chemically pure nitric acid. When the solution is effected, carefully evaporate to the consistence of syrup. Remove from the heat and add distilled water enough to make the whole measure exactly 1,000 c. c , or a litre. Should there be a precipitate of any basic salt, add a few drops more of nitric acid until it becomes redissolved. One c. c. of this solution precipitates 0.01 gramme of urea.

Methæmoglobulin. *Methæmoglobin.* This is an intermediate condition between hemoglobin and hematin, and may be observed when the former is subjected to the action of carbonic acid gas, when deprived of oxygen, or when it ceases to circulate, as, in extravasations of blood. In hemorrhages from large vessels the urine is brighter and contains more hemoglobin ; in hemorrhages from the capillary vessels, it is darker, of a brownish-red color, and contains more methemoglobin. The determination of this substance as being derived from blood, or, as being one of the steps in the progress of its dissolution, is the same as for *hematin, hemin crystals.*

Microscope. At the present day the microscope forms an important and valuable instrument for the physician and student; and he who does not avail himself of its utility in the diagnosis and treatment of disease, especially when of an obscure or obstinate character, is greatly behind the present era of his profession. Indeed no practitioner, who is not conversant with the revelations made by this instrument concerning the structure of the various parts of the system in health, and the changes effected upon these structures by disease, can lay claim to the title of a " respectable physician." It has often occurred that during the initial stages of severe and fatal maladies, the microscope has afforded information which, even before any appreciable symptoms had become developed, revealed to the practitioner the alarming condition towards which the patient was more or less rapidly hastening, and thereby led him to adopt an efficacious course of treatment long before the pathological lesions had become serious or of a permanent nature. But the value of the microscope does not cease with medicine, it has proved of immense service in geology, botany, mineralogy, chemistry, and indeed in nearly every department of science, besides having proven exceedingly advantageous in the detection of adulterations in food and drugs, as well as of filth and impurities. Unfortunately some have conceived the microscope to be a mysterious instrument, capable of being managed or understood only by certain particular versons; this is a great mistake; the instrument is intended solely as an aid or improvement to our sense of sight. Objects which can be seen well by the naked eye, do not require its assistance, save for the investigation of their minute structure ; but with those which are too small to be thus seen, the power of vision is greatly aided by its employment, and every individual possessed of sight can readily avail himself of such assistance. Parties with imperfect vision employ lenses or spectacles to improve this sense ; and the microscope, a combination of lenses, improves the magnifying and defining powers of the eye, almost enabling it, as it were, to observe and scan the invisible. Many parties possessing a microscope, instead of perfecting themselves in its use and its revelations, allow it to remain peaceably in its case, until they are suddenly aroused to demand some information from it ; a satisfactory response is not obtained, and the conviction is then had that the instrument is of no service. This al-

Fig. 20.

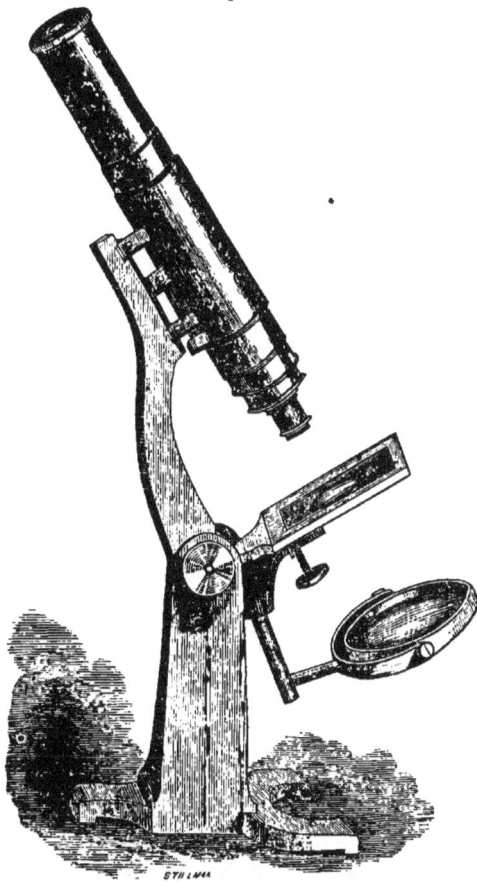

Students' Microscope.

most always happens with those who have not, by practice at leisure moments, made themselves capable of interrogating it by proper management— a mortifying and painful acknowledgment ! And yet, no better results are obtained from the ophthalmoscope, the laryngoscope, or the aural mirror, instruments of undeniable ability, occupying definite places among our means of diagnosis, if, perchance, one has never learned their management. It does not require a person to be an eminent microscopist or an accomplished chemist, in order to be a successful diagnostician and therapeutist.

A great obstacle to the more common use of the compound achromatic microscope, heretofore, has been its expensiveness; but excellent instruments are now made by our best opticians, termed "students' microscopes," which will accomplish all that a practitioner can desire. The value and usefulness of a microscope does not lie so much in the beauty or workmanship of its brass mountings and other metallic accompaniments, as in the quality of its objectives and eye-glasses. However great may be the magnifying powers of a microscope, it is useless unless it likewise possesses penetrating and defining powers. And in the purchase of one of these instruments, the name of its manufacturer should always be learned, as our best opticians never permit poor or imperfect glasses to leave their workshops.

In selecting an instrument the error is frequently made of purchasing a very complete and expensive one, in the hope that it may be more easy to manage, and will give superior results; but the contrary more generally happens. For if an individual does not apply himself to regular histological researches, he will soon ascertain that he has needlessly expended several hundreds of dollars, unless, indeed, his pecuniary resources are large and abundant. Movable stages, condensers, extra rack movements, diaphragm holders, etc., are absolutely useless for daily and practical observations. It is at once more economical and more convenient to purchase a simple "students' microscope," the optical apparatus of which is unobjectionable; to which may be added, from time to time, as may be desired, a micrometer eye-piece, of which we should calculate the value of each division for the series of powers possessed by the microscope, and the figures indicating such values should be arranged in tabular form on a card, which should be kept suspended near the table at which we work. This will enable us at once to determine the dimensions of any microscopic element submitted to observation.—The employment of the camera lucida, adapted to the microscope, enables us to draw that which we observe; a sheet of strong and smooth

paper, or Bristol paper, may be used, and the best pencils for this purpose are Faber's (graphite of Siberia), No. 5, or H. Then a sketch may be made of all known or unknown bodies, and the action of reagents, acetic acid, alkalis, coloring agents, etc., noted down; and at a later period these indications, together with other information subsequently acquired, may enable us to correctly determine them. A single personal observation is a thousand times preferable to the most detailed descriptions, and multiplied drawings. Every compound microscope should be furnished with a camera lucida that investigators may, in addition to what has already been stated, estimate the magnify-

Fig. 21.

Dissecting Microscope.

ing power of their lenses, as well as the diameters of objects under examination.—A microscope giving magnifying powers of 100 and 300 diameters is suitable for the greater part of observations; but where it can be afforded, powers of 100, 250 and 500 diameters are to be preferred. A good work should also be procured to serve as a guide in the management of the instrument, and in the microchemical manipulations.

It has been frequently objected that the employment of the microscope in medical practice occasions a loss of time which is not sufficiently compensated by the results obtained. This inconvenience has been greatly exaggerated, and is owing to want of skill, and often to an excess of care. Many suppose that after each and every observation, the instrument must be returned to its case, for fear that it may become injured. This manœuvre, very annoying moreover and needless, requires, in fact, triple more time than would be necessary to make a genuine and simple examination. There should be no hesitation in dispensing with it. By leaving the instrument properly equipped before a window, upon a small table furnished with a drawer or two, and upon which is kept the ordinary reagents, a flask of distilled water, a drop pipette or two, several glass slides, a piece of soft old linen, and a pasteboard or bell-glass cover to protect the instrument from dust, there will be constantly on hand,—all that is necessary for a rapid observation. There is nothing to fear from this apparent abandon. Any dust on the lenses may be removed by means of a soft camel's hair pencil; if they become spotted or finger-stained, they may be gently rubbed with a piece of moistened soft old kid glove or linen, and the use of a similar dry article to dry the surfaces. Mineral acids, the vapors of which rapidly attack the brass mountings, should not be kept near the instrument. Indeed, strong acids and alkalis, volatile caustic fluids, sulphureted hydrogen, and all agents that may injure the metallic portion of the instrument, or that may act upon the lead of the flint glass of the lenses, should be avoided as much as possible. And should a lens unfortunately come in contact with any such agents or their vapors, it should at once be dipped into distilled water. To a certain extent any injury to the metallic part, from the above causes, may be avoided by giving the brass, etc., one or two coats of a mixture of 1 part of paraffin dissolved in 4 parts of benzine; and to protect the surface of the objective lens, a thin glass may be placed over it by means of a layer of the above mixture carefully applied upon the metal surrounding it. During cold and excessively damp weather the room containing the microscope should be kept heated to prevent moisture from forming, especially upon the lenses. Finally, a suggestion which may appear singular, but the propriety of which will be sooner or later appreciated, is, to prohibit the person who attends to the apartment from dusting or wiping the table and the microscope; this is the only method of avoiding any clumsiness or excess of zeal that may soon place the instrument out of use.

It has been frequently remarked that it is very unfortunate the physician can not, while making professional visits, have a microscope with him; but this is not our opinion. A practitioner can always very conveniently carry, in a small sample tube, the urine it is desired to submit to a microscopic examination. But portable microscopes are now manufactured, by means of which examinations may be made in the sick chamber; and, if a physician's pecuniary condition will not permit him to possess one of these, he can find in nearly every philosophical instrument store a small instrument of very moderate price, known as the "Coddington Lens" (*Fig.* 1), by means of which the greater part of urinary sediments and crystals can be perfectly recognized; while with the microscope at home, he can subsequently complete the investigation. With a little careful and attentive practice, a person will soon become thoroughly conversant with the management and employment of the microscope, as well as with its revelations.

The best microscopic objectives of high powers have their chromatic and spherical aberrations so perfectly corrected, that the thinnest glass cover upon the object renders its image indistinct to view. To obviate this defect, they are furnished with a compensating arrangement, by means of which distinct images of objects examined may be seen, from uncovered to a covering with quite a thick glass cover; some mark is generally made upon the objective indicating how the glass should be corrected for uncovered objects, and by turning a collar so that the graduated line is carried towards "covered," the glass may be corrected for any thickness of thin glass cover, and which correction simply consists in bringing the anterior lenses closer to, or more distant from, the posterior combination, by means of the regulating collar. In adjusting the objective for a covered object, it must first be arranged for "uncovered," and then we carefully focus for as distinct a view of the image of the object as can be obtained; this done, turn the regulating collar around until a view is had of dust that may have settled upon the upper surface of the thin glass cover. Now carefully focus for the object by means of the fine adjustment. Note what the graduation is on the collar of the objective and mark it on the slide for future use, as it will always be the same for the same objective and eye-piece.—The best microscope makers in this country, are, Chas. A. Spencer & Sons, Geneva, N. Y.; R. B. Tolles, Boston, Mass.; J. Grunow, New York city; Wm. Wales, Fort Lee, N. J.; J. Zentmayer, Philadelphia Pa. Among the best works on the use of the microscope may be named Beale, Carpenter; or, a very cheap and useful little work by W. L. Notcutt.

Microcosmic Salt. *Sal Microcosmicum.* A triple salt obtained from the urine, composed of ammonia, soda, and phosphoric acid. See *Phosphate of Soda, Neutral.*

Micro-spectroscopy. This refers to a mode of examination of substances too minute to be observed under an ordinary spectroscope; a microscope, as well as a spectroscope especially adapted for the purpose, are employed in combination. The Sorby-Browning micro-spectroscope is an ocular spectroscope designed for use with a microscope, and is probably the best one now made. Among other valuable determinations, the micro-spectroscope enables us to recognize the state of oxygenation of the blood, its richness in red corpuscles, and the presence of oxide of carbon in these corpuscles. The manner of using the instrument is as follows :—Place a glass slide upon which is a drop of blood, upon the stage of the microscope, and examine it, with any power whatever, in the usual manner. When a clear and distinct image of the blood corpuscles is obtained, remove the eye-glass and replace it with the ocular spectroscope. We now do not see the image of the blood elements, but a spectrum according to the condition of the blood, oxygenated or deoxgenated, etc.

Oxygenated hemoglobin furnishes a spectrum presenting two bands of absorption in the yellow-green part of the ordinary luminous spectrum ; the one to the left or yellow part being smaller than the one to the right or green part. To obtain these lines or bands the fluid examined must be properly diluted.

Reduced or deoxygenated hemoglobin furnishes but one band of absorption, large and diffuse, and occupying the same part of the spectrum as the preceding. The red of the spectrum is darker, and the blue clearer than is observed with the oxyhemoglobin.

Oxygenated hematin furnishes one large black band on the limits of the red and yellow part of the spectrum, if it be acid ; and a larger band more to the right or yellow part of the spectrum, if it be alkaline.

Reduced or deoxygenated hematin furnishes two bands, one, observed first, is large and intense, and located on the left in the yellow-green part of the spectrum ; the other does not appear until the first is completely formed, it is smaller and less distinct, and, if the quantity of blood employed in the experiment be too slight, it may be invisible,—it is situated on the right and in about the center of the green part of the spectrum. By agitating the solution in the air, these two bands disappear, and are not replaced by any others, unless hemoglobin is present.—In these experiments, as 100 parts of hemoglobin contain only 4 parts of hematin, we should experiment with concentrated solutions of hemoglobin.

Methemoglobin, furnishes three bands of absorption similar to those of oxygenated hemoglobin, and occupying exactly the same position in the spectrum, with the exception of the first band to the left being diffused and placed nearer the red of the spectrum ; the two rays in the green part of the spectrum are fainter.

Oxygenated Methemoglobin differs from the above as follows : The band or line at the red of the spectrum disappears ; the two bands in the green become darker, and a faint band appears in the orange. On deoxydizing this solution, we obtain the spectrum of oxygenated hemoglobin.—Sorby is probably correct, when he considers methemoglobin as a simple peculiar state of oxidation of hemoglobin.

When the blood is poisoned by the *oxide of carbon* the hemoglobin assumes a blueish-red color. The spectrum is very analogous to that of oxygenized hemoglobin, the difference being that the two bands of absorption are placed a little nearer to the right or green part of the spectrum. The best characteristic of the presence of this oxide is that its spectrum undergoes no change from the action of reducing agents, as hydrosulphate of ammonia, tartaric acid and protochloride of tin, ammoniacal solution of tartaric acid and sulphate of protoxide of iron, etc.

The most important characteristic spectra of blood are given in the following table, together with the means for obtaining them, when the blood is under micro-spectroscopic examination. By pursuing these experiments according to the order indicated by the figures attached to the names, and to the experiments, even a novice may be enabled to make a successful examination.

Table VIII.

The substance to examine, blood, is recognized *by the spectrum of:*

1. Soluble in water,....................................... $\left\{ \begin{array}{l} \text{Hemoglobin, 3.} \\ \text{Methemoglobin, 9.} \end{array} \right.$

2. Insoluble in water, but completely soluble in a weak solution of citric acid, or of ammonia,........................... ....Hematin, 13.

3. Add ammonia, double tartrate, and iron salt,............ ....Deoxidized Hemoglobin, 4.

4. Stir, in the air,.. Oxygenated Hemoglobin, 5.

5. Add citric acid,... Hematin, 6.

6. Add ammonia,...Oxygenated Hematin, 7.

7. Add iron salt,.................................... ....Deoxidized Hematin, 8.

8. Stir thoroughly,....................................Oxidized Hemoglobin.

9. Add ammonia,........................... ....Oxidized Methemoglobin, 10.

10. Add double tartrate, and iron salt,.......... .Oxidized Hemoglobin, 11.

11. Allow it to rest,.............................. ....Deoxidized Hemoglobin, 12.

12. Stir in the air,.................................. ...Oxidized Hemoglobin.

13. Add double tartrate and iron salt,...............Deoxidized Hematin.

The usual reagents in these investigations are, 1, a dilute solution of ammonia; 2, a dilute solution of citric acid; 3, double tartrate of potassa and soda in solution, employed to prevent the precipitation of the oxide of iron; 4, solution of sulphate of protoxide of iron and ammonia, which serves to deoxidize; 5, dilute hydrochloric acid; 6, refined boracic acid; 7, sulphite of soda; 8, a small platinum wire to stir with; 9, a glass cell slide to hold the fluid under examination.

In all cases where the life of a human being may depend upon the result of the examination, it is always indispensable that the result be not recorded until it has been controlled by the superposition of the normal spectrum with that of the fluid under investigation; this is readily effected with all good modern instruments, which are furnished with a special prism for this purpose. For further information the reader is referred to Proceedings of the Royal Society, Vol. XV, page 53, "*On a Definite Method of Qualitative Analysis of Animal and Vegetable Coloring Matters,*" by H. C. Sorby; also to a work entitled: "*On Spectrum Analysis as Applied to Microscopic Observation,*" by

W. T. Suffolk, and sold only by John Browning, optician, No. 63 Strand, London, Eng.

Mictio. *Micturition.* The act of voiding urine.

Mitscherlitch's Saccharimeter. See *Polarizing Apparatus.*

Monad. Naked bodies of roundish or oblong form, without any appendages or variable expansions, having a single flagelliform filament, and a slightly vacillating motion. These corpuscles are very small, much smaller than a blood globule, being from $\frac{1}{12000}$th to $\frac{1}{16000}$th of an inch in diameter, and requiring a microscope of $\times 500$ in order to see them accurately. They are colorless and transparent; their filament or cilium is difficult to perceive, and is frequently wanting. They are met with in all fluids commencing to change, mingled with vibrios and bacteria. They are very abundant in urine containing a great amount of mucus and albumen, and may be observed shortly after its discharge; they also exist in the bladder of persons laboring under chronic cystitis.

Morphia. This substance passes into the urine partly unchanged, whether it be derived from opium or from the morphia salts. It does not appear to have any important influence upon the water of the urine, the quantity of which is variable according to the temperature of the body and other circumstances, though it diminishes the amount of urinary solids, especially the earthy phosphates. Micturition usually becomes less frequent, and the urine more highly colored, especially when the action of the skin is augmented, and there is apt to be a deposit of urates. It is very probable that morphia undergoes a partial or complete decomposition in the organism; but when it is present in urine it may be detected by two or more of the following tests, first rendering 10 or 20 c. c. of the urine, pale, if required, by the addition of sufficient distilled water, and from which samples may be taken for testing:—
1. To a small quantity of the diluted urine, slightly acidulated, add 2 or 3 drops of a solution of crystallized perchloride of iron 15 grains in distilled water 2 drachms; agitate, and then add 1 or 2 drops of a solution of red prussiate of potassium 1 grain in 2 drachms of distilled water. A light or deep blue color will at once appear, according to the amount of morphia present. *Kalbruner.*—2. To a small quantity of the diluted urine, add a drop or two of pure hydrochloric acid, and gently heat; then add a very small quantity of pure perchlorate of potassium (entirely free from chlorate); the fluid immediately surrounding the perchlorate will at once assume a dark-brown color, if morphia be present, which will soon spread and extend. *L. Siebold.*—3. A few drops of sulphomolybdate of ammonia produces, in a sample of the urine, a dark-red color, which, on standing, changes to purple, and finally dark blue. The test fluid is prepared by carefully heating 2 grains of molybdate of ammonia with 1 drachm of chemically pure sulphuric acid. This test should be prepared fresh as wanted. *J. Buckingham.* Frœhde's reagent, of which the above is a modification, gives a violet color with morphia, and is considered superior to Buckingham's; it is made of 1 part of sodium molybdate dissolved in 368 parts of pure concentrated sulphuric acid, and to be used

while fresh. Dragendorff found 1 part of the sodium salt to 1,840 parts of the acid, to be a still more reliable test.—4. Add a drop or two of concentrated sulphuric acid to the diluted urine, and then a small quantity of oxide of cerium, on agitating the mixture an olive-brown color, becoming brown, will be produced. *Sonneschein.*—5. Mix 3 drops of nitric acid of 1.25 sp. gr. with 50 c. c of distilled water, and add 5 drops of this mixture to 10 grammes of pure concentrated sulphuric acid. Now to a sample of the urine add 2 or 3 drops of pure concentrated sulphuric acid; to the colorless solution add drop by drop, from 5 to 15 drops, as may be required, of the preceding solution of sulphuric acid mixed with nitric acid, and then 2 or 3 drops of water; in 15 or 30 minutes a violet-red color is produced. Upon adding a few small fragments of binoxide of manganese, or of chromate of potassa, an intense mahogany-brown color is produced. To this brown fluid add 4 times its volume of distilled water, and then render it *almost* neutral by addition of ammonia, a dirty-yellow color appears, changing to brownish red upon supersaturation with ammonia, without an appreciable precipitate. *I. Erdmann.*—6. To the clear urine add iodic acid, and then a little of a mixture of bisulphide of carbon and chloroform; a beautiful violet color appears. This will detect $\frac{1}{100000}$th part of morphia. See *Husemann's Test; Quinia.*

Mucin. *Mucosin.* An albuminoid substance which is the principal organic constituent of mucus. It is entirely soluble in water in some instances, in others, it swells up; it does not coagulate by heat; is precipitated by dilute, and redissolved by concentrated, acids. Alcohol separates it in flocculi or threads, which resume their original properties on being washed with water. When the mucous membrane is attacked with inflammation, an increase of mucin renders the discharge very tenacious. When mucus is very watery, its mucin is readily filtered; when it is very thick, the mucin almost entirely remains upon the filter. Mucin likewise appears to act differently with the same reagents, when the mucus, in which it enters, is collected from different mucous membranes. This substance may be readily studied in the white of egg. Mucin may be determined from fibrin, by the addition of dilute acetic acid, which causes the former to retract and present a very distinct striated appearance, becoming more and more concrete, while the latter, as well as its connectives, swell up, and take a homogeneous and gelatinous aspect, and are finally entirely dissolved. See *Mucus.*

Muco-pus. Muco-pus in urine is thready and gelatinous, from the action of carbonate of ammonia in this fluid. When this substance is from the mucous membrane, renal or vesical epithelia, dotted with fat, are present; when the pus proceeds from rupture of abscesses in the neighborhood of the urinary passages, these epithelia are absent, or very few in number. Alkaline urine dissolves the purulent globules.

Mucus. All surfaces covered with epithelium give rise to a more or less fluid product, possessing the same general properties in all regions, and which is termed "mucus." It is to mucous surfaces what the furfuraceous desquamation of the epidermis is to the cutaneous surface. Mucus is composed of mucin, mineral salts in solution, especially chloride of sodium, carbonated alkalies, phosphates, sulphates, etc., and certain principles of organic origin which may form, by precipitation, crystals, or amorphous grumous masses. Under the microscope may be seen (in mucus) epithelial cells, and debris of such cells, leucocytes, and droplets of fat. The cells and debris are characteristic of each mucus discharge, and aid us in determining the nature of the mucous membrane that has furnished this product. Leucocytes (mucous corpuscles) are identical in appearance, size, and anatomical characters, with the white blood globules, or with the elements of lymph and chyle; they invariably exist in every mucus, but they may be very rare in it, and which is a sign that the mucous membrane is in a perfectly healthy condition; and as this membrane becomes more and more irritated, so will the production of the leucocytes become more abundant; and thus we may observe all the transitions between physiological mucus and *muco-pus*,—a substance of a whitish appearance, containing an abundance of leucocytes, and which may at last undergo certain degenerations. The mucus of the several mucous membranes varies as to reaction, from some membranes giving an acid, and from others an alkaline, reaction.

Normal urine presents only a few epithelial cells proceeding from the desquamation of the bladder, and of the urethral canal, some leucocytes, remarkable for their lack of dimensions relatively to the white globules found in other liquids, and traces of mucus in solution which give to the urine the property of frothing upon agitation. (But when the urine becomes ammoniacal, from decomposition of urea, the leucocytes are more voluminous, turgescent, and swollen, the same as the leucocytes in buccal mucus.) In the normal state, the mucus is so mixed with the urine, as not to interfere with its transparency; but as it is not dissolved in this liquid it becomes soon deposited in or upon the urine as this fluid cools, forming a light cloud, *enœorema*, which holds some urates and epithelia in suspension, and it is so tenuous and swollen by the water of the urine, that, with acetic acid, it shows with difficulty the characteristic striæ of the mucin. The shining, varnish-like, sometimes scaly, layer, observed upon the dry filter, after having filtered urine, is its mucus. This mucus is normally secreted, but in very small amount, by the mucous membrane lining the urinary passages, and is carried off by the urine. And whenever it is present in abundance, quite visible, and very readily characterized, it indicates a pathological condition of the urinary passages. In health, from 4.50 to 1.95 grammes of mucus are passed with the urine in 24 hours.—Mucus appears to be an important factor in the fermentative process of urine.

Chemical Characters of Mucus. ‡ The characteristic element of mucus is *mucin*. Dissolved in a fluid, it renders this viscous and thready, but it does not coagulate by heat, which distinguishes it from albuminous fluid. Mucin is precipitated from its solutions, by acetic acid, in the form of filamentous flakes, somewhat resembling coagulated fibrin, and which are insoluble in an excess of the acid. The addition of a solution of iodine and iodide of potassium, renders these flakes or threads much more distinct. A solution of alum precipitates it, so do mineral acids, but the precipitate formed by the acids is readily resoluble in a slight excess of the acid; if a concentrated acid is employed, the solution is effected at once without any precipitate. Basic acetate of lead and alcohol precipitate mucin in dense fibrous flakes. Alkalies dissolve it. Hydrochloric acid dissolves mucus and clears up urine rendered turbid by its presence, when albumen is present. Dilute nitric acid dissolves mucus, thus distinguishing it from albumen.

Microscopic Characters of Urinary Mucus. If, by means of a pipette, a small portion of the flocculent cloud held in suspension in the urine, be placed upon a glass slide, and then examined under a microscope of 450 or 500 diameters, leucocytes (mucous globules) will be observed, epithelia, some fatty droplets, and fine granulations of urates, the whole being included within an extremely pale substance, disposed in streaks, frequently anastomosed. If now, acetic acid be added, or, which is better, if a new drop of the mucus cloud be taken, and treated with a drop of acetic acid, and then be submitted to microscopic examination, the same elements will be observed as at first, but modified. Two or three nuclei will be seen in the mucous globules (leucocytes), the epithelia will be paler and their nuclei will be more distinct; finally, the

Fig. 22.

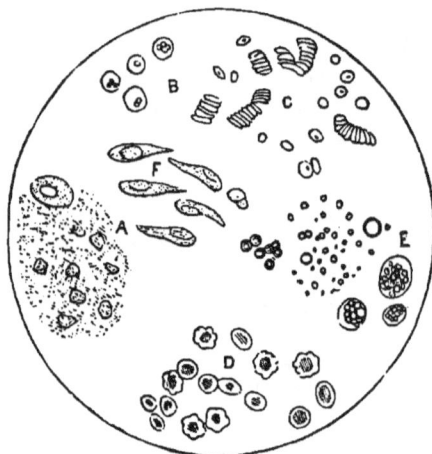

A. Mucus found in urine.
B. Mucus, acted upon by acetic acid.
C. Blood corpuscles, cohering.
D. Blood corpuscles, separate.
E. Oil globules.
F. Epithelial cells of the urethra.

pale substance will show itself fibrillary, or finely punctated. This latter substance is the mucus, the mucin of which has been precipitated by the acid. After a short time, small, square, or lozenge-like crystals of uric acid will appear, arranged in series along the filaments of the mucus. This reaction is important, as it enables us to determine fibrin from mucus, for, as already stated, an inverse action is effected upon fibrin by this acid, which renders it paler, less distinct, and eventually, by dissolving it, causes its disappearance.

Mucus is slightly colored by solution of carmine, which should not be of too dark a color, nor contain any ammonia; the advantage of this reagent is to render the *epithelia* and *leucocytes* more conspicuous. Vogel recommends

9

the tincture of iodine, which both precipitates and colors the mucus.—Beside urates, the mucus may hold crystals of oxalate of lime, or, of ammonio-magnesian phosphate. However, these are exceptional cases, and are only met with in certain pathological conditions, referred to under *Clinical Import*. —Nasal, or expectorated mucus, possesses the same characters as just described, but without the presence of any crystals.—It may be proper to state here, that C. Méhu has determined that the substance generally considered to be urinary mucus, is not mucus in reality since it contains no mucin, but is merely a normal or pathological urinary sediment consisting of epithelial or organic detritus, pus, sperm, phosphates, urates, etc.; and that its appearance varies with the elements composing it, and with the acidity or alkalinity of the urine. Query. Does urine exert a decomposing action upon mucin?

Clinical Import. When an abnormal amount of mucus exists in urine, it forms a more or less considerable glairy deposit, adhering to the bottom of the urinal, as, in vesical catarrh. But to be exact, it must be added that it is mixed with pus, rendered glairy by the action of the carbonate of ammonia, which is so rapidly produced in the urine of patients laboring under this affection.—Again, the presence of an abnormal amount of mucus in the urine, indicates, to the physician, the existence of a slight irritation at some point of the urinary mucous membrane, and which point may frequently be determined by the form of the accompanying epithelia. With females, this irritation may also be seated in the genital mucous membrane. At other times the excess of mucus is due to a general condition, as, fevers, typhus, pneumonia, etc.—Finally, there is a particular form of mucus in the urine, which should be described. We have encountered it very frequently, and have referred to it elsewhere. See *Pus.* This form presents more or less transparent, whitish filaments, which, with many persons, are voided with the first jet of urine. They are 1, 2, or 3 centimetres long, and from $\frac{1}{15}$th to $\frac{1}{5}$ths millimetre thick, and when left at rest slowly fall to the bottom of the urinal. Beside, the microscope (at least 400 diameters), shows smaller ones, invisible to the naked eye, and almost imperceptible in the deposit, or in the nubecula of normal urine. They may likewise be found in various morbid deposits. They consist of a quite firm microscopic filament of mucus, gradually swelling in the urine, ordinarily striated lengthwise, sometimes with no striæ or hardly visible, but becoming manifest on contact with acetic acid. This mucus includes within it fine granulations; more or less deformed and elongated leucocytes, recognizable under the action of acetic acid; spherical leucocytes; at times, pavement epithelial cells; and sometimes, with man, dead spermatozoids, or living, when the urine is fresh. These filaments are rectilinear or diversely flexuous, according to the fortuitous circumstances of the preparation; they contain leucocytes in variable quantity in different subjects, and even in the different filaments. These elements are more numerous among persons who have had gonorrhea, and the filaments are then whiter and more abundant.

Authors have variously described these filaments, as being:—1, casts from the renal tubes; 2, more frequently as coming from the prostatic canals; and, 3, as coming from the testicular and epididymary tubes. But in the normal and even in the morbid state, none of these organs produce any *mucus*. On the contrary, it is certain that these filaments, simple or ramified, are formed from the mucus accumulated in the folds of the urethral mucous membrane, in which it includes leucocytes, epithelial cells, etc. In man, they are found especially in the folds of the sinus of the urethra (of Lecat, Fr.), at the junction of the membranous and bulbar parts of the canal; at this point they receive spermatozoids which escape in small amount, normally and independently of any seminal loss, with individuals who, for any cause whatever, abstain from sexual contact for several weeks. The filaments formed in the period following coition also contain them.—The presence of these filaments with leucocytes in the urine of women, in the unimpregnated as well as in the impregnated condition, shows that they do not come from the prostatic nor the seminal tubes.

Mulberry Calculus. *Mural.* Names given to the rough oxalate of lime calculi, having tuberculated surfaces, resembling those of mulberries.

Mulder's Test. For *Grape Sugar.* Place a little pure sulphate of indigo in a test tube, add the urine to be tested, and boil the mixture. Add to it a solution of carbonate of potassa or of soda; if glucose be present the mixture becomes decolorized; if not, it remains blue. This determines the presence of a very minute quantity of sugar; and if traces of it only be present, a very small amount of the indigo salt should be added to the urine. The solution of soda has to be in excess, so as to render the indigo alkaline.

Murexide Test. See *Purpurate of Ammonia. Uramile.*

Muriate of Ammonia. See *Hydrochlorate of Ammonia.*

Mycoderma Cerevisæ. See *Torula C.*

N.

Nepheloid. Nebulous, or cloudy.

Nephranuria. A non-secretion, or very diminished secretion, of urine by the kidneys. Renal ischuria.

Nephritic. Of, or belonging to, the kidney.

Nephritis. Inflammation of the kidney, of which there are several varieties.

Nephrochalazosis. A name for Bright's disease.

Nephropyosis. Renal suppuration.

Nessler's Test for Ammonia. To 10 or 15 drops of a cold concentrated solution of chloride of mercury, add solution of iodide of potassium (7 Tr. grains to 6 f℥ distilled water), drop by drop, agitating the mixture so as to dissolve the red precipitate of mercuric iodide as fast as it forms, and continuing the operation until this precipitate is *exactly* dissolved, or, at least, until a very minute amount of it remains undissolved. Filter. Then make the solution strongly alkaline with hydrate of potassa (20 Tr. grains caustic potassa to 10 f$\mathfrak{z}$ distilled water), and add enough distilled water to make the whole measure 25 fluidrachms; filter if necessary.

If this solution (Hg I$_2$ 2 K I + K) be added to water or diluted urine, containing .03 of a grain of ammonia to the gallon, a yellow color will be produced; a greater quantity of ammonia will give a brownish-yellow color, or a reddish-brown precipitate, supposed to be tetrahydrargyro-iodide of ammonium (N Hg$_2$ I). This is a very delicate test, interfered with, however, by the presence of cyanide of potassium, or, of sulphide of potassium.

Another method of using it is, to place the urine in one vessel, and immediately, above this, place another vessel containing a mixture of 1 part sulphuric acid and 9 parts distilled water. Now cover the whole with a bell glass. (An ordinary drying apparatus will answer for this purpose.) In 20 or 24 hours add 1 drop of the mercurial solution to the acid fluid; if it contains ammonia, absorbed from the ammoniacal vapor evolved from the urine, a precipitate will occur of a brownish tint. See *Rabuteau's Method.*

Neutral Urine. Urine which gives neither an acid nor an alkaline reaction is termed neutral. This is not owing to the positive loss of acid, but to its neutralization by bases, fixed or volatile alkalies. Even when voided acid, urine may become, in a short time after its discharge, neutral, or alkaline, from decomposition of its urea into ammonia and carbonic acid. But urine may be neutral when passed, and which may be due to decomposition of urea in the bladder, or to the use of certain agents, as, alkaline carbonates, carbonate of lime or magnesia, etc. In anemia there may be a diminution of the phosphoric or sulphuric acid of the urine. As neutral urine most commonly becomes alkaline in a short time, being as it were the first step towards alkalinity, in urinary investigations neutral and alkaline urine are considered as equivalents. See *Alkaline Urine.* Prof. Loreta, of Bologna, has related several instances in which concussion of the brain was followed by neutral urine during a period of from 8 to 24 hours after the shock; M. Testi explains this by stating, that from the abatement of the velocity of the circulation, the phosphate of soda contained in the blood does not encounter uric acid in the kidney, hence this salt remains neutral instead of becoming acid, and the urine is thus left neutral.

New Constituent of Urine. This was first separated from the urine of a dog, and finally from human urine, by F. Baumstark. It forms white, hippuric acid-like columnar crystals, of the formula C$_3$ H$_8$ N$_2$ O, which are sparingly soluble in cold water, freely soluble in boiling water, insoluble in ether or absolute alcohol, and forms soluble salts with acids. Heated in a tube it evolves dense white vapors having the odor of ethylamin, and a combustible gas of an alkaline reaction.

Nickel. A salt of nickel in solution is mixed with solution of acetate of soda, to which solution of hypochlorate of soda is added, and the whole heated to the boiling point, a dark blue precipitate of peroxide of nickel occurs, which is soluble in nitric acid.

Nitrate of Silver. *Argentic*, or *Silver Nitrate*. Indicates the presence of *hydrochloric acid*, whether free or combined, in the fluid under examination, by causing an instantaneous white cloud of chloride of silver, which precipitates in a caseous magma, and which, by exposure to the air, changes to a hyacinth, or more or less, violet color. Under the microscope the magma preserves its curdled form. This precipitate is insoluble in dilute nitric or hydrochloric acids; in the cold it is soluble in ammonia, and in very concentrated hydrochloric acid by heat. Solution of nitrate of silver gives a yellow precipitate with phosphate of soda; a brick-red with arsenic acid; olive-brown with lime water or the fixed alkalies; and yellow with the alkaline arsenites. Metallic copper or phosphorus added to a solution of nitrate of silver, occasions a precipitate of metallic silver. The yellow precipitate of tribasic phosphate of silver, formed when nitrate of silver is added to a phosphate, is soluble in excess of ammonia, and in excess of nitric acid. In urinary investigations the argentic nitrate is employed for the detection of chlorides and phosphates.

No. 1. *Nitrate of Silver Solution.* Take of nitrate of silver, in crystals, 1 gramme, distilled water 10 grammes; mix. Preserve in a well stopped vial and label, "*Solution of Nitrate of Silver at the $\frac{1}{10}$th. For Detecting Chloride of Sodium.*"—No. 2. *Nitrate of Silver Solution.* Take of pure fused nitrate of silver 11 grms .63, distilled water 400 grammes; mix. Preserve in a well stopped vial, and label, "*Solution of Nitrate of Silver. Exact Estimation of Chloride of Sodium. 1 c. c. of the Solution = 1 centigramme of Chloride of Soda.*" The purity of this reagent should be ascertained as follows: The solution must be perfectly neutral; on precipitating a portion by hydrochloric acid and filtering, the filtered liquid should leave no residue when evaporated on a watch glass, and must neither be colored nor precipitated by hydrosulphuric acid. See *Stains, etc.*

Nitrate of Urea. See *Urea.*

Nitric Acid. *Azotic Acid.* This acid colors fresh albumen yellow, especially when heated; it also colors the albuminoid substances of animal tissues yellow, as, nails, horn, etc., often causing the outlines of the individual cells to become very distinct. When vapors of ammonia are evolved, by subjecting them to the action of the vapor from nitric acid they become white; this may be effected by holding a glass rod, moistened with the acid, near to the point from which the ammoniacal evolution is occurring. Nitric acid dissolves cystine, oxalate of lime, urea, and triple phosphates, converts hippuric acid into benzoic, and precipitates albumen. The remarks concerning the care required in employing *hydrochloric acid* with the microscope, are equally applicable to nitric acid. Pure colorless nitric acid is used in urinary investigations; also, a solution of nitric acid 1 part, distilled water 10 parts. See *Stains, etc.*

Nitroso-Nitric Acid. *Fuming Nitric Acid.* This is a common commercial acid containing nitric and nitrous acids with water, and is generally sold as nitrous acid. It fumes in the air, has a brown or yellowish-red color, and is more corrosive than concentrated nitric acid. Its formula has also been given as $H\ N\ O_3 + N_2\ O_4$, being nitric acid containing tetroxide of nitrogen. It constitutes Gmelin's test for determining the presence of *bile pigment* in the urine.

O.

Odor of Urine. Odor is a characteristic reaction of certain bodies; but it can not describe them, it can only compare them. The alliaceous odor denotes the presence of arsenic in an inorganic substance when thrown on burning charcoal. The odor of chlorine is characteristic, resembling somewhat that of nitrous and sulphuric acids; in this case, practice alone will enable one to detect it. The odor of iodine is somewhat similar to that of saffron. Sulphuric acid, when heated, and simply mixed with water, evolves a distinct odor. Ammonia in a state of evolution is known by its odor, which is very similar to that of carbonate of ammonia; its vapor causes the tears to flow. It is similar with acetic acid. The addition of a drop of hydrochloric acid, gives an odor different from that of any animal or vegetable; the most fetid are often made to assume a bouquet which approximates the odor of violets, apples, etc. The acid phosphates of ammonia give a very repulsive odor to the breath of certain persons. Nevertheless, the organs of smell and taste are two reagents whose indications vary according to the individuals. On this account, every one should make from practice and memory a table of indications for his own use; these reactions being, after all, only signs which we can not transmit positively to others, and which only serve to point out the method that led to the indication.

The odor of healthy urine is characteristic, and of a peculiar kind, due to the presence of certain volatile acids, determined by Stadeler, as carbolic, taurylic, damaluric, and damolic, and it is so indestructible that an analyst can not remove it by any of the processes pursued in urinary investigation. When a mineral acid is added to urine, its odor becomes modified and stronger. The urinous odor may be produced by scorching a drop or two of urine on a piece of linen, or, by treating it with sulphuric acid; by this method urine can be determined from any other fluid. The urine of animals varies in odor, each one having an odor characteristic of the animal, and usually smelling like the fat of the animal from which it is obtained. The strong urinous odor of concentrated urines is due to the presence of urea, and when this undergoes decomposition, an ammoniacal odor, more or less pungent, is emitted from the carbonate of ammonia formed, and, should the organic matters, as, mucus, etc., undergo decomposition at the same time, a putrid odor is also observed, and which is more common with urine during some destructive renal or cystic malady, when the organic substances are in greater amount; if sulphur be formed during the process of decomposition, sulphureted hydrogen may be emitted. If a fixed alkali be present the urine has a sweetish aromatic odor. Fresh diabetic urine has a sweetish or faint whey-like odor, which changes to that of sour milk on fermenting. Blood, pus, or sanious discharges also give a stale, offensive odor, resembling that of tainted flesh. With these exceptions, the urine in all diseases, persistently retains its characteristic urinous odor, and hence, in a clinical point of view, the smell of the urine is of minor importance, being simply suggestive.

However, many articles of diet, medicine, etc., communicate certain peculiar odors to the urine: As, asparagus, valerian, assafetida, castor, coffee, garlic, onions, turpentine, cubebs, copaiba, sandal-wood oil, saffron, etc. Turpentine imparts a violet odor, whether it be swallowed or inhaled. It has been stated that in gout, and in organic diseases of the kidneys, these odorous substances, when taken internally, especially asparagus, turpentine, etc., can not be recognized in the urine by their smell, but observations by Vogel and others have disproved this statement.

Oil Globules. See *Fatty Matters.*

Oliguresia. *Oliguria.* Want or deficiency of urine.

Omichesis. Voiding urine.

Omichma. The urinary fluid.

Onions. See *Garlic.*

Organic Globules. Described by Dr. Golding Bird. The "large organic globules" have been observed in cases of ardor urinæ, in the urine passed during the latter months of pregnancy, but more abundantly in confirmed cases of Bright's disease. They consist of a cell wall, containing more or less fat globules, diffused throughout a urine in which there is no albuminous or viscid element, as observed in urine containing pus or mucus. Acetic acid added to these globules, develops internal nuclei. They greatly resemble, what has been termed, "exudation cells," or "compound granular cells."

The "small organic globules" differ from the preceding in not having a granular surface, and presenting no nuclei or granular structure, even under the action of acetic acid. They are not so commonly observed as the large variety, and are supposed to be the escaped nuclei of certain cells. See Fig. 30.

Ourema. *Ouron.* Same as *Omichma.*

Oxalate of Lime. *Calcic* or *Calcium Oxalate.* Urine may normally contain oxalate of lime, in a state of solution at the time of emission, and which in a few hours becomes deposited in crystalline form. It never exists in such quantity as to form a deposit by itself, but is found mixed with other salts, generally urates or uric acid; however, it is by no means uncommon to find it accompanied with an abnormal deposition of earthy phosphates, the urine, in such cases, being either feebly acid at the time of its discharge, neutral, or alkaline. In these instances, the addition of acetic acid will dissolve the phosphates and leave the oxalate crystals intact. I will remark, however, that in certain cases of hepatic and digestive derangement, and in a few cases of functional cardiac affections following sexual excesses, in which, during the flow of the last drops of urine, from half a teaspoonful to a teaspoonful of a white deposit was passed, I found this deposit to consist wholly of crystals of oxalate of lime. The urine was more generally that passed first in the morning, and in several of these cases, physicians who had been previously consulted (without a microscopic examination), pronounced this white substance, semen. Whenever, a urine within 24 hours after its emission, deposits oxalate of lime crystals, it is a certain indication of an unhealthy condition of the system, whether this be temporary or persistent. Two forms of crystals of oxalate of lime occur in urine, the octohedral, and the dumb bell, and either may be readily recognized under the microscope with a power of 350 or 400 diameters. It is impossible to determine them by the naked eye; the crystals may fall to the bottom of the tube containing the urine, or they may be entangled in the mucus flocculi and float in the liquid with it. The urine containing oxalate, is generally of decided acid reaction (rarely neutral or alkaline), of dark amber color, of sp. gr. 1.015 to 1.025, with a greater or less abundance of epithelia; if the urine be greenish, the coloring matter of blood may be present.

Microscopic Characters. Crystals of oxalate of lime formed in urine, are shining, well defined, perfectly transparent, square octohedra, having a strongly refractive power, and, when very regular, the intercrossing of their axes gives to them somewhat the appearance of a letter envelop. They are generally very small; smaller than a leucocyte. However, some have been seen very regular, and having a diameter of 0mm .011; these being accompanied by a host of others of much smaller dimensions. Their small size permits then, *a priori*, of their separation from certain crystals of ammonio-magnesian phosphate which greatly resemble them in form, but are of larger size. Sometimes they present the appearance of a square, crossed obliquely by two bright lines; and when exceedingly small

Fig. 23.

A. Octohedral crystals of oxalate of lime.
B. The same, when dry.
C. Dodecahedral crystals of oxalate of lime.
D. Dumb-bells—oxalurate of lime.
E. Oval forms of oxalurate of lime.

they look like dark squares having a transparent point in their center. Other more or less singular forms have been observed; G. Bird has described some under the name of hour glass or dumb bell crystals, that resemble two kidneys united with their concavities opposite, or a bottle gourd; these crystals are more frequently mingled with the octohedral, which will enable one to determine them from the dumb bells sometimes formed by urates, or uric acid. He has likewise observed oval grains, presenting a kind of nucleus, and perfectly analogous with certain forms of carbonate of lime, but which he chemically recognized as oxalate. Other varieties have also been described, the irregular disc, the well defined diamond shaped, etc.; and all are almost invariably mixed with the ordinary octohedral crystals, the appearance of which is so well known, that but few persons hesitate to pronounce them oxalate of lime, when observed under the microscope. This is wrong, however; the chemical character of these crystals, as well as that of all others, should always be ascertained, because their appearance under the miscroscope may deceive the most experienced. Urates have been found presenting the dumb bell form; carbonate of lime, that of discs; uric acid, of small diamond shaped crystals; and chloride of sodium from presence of urea, of octohedrons.

Chemical Characters. Oxalate of lime may be confounded with crystals of ammonio-magnesian phosphate, of carbonate of lime, and of uric acid. A drop of acetic acid passed under the thin glass cover, dissolves the first two (giving effervescence with the carbonate), and a drop of liquor potassa will dissolve the uric acid,—in each instance leaving the oxalic acid intact. As to chloride of sodium crystals, they are never found in unconcentrated urine, being soluble in water.—Oxalate of lime is insoluble in water, in acetic acid, and in liquor potassa, and ammonia, but is soluble in nitric or hydrochloric acid without effervescence. The dumb bells, kept in liquid for any length of time, gradually pass into octohedra; they give splendid colors under polarized light,—the octohedra very faintly under peculiar management. When remaining in acetic acid for some time, the dumb bells lose their crystalline substance, leaving a skeleton, as it were, of the original shape of the crystal. If oxalate of lime be calcined in a platinum capsule, a drop of acetic or hydrochloric acid, added to the residue (carbonate of lime), gives rise to effervescence from disengagement of carbonic acid gas; and if the remaining lime be treated with oxalate of ammonia, crystals of oxalate of lime will be formed anew. See *Phosphates, Micro-chemical Diagnosis of Earthy Phosphates, etc.*

Clinical Import. When the presence of oxalate of lime crystals is persistent, it is due to some abnormal condition, not as a cause but as an effect, and this is termed *oxaluria.* Thus, it often accompanies dyspepsia, affections of the respiratory organs, and other pathological conditions which occasion a diminution of oxygen in the system, or, in which there is mal-assimilation of food in primary digestion, and, hence, its presence will be an indication of the existence of such conditions, to be determined by other investigations than those solely derived from the urine. When the presence of these crystals in the urine is temporary, an explanation must be sought for in one or or more of the causes hereafter referred to, thus :—1. It may proceed from certain vegetables, etc., used as diet, as, cauliflower, asparagus, garlic, gooseberries, bananas, turnips, tomatoes, water-cresses, carrots, onion, sorrel, parsnips, apples, sugar in excess, rhubarb plant, etc.; or, from medicines, as, gentian, rhubarb, squills, valerian, canella, elder, etc. It is not always crystalline at the time of voiding, but becomes so when the urine cools.—2. It may proceed from the metamorphosis of vegetable, animal, or mineral substances; for example, by the interrupted oxidation of uric acid, and the imperfect oxidation of saccharine and starchy bodies, as well as of salts of organic acids, which instead of being completely transformed into carbonic acid and urea, pass partly into the state of oxalates.—3. It likewise appears after drinking alkaline waters, carbonated drinks, fermented liquors, sparkling wines, lime in drinking water, etc., especially with dyspeptics.—4. Dalton observes that when oxalate of lime is observed in urine that has been exposed for 24 or 48 hours to the action of the air, it may be due to *acid fermentation.* May not the same cause occurring in the kidneys, ureters, or bladder, exist, when this oxalate is found in urine just passed without any constitutional symptoms being manifested?

As oxalate of lime is insoluble in water, how does it pass through the walls of the renal capillaries into the urine? The researches of Neubauer, Modderman, and others, have proven that this salt is soluble to a certain degree in the acid phosphate of soda, and that its solubility is favored by chloride of sodium and urea. And the acid phosphate of soda has been advised as an internal remedy, in cases where mineral acids and tonics have failed.—Oxalate of lime is often present in the urine of patients convalescing from acute diseases.—Crystals of oxalate of lime may be procured by carefully adding a dilute solution of oxalate of ammonia to normal urine just voided, and allowing it to stand for a short time ; the precipitate will contain various forms. They may frequently be procured after eating some sorrel, watercress, or rhubarb plant, during a meal, and subsequently collecting the urine. In the nubecula which forms during cooling, many beautiful octohedral crystals of the oxalate will be found imprisoned in a small quantity of mucus. See *Oxaluria.*

Oxalate of Urea. See *Urea.*

Oxalic Acid. This poisonous acid is one of the most powerful among the organic acids, and appears to be constantly present in the blood in minute quantity, being excreted from the kidneys in certain unhealthy conditions, or after the ingestion of medicines or articles of diet containing it, as, for instance, rhubarb, onions, etc. When oxalic acids or oxalates have been taken in considerable quantity, or for a period of time, free oxalic acid may exist in the urine, but, more generally, it occurs in combination with lime, forming oxalate of lime. Free oxalic acid may be detected in urine, by adding a few drops of acetic acid to a drachm or two of the suspected urine, previously neutralized by ammonia, and then adding a solution of chloride of lime; a

white precipitate occurs, which after a few hours' standing will be deposited
in the form of crystals of oxalate of lime.—A quantitative estimation of
oxalic acid present in urine is not always so easily effected. Lehman advises
to take, from the collected urine of 24 hours, 200 c. c., add a little milk of
lime to it, filter, and evaporate to dryness. Acidulate the residue with acetic
acid, and treat it with moderately concentrated alcohol; agitate this alcoholic
solution with about 50 per cent. of ether. The insoluble precipitate formed
of oxalic acid is collected on a filter, washed, dried, and weighed. From
this may be estimated the whole amount passed in the 24 hours; thus, if 200
c. c. of urine contain 5 grammes of oxalate of lime, what will the amount
(1,000 c. c.) passed in the 24 hours contain? $\frac{5 \times 1,000}{200} = 25$ grammes. From
this, the amount of oxalic acid is to be determined.

When oxalic acid exists in the urine, free or as a salt of lime, it may be
derived from food or medicines; but when it exists as the result of disease it
is due to an interrupted retrogade metamorphosis of uric acid (creatin, leucin, tyrosin, lactic, acid, etc.), and of the albuminoid group of foods. Perfect
oxidation of uric acid changes it into urea and carbonic acid; imperfect
oxidation into urea, carbonic acid, and oxalic acid, in proportion to the
degree of oxidation. Schunk, who has devoted considerable attention to this
subject, found oxaluric acid in urine; it being the result of oxidation of uric
acid. This acid may readily be converted into oxalic acid, in any portion of
the urinary passages, or, even after the urine has been discharged from the
bladder, and thus ultimately give rise to oxalate of lime formations.

Oxaluria. Also erroneously termed *oxalic acid diathesis*, is that condition
of the system in which there is a deposit of oxalate of lime in the urine
within 24 hours after its discharge, and which deposit persists daily for weeks
or months. It is common in nervous and dyspeptic diseases, where impaired
nutrition exists, as well as in bronchitis, phthisis, catarrh, pneumonia, emphysema, also heart and liver affections, in which there is an imperfect
oxygenization of the blood. The persistence of this oxalate in the urine indicates the existence of oxalic acid in the blood, which slowly poisons the
brain and spinal cord. These crystals are almost invariably present in the
urine of persons affected with spermatorrhea. The symptoms accompanying
oxaluria are variable, being generally of a nervous and dyspeptic character;
as, more or less depression of the vital forces, melancholy, excitable disposition, inability for mental or physical exertion, distaste for society, deficient
or absent sexual power, pain or weight in the lumbar region, irritability of
the neck of the bladder, gradual derangement of the general health, and
confirmed hypochondriasis. When the urine is alkaline, or neutral, calculus,
or disease of the bladder, may be present; when the oxalate exists in an
albuminous urine, or in urine containing renal casts, there may be some
disease of the kidneys, or, these conditions in the urine may be the result of
the mechanical irritation produced in the renal apparatus by the oxalate
crystals. Blows across the loins, and a rough catheterism, are stated to have

also occasioned oxalate deposits in the urine. It must, finally, be observed, that cases of oxaluria frequently exist, in which no symptoms or other manifestations of disease can be detected, the persons being to all appearance, and as far as feelings are concerned, quite healthy; and which cases are very probably due to the formation of oxalate of lime crystals in the kidneys, ureters, or bladder, from the process of fermentation.

In instances where oxalates and phosphates are found in the urine together, or where they appear in successive alternations, the formation of a calculus is more especially to be feared. The remedies are nitro-muriatic acid, bitter vegetable tonics, nitrate of ammonia, inhalation of oxygen, cold bath with frictions, attention to bowels, digestive functions, and skin, acid phosphate of soda, plain digestible diet, avoidance of all the articles named under *oxalate of lime*, as well as of fatigue, sexual excesses, and excessive mental application; if *anemia* be present, tincture of chloride of iron, pyrophosphate of iron, ammonio-tartrate of iron, etc., and for stimulants, when necessary, whisky, brandy, gin.

Oxidum Uricum. *Oxidum Uranicum.* See *Xanthine.*

Ozonic Ether? (*Antozonic Ether.*) See *Ethereal Solution of Peroxide of Hydrogen.*

P.

Parabanic Acid. This acid is one of the derivatives from uric acid, and like all the other derivates of this acid may be converted into urea and oxalic acid. It is procured from uric acid by dissolving this in nitric acid, and then evaporating the solution until no more gas is evolved, and until the alloxan originally produced is decomposed.

Pavy's Test, or **Solution.** (See *Copper.*) Take of sulphate of copper 320 grains, neutral tartrate of potassa 640 grains, caustic potassa 1,280 grains, distilled water 20 fluid ounces. Dissolve the neutral tartrate and the potassa fusa in one portion of the water, and the sulphate of copper in the other; afterwards mix the two solutions. Label, "*Pavy's Cupro-Potassic Solution. Detection of Sugar.*" This differs chiefly from Fehling's in the substitution of caustic potassa for the caustic soda, and requires similar care and management in its preparation and preservation. If at any time this solution gives any precipitate upon being boiled, it is not fit for use, but may be made as good as ever by a fresh addition of caustic potassa.

This is an excellent modification of Fehling's test for sugar. But as it gradually decomposes upon standing, it is better to keep the several articles in solution separately and mix them when required, as follows:—1. Dissolve pure sulphate of copper 160 grains, in distilled water 5 fluid ounces. Label, "*Solution of Copper. Pavy.*"—2 Dissolve neutral tartrate of potassa 320 grains in distilled water 2½ fluid ounces. Label, "*Solution Tartrate Potassa.*

Pavy."—3. Dissolve pure caustic potassa 640 grains in distilled water 2½ fluid ounces. Sp. gr. 1 12. Label, "*Solution Potassa. Pavy.*" When required for use mix 2 fluidrachms of the sulph. copper solution with 1 fluidrachm of the tart. potass. solution, and then add 1 fluidrachm of the caustic potass. solution. 100 minims of this test solution are decolorized by half a grain of sugar; that is, in other words, half a grain of sugar will convert the whole of the oxide of copper in 100 minims of the test liquid into the state of suboxide. As diabetic urine undergoes rapid decomposition, especially in warm weather, the analysis should be made, if possible, immediately after the urination.

To use this solution, dilute the diabetic urine with four times its bulk of distilled water; place it in a graduated pipette, and let it fall drop by drop into a porcelain capsule containing exactly 100 minims of the above test solution along with a piece of caustic potassa, about the average size of a pea, which solution must be kept gently boiling all the time. As soon as the blue color has disappeared and an orange color appears, read off the amount of diluted urine employed. The number of minims employed will give the amount of sugar per fluid ounce, as in the table below. But, as the urine is diluted, and only represents the fifth part of an ounce, the amount given in the table must be multiplied by 5, to get the amount of sugar in the ounce of urine. Thus, if 28 minims of diluted urine were required to decolorize the 100 minims of test solution, according to the table this would give 8.57 grains of sugar to the fluid ounce; but this is only ⅕th of an ounce, the urine being diluted, and must be multiplied by 5, which gives 42.85 grains of sugar in an undiluted fluid ounce of urine. Three gallons of urine per day, are 384 fluid ounces, yielding 16,454.40 grains of sugar.

If there are only traces of sugar in the urine, concentrate by evaporation, and then treat with acetate of lead to get rid of coloring matter, urates, and phosphates, and then filter. Under ordinary circumstances no preparation of the urine is required for the application of this test.—Should a small quantity of the urine be boiled with some crystals of sulphate of soda, q. s., to insure a total separation of all that is vegetable, and then filtered, the soda will not interfere with the test.

IX. Table,

*Showing the Quantity, in Troy Grains, of Sugar per Fluid Ounce, for Minims
(from 15 to 100) Required to Decolorize 100 Minims of F. W. Pavy's Test-fluid.*

Minims to Decolorize.	Sugar per Fluid-ounce in Grains.	Minims to Decolorize.	Sugar per Fluid-ounce in Grains.	Minims to Decolorize.	Sugar perFluid-ounce in Grains.
15	16.	44	5.45	73	3.28
16	15.	45	5.33	74	3.24
17	14.11	46	5.21	75	3.20
18	13.33	47	5.10	76	3.15
19	12.63	48	5.	77	3.11
20	12.	49	4.89	78	3.07
21	11.42	50	4.80	79	3.03
22	10.90	51	4.70	80	3.
23	10.43	52	4.61	81	2.96
24	10.	53	4.52	82	2.92
25	9.60	54	4.44	83	2.89
26	9.23	55	4.36	84	2.85
27	8.88	56	4.28	85	2.82
28	8.57	57	4.21	86	2.79
29	8.27	58	4.13	87	2.75
30	8.	59	4.06	88	2.72
31	7.74	60	4.	89	2.69
32	7.50	61	3.93	90	2.66
33	7.27	62	3.87	91	2.63
34	7.05	63	3.80	92	2.60
35	6.85	64	3.75	93	2.58
36	6.66	65	3.69	94	2.55
37	6.48	66	3.63	95	2.52
38	6.31	67·	3.58	96	2.50
39	6.15	68	3.52	97	2.47
40	6.	69	3.47	98	2.44
41	5.85	70	3.42	99	2.42
42	5.71	71	3.38	100	2.40
43	5.58	72	3.33		

Penicilium Glaucum. *Mould Fungus.* See *Fig.* 31, page 178.   This
fungus is one of the most common met with on decaying vegetable substances, and especially on vinegar, albuminous, and semi-fluid matters. It
consists of pedicels or partitioned tubes, terminating in a repeatedly bifurcated pencil, each branch of which bears a moniliform row of spores. It is
common in acid urine containing organic substances, and may be seen in its
various phases of development, as, spores, round or oval cells; thallus,
branches or interlacing fibres; and aerial fructification or mould, in which
a downy pile of threads grow out into the air. The spores show a nucleus,
tend to an oval or elongated form, are green or blueish,and after fructification may be found in the bottom of the urine. The fungus should be studied on preserves, and certain kinds of cheese; it may be recognized by its

partitioned tubes terminating in a pencil of filaments carrying the spores arranged in a bead-like manner. Solution of iodine renders them more distinct, colors them yellowish, and arrests any movements. See *Fungi; Torula; Vegetable Organisms.*

Pettenkofer's Test. This test is for the determination of the bile acids in urine, and is based upon the action of sulphuric acid upon the bile acids in presence of sugar, producing a purple-violet color. The original process has been simplified by Neukomm, and others, which see under *Bile Acids*. But all these are frequently impracticable clinically, and may even fail. Probably the best process is the following, which, however, requires considerable time and some careful management: Evaporate on a water bath, 200 or 250 c.c. of the urine to be investigated. Add an excess of absolute alcohol to the dry residue, filter, and to the filtered liquid add a large excess of ether. The precipitated bile acids are separated by filtration, dissolved in distilled water, and decolorized by filtering through animal charcoal. From 4 to 6 c.c. of the colorless liquid are then placed in a porcelain capsule or test tube, and one drop of a very dilute syrup of cane sugar added. Now gradually drop sulphuric acid into the mixture, keeping down an excess of temperature by holding the tube or capsule in cold water, and when an amount of acid has been added nearly equal to that of the fluid under investigation, the characteristic purple-violet color will appear if bile acid be present.

Phosphates. The phosphates existing in the urine are those with the alkaline bases, soda and potassa, and those with the earthy bases, lime and magnesia; in healthy urine these are never spontaneously deposited, but are held in solution, owing to the presence of certain salts and the acidity of this fluid. But as soon as the urine undergoes alkaline fermentation, these phosphates, becoming insoluble, are deposited in the form of amorphous or crystalline precipitates. They are likewise met with in neutral urine, or urine rendered slightly alkaline from the use of mineral waters, etc. In normal urine the alkaline phosphates are in greater amount than the earthy, the latter varying according to circumstances, sometimes lime preponderating, at others, magnesia. It must likewise not be forgotten that the phosphatic salts are constantly subject to changes from the condition and changes occurring in the urine; a basic salt, with an alkaline reaction, from a very slight chemical change may be converted into a neutral, and even into an acid salt; this is due to the changes occurring in the organic acids of the urine.— Earthy phosphates are soluble only in acid solutions; alkaline phosphates are very soluble, and do not precipitate with ammonia. The earthy phosphates are the only ones of these several varieties, met with in the sediment of urine, they are daily observed, and are the most important among the phosphates; we will, therefore, in order to facilitate investigation, first give their common and distinctive characters, etc., and then the determination of *phosphates in general.*

‡ I. Common Characters.

Phosphate of Lime and Ammonio-magnesian.

1. Insoluble in water.

2. Insoluble in alkaline solutions, which consequently precipitate them from their solutions.

3. Soluble in acids, even in *acetic acid.*

‡ II. Distinctive Characters.

Phosphate of Lime.	*Ammonio-magnesian Phosphate.*
Nothing.	Heated in a glass tube with a solution of caustic alkali, it disengages ammonia, which may be recognized by its odor, by litmus paper, and by the glass rod dipped in hydrochloric acid.
Dissolved in acetic acid, its solution neutralized with carbonate of soda, gives a precipitate of *oxalate of lime*, when solution of oxalate of ammonia is added to it.	Nothing, especially if care has been taken to add a few drops of solution of hydrochlorate of ammonia to the acetic solution.
Affects the amorphous condition; rarely crystalline.	Always crystallized.

III. Characters of the Phosphatic Sediments.

Deposits of the phosphates are always *white*, unless they are colored by abnormal coloring matters in the urine, as blood, senna, bile pigment, etc. They are soluble in hydrochloric and acetic acids, insoluble in ammonia and caustic alkaline solutions. Upon heating the urine, this deposit undergoes no change, unless it be an agglomeration in small masses. *Mucus, pus,* or *blood,* may at the same time be present in the urine, and mask the deposit, in which case we must proceed as follows:—Treat the sediment with hydrochloric acid (which dissolves the phosphates more rapidly than acetic acid, without decomposing them) until the salts are completely dissolved; then filter. The filtered liquor contains the phosphates in solution, while the elements of organic origin, pus, blood, mucus, albumen, remain on the filter. The phosphates may now be precipitated by ammonia and then examined under the microscope. The phosphate of lime will be found in very pale amorphous grains, united by patches, or in flaky groups, and the ammonio-magnesian phosphate will be in stellar or fern-like foliaceous crystals. If we wait a little time, the ordinary prismatic crystals will also be observed.— If it be desired to make a chemical analysis of the precipitate thus obtained, dissolve it in acetic acid, and proceed as named hereafter, under *Phosphates in General; Detection, Separation,* and *Estimation.*

Microscopic Characters. The two phosphates are more generally together in the deposits; rarely the one without the other. The distinctive characters of each phosphate seen under the microscope, are as follows:—

1. *Phosphate of Lime. Basic Phosphate of Lime; Basic Calcic, or Calcium Phosphate.* (See *Fig.* 24, page 146.) This is present in an amorphous state in the form of extremely transparent, irregular flakes or patches. A certain amount of experience is required to recognize them under the microscope; their transparency likewise enables us to distinguish them from certain very neighboring forms of urates, the granules of which are somewhat larger, and especially darker.—This is the usual aspect of phosphate of lime in the sediments. But it may be met with under many other forms, which may be simply referred to, as they have been hardly studied, and are not well known: The most common in the sediments are the forms in wand. in small crosses, or in beads. These three varieties are observed together, and result from the grouping of small grains of phosphate of lime. They have been seen in a deposit of bilious urine.—The following are the several varieties under which phosphate of lime may appear:

Forms Assumed by Phosphate of Lime.

Amorphous or in badly defined crystals.	Small pale granules grouped in irregular patches. Small spherules with dark outlines, isolated either in		
	Agglomerated.	Sausage form, or in beads (3 or 4). In cross. In hour glass or dumb bells.	
Crystallized.	In needles.	Isolated.	
		Grouped.	In fasciculi, tufts, sheaf-like bundles. In broom form. In fan shape. In stellar form.
	In acicular prisms.	Isolated. Grouped.	

The amorphous forms are observed in the sediments. The crystalline forms have been obtained by precipitating the phosphate of lime in fresh urine with carbonate of soda. By this method an amorphous precipitate is obtained, in which, after several hours, a variable quantity of crystals will be mixed. Hassal states that, though rapidly precipitated, phosphate of lime forms an amorphous powder, when precipitated slowly, prismatic crystals of this salt are formed, $2\,CaO, HO, PO_5 + 3\,HO$. Chemical analysis of this precipitate has proven that it never contains magnesia; operating according to the process named on page 152.—However, Vogel has found crystalline sediments of phosphate of lime sometimes alone (the most frequent), sometimes mixed with ammonio-magnesian phosphate. The size, the form, and the grouping of the crystals in the sediment are, he states, extremely variable. Sometimes they are isolated, at other times aggregated. At times they are delicate, in needle form, and then in crossing each other at right angle, being placed one upon the other, they often form masses of globular crystals; at other times they are thin and with perfectly smooth surfaces, their extremities termin-

10

ating in acute points. Very often the crystals are thick, more or less cunei-
form, and adhering together by their pointed extremities so as to describe a
more or less considerable portion of a circle (rosettes). The urine depositing
crystalline phosphate of lime in large amount has ordinarily a pale color, is
abundant, with a feebly acid reaction, but it readily becomes alkaline under
the influence of the mucus with which it is mixed. According to Bence
Jones, this deposit can be produced at will by administering lime water, or
the acetate of lime.—Two general remarks may be made concerning the forms
which phosphate of lime may assume:

  a. When it is amorphous, it is transparent, or, if its granules are dark they
are in groups of two or three, never more than four or five together, and this
separates them from the granules of urates, which are yellowish and much
larger when they are isolated, and ramified like sprigs of moss when they are
agglomerated.

  b. When crystallized, its crystals manifest a great tendency to grouping,
which distinguishes them from the habitual solitary ones of ammonio-magnesian phosphate. This salt is of greater pathological importance than the
triple phosphates.

  These rules, be it understood, are not absolute, and their exactness must be
ascertained by micro-chemi-
cal analysis, as indicated
hereafter.

Fig. 24.

A. Prismatic crystals of triple phosphate.
B. Penniform crystals of triple phosphate.
C. Stellar and foliaceous crystals of triple phosphate.
D. Mixed phosphates; amorphous phosphate of lime.
    A few drops of solution of sesquicarbonate of am-
    monia (one drachm of the salt to one fluid ounce of
    distilled water), added to urine passed after the
    digestion of a meal, will precipitate the neutral
    triple phosphate.

  *Phosphate of lime is the pre-
cipitate formed on heating the
urine,* or, *on treating it with
an alkaline solution* (soda or
potassa). However the precipitation by heat alone occurs only when the urine
contains an excess of phosphate; it may be distin-
guished from albumen by
the addition of a little nitric
acid, which dissolves the
phosphate leaving the albumen intact.

  2. *Ammonio - magnesian
Phosphate. Triple Phosphate.*
This deposit in small
amount may be normally
present in slightly acid or
neutral urine. But it al-
ways appears when the
urine becomes alkaline. In
some diseases of the spinal

cord (paraplegia), and especially in chronic affections of the bladder, sediments are observed consisting solely of crystals of ammonio-magnesian
phosphate. These crystals are remarkable for their glistening, glass-like
aspect, the regularity of their form, and their large size. The *coffin-lid form*
is the more frequent; it may be termed the standard form. These are varieties easy to recognize and which every one may observe. See *Fig.* 24.

When this phosphate is suddenly precipitated by the addition of an excess
of ammonia to the urine, it assumes a different, but no less characteristic,
appearance; the penniform or foliaceous crystals thus obtained are generally
known as *fern-leaf crystals*, which gives a good idea of their appearance. See
Fig. 24.—Sediments of ammonio-magnesian phosphate are almost always
accompanied with phosphate of lime, either amorphous or in granules; they
may likewise be mixed with finely granulated deposits of whitish alkaline
urates; roundish grains of carbonate of lime; and octohedral crystals of
oxalate of lime. The mode of distinguishing these deposits from each other,
chemically under the microscope is as follows:

IV. Micro-chemical Diagnosis of Earthy Phosphates, and of other Crys-
talline Deposits.

Add a drop of *acetic acid* to the drop of deposit on the glass slide.	The *earthy phosphates* are dissolved; the phosphate of lime more slowly than the ammonio-magnesian phosphate.
	The *urates* equally disappear, but are replaced by lozenge shaped, or square crystals of uric acid.
	Carbonate of lime, if present, which is rare, is dissolved, giving out gas bubbles.*
	Remain undissolved in the preparation. { *Oxalate of lime.* *Uric acid.*
Add a drop of *carbonate of soda* to another drop of the deposit.	Are dissolved. { *Urates.* *Uric acid.*
	Remain undissolved. { *Earthy phosphate.* *Oxalate of lime.* *Carbonate of lime.*

* (The bubbles of gas that rise in the liquid of the preparation, and which
are due to the decomposition of the carbonate of ammonia of the alkaline
urine by the acetic acid, must not be confounded with those disengaged by
the carbonate of lime. See *Urinary Sediments. Action of Acetic Acid.*)

Each operation of the above table, must be made upon separate drops of
the deposit, and no two successively upon the same drop.

*Separation of Phosphate of Lime from the Ammonio-magnesian Phosphate, under
the Microscope.* In certain instances this separation is very difficult to effect.
The best method is to add a drop of solution of oxalate of ammonia to a drop
of the urine containing the mixture of phosphate of lime and ammonio-magnesian phosphate. The amorphous mass of phosphate of lime will disappear,
and be replaced by small squares of oxalate of lime, while the crystals of
ammonio-magnesian phosphate will remain unaffected.

Study of the Earthy Phosphates. In hospitals this study can be made very readily with the urine of patients laboring under vesical catarrh. Normal urine will also yield them if permitted to undergo putrefaction.—In order to examine each phosphate separately, treat fresh normal urine with carbonate of soda; a precipitate of phosphate of lime will be obtained, which permit to become deposited. Then treat the clear supernatant liquid by an excess of ammonia, and allow it to rest for several hours; a light precipitate will be obtained of ammonio-magnesian phosphate, in the form of foliaceous, stellated, or penniform, and sometimes in isolated, crystals.—When fresh urine is treated by ammonia, the precipitate formed will consist of a mixture of the triple phosphate and phosphate of lime.—It is very difficult, if not impossible, to determine an excess of earthy phosphates, except by quantitative analysis. A normal or diminished amount of these phosphates may be precipitated in an alkaline urine; while on the other hand an excessive quantity may be held in solution if the urine be very acid. The following quantitative analysis may be made: Take 50 c. c. of urine and add ammonia to it; filter to remove the earthy phosphates formed. To the filtered liquor add a little carbonate of ammonia, and then some sulphate of magnesia; allow this to stand for 12 or 24 hours, and filter to remove the precipitated penniform crystals of triple phosphate. Dry and weigh each precipitate; the first precipitate obtained will give the amount of the earthy phosphates of lime and magnesia in the urine employed; the second, the amount of the alkaline phosphates of soda and potassa. And from this, the amount of each in the urine of 24 hours may be calculated. It must not be forgotten in the analyses for phosphates, that certain salts of magnesia taken internally temporarily increase this substance, and, consequently occasion a phosphatic deposit, in the urine.

Quantity of Earthy Phosphates Eliminated Daily. Acccording to Bencke, a man eliminates by urine in 24 hours, 1 grm .20 of earthy phosphates. According to Neubauer, in 100 parts of earthy phosphates there are 33 parts of phosphate of lime, and 67 parts of phosphate of magnesia. It must be borne in mind that a considerable quantity of earthy phosphates is eliminated with the feces. It is probable that the excess of phosphates in the food eaten, is rejected by this route. For, according to Neubauer, the ingested salts of lime do not pass through the urine, although Roberts sustains the contrary.—As the rule, phosphates exist in smaller proportion in young children than in adults; old persons have an abundance of phosphates in their urine. Diet, and exercise, both mental and physical, exert, however, an influence upon this amount. Beef, milk, potatoes, and bread, are exceedingly rich in phosphates; while starchy, saccharine, and fatty foods contain but small amounts, if any.

Clinical Import. The presence of earthy phosphatic deposits in the urine apprises the physician of an alkaline, or neutral, condition of this fluid, the resulting consequences of which he should be aware, and should, therefore, correctly ascertain the cause of this condition, and promptly employ the appropriate remedy.—If the urine contains a sediment of earthy phosphates at the moment it is voided, it is evident that this sediment must originate in the inner urinary passages, and then the formation of vesical calculi is to be feared. An increase of the phosphates may be present in cases of phrenitis, injuries to the head; paralysis, especially when the result of some malady of the spinal cord, rickets, etc.; but in such cases, it is due to the disease, which must be removed by appropriate treatment in order to remove the abnormal amount of the phosphates. In addition to which, measures must be taken to remove any local cause that may exist favoring the production of gravel or stone, as the presence of mucus, fibrin, blood, etc.

If the precipitate of earthy phosphates is entirely amorphous, it is evident that the alkali causing it is not ammonia, and calculous formations are not so much to be feared. But should the deposit contain crystals of triple phosphate, the presence of ammonia, probably from decomposition of urea, is indicated, and the more persistent this deposit, and the nearer the period of emission at which it occurs, the greater reason for apprehending the formation of calculi, irritation and ulceration of the lining membrane of the bladder, or, some affection of the spinal cord. See *Alkaline Urine.*—It has been supposed by many that the excretion of phosphate of lime by the urine is increased in certain osseous diseases, as, osteomalacia; but there are very few researches upon this question. In a case of extensive burn, a considerable increase of earthy phosphates in the urine was observed.—Phosphates are found in abundance in the urine of maniacs, but are diminished in renal and intestinal affections, as well as in grave pneumonia,—their reappearance is a favorable prognosis. Phosphate of magnesia increases in meningitis, and diminishes in typhus and grave fevers, but reappears as convalescence comes on. Phosphate of magnesia, as well as *creatine*, abounds in the urine of those attacked with progressive atrophic muscular paralysis. Phosphate of lime augments in mollities ossium; it is also abundant with diabetics who satiate their thirst,—and in meningitis, cerebral tumors, tumors of the spinal cord, caries, osseous tumors, and tertiary syphilis.

It must not be forgotten, however, that amorphous phosphate of lime is frequently present in the urine of those who are severe students, who lessen their hours of sleep, who exhaust their nervous systems, who use certain articles of diet, citrates, tartrates, carbonates, etc. On the contrary, food or drink containing lime, as well as certain maladies, cancer, diabetes, etc., are very apt to give rise to the crystalline forms of phosphate of lime. The presence of ammonio-magnesian phosphate indicates an *alkaline fermentation* of the urine, due to retention of urine, vesical calculus, or vesical paralysis.

Phosphate of Lime, Acid. *Acid Calcic, or Calcium Phosphate. Biphosphate of Lime.* This salt exists in the urine in variable quantity in a state of solution. It is deposited when the urine has been evaporated to one-third, or one-half, its volume. The crystals are both large and very small, have the form of elongated hemioctohedrons, are colorless, transparent, with clear edges, and hardly refracting light. They may be isolated, but are more commonly united in twos or in more voluminous groups. This salt does not always crystallize, but frequently forms an amorphous layer on the surface of the evaporated fluid which unites into groups with the blackish spherical masses of urate of soda. When crystallized it is always accompanied by this amorphous matter which holds it in groups with the soda urate. This salt is very soluble in acetic acid. It must be recollected that the composition of the phosphatic salts is constantly undergoing change, from the presence and action of the acids and bases in the urine.

Phosphate of Magnesia. *Magnesic,* or *Magnesium Phosphate.* This salt is met with in all the fluids and solids of the body, especially in the muscular flesh; it has likewise been detected in pus, in the serosity of several cysts, in the serosity and pus of the pleura and of the peritoneum in chronic pleurisy, and in the fluid of ovarian cancer, etc. It has also been found in uric acid, and phosphate of lime, calculi. It is generally in liquid form, though it may readily pass to a solid and crystalline state. From its diffusion throughout the system, it must perform an important part in the constitution of the solids and tissues, not yet determined by physiologists. The surplus amount of this earthy salt passes partly by stool, and partly by the urine. To determine its presence, reduce the solid or liquid to an ash; dissolve this in very dilute hydrochloric acid; then add to the solution hydrochlorate of ammonia, and ammonia, which occasions a precipitate of ammonio-magnesian phosphate. When the urine of herbivorous animals is evaporated, we obtain the phosphate of magnesia in the form of brilliant crystals, oblique prisms with rhomboidal base, having unequal growths on two of the vertical edges. However, there will often be slight differences in the aspect of these prisms.

Phosphate of Potassa. *Potassic,* or *Potassium Phosphate.* This salt, although it has been found in the substance of the nerve centers, has been detected rarely, if at all, in human urine.

Phosphate of Soda, Acid. *Biphosphate of Soda. Acid Sodic,* or *Sodium Phosphate.* This salt is met with in urine, and is considered one of the principles from which this fluid derives its acidity. In fresh and acid urine it is almost constantly present. It may be obtained by pursuing the same process as given below for procuring the neutral salt. Three or four days after the neutral salt has crystallized, transparent, rectangular, prismatic crystals will be observed deposited chiefly upon the sides of the vessels, and which are more soluble in water than the neutral phosphate of soda. Their formation may be hastened by adding ether to the liquid already diluted with absolute alcohol.

Phosphate of Soda, Neutral. *Neutral Sodic,* or *Sodium Phosphate. Microcosmic Salt. Essential Salt of Urine.* This neutral phosphate is found in all the solids and fluids of the system; being very soluble in water, it is almost always met with in a fluid state in the economy. Like the other phosphates in urine it may change into an acid or an alkaline salt, according to the chemical composition of the urine, with regard to its acid or alkaline condition. It may be obtained from the urine by first depriving it of any fat or albumen it may contain; then strongly concentrate the urine, decant or filter to separate it from the saline deposit, and to the clear liquid add absolute alcohol. Crystals of neutral phosphate of soda slowly form upon the sides of the vessel. They are in table forms derived from the rectangular prism, with unequal growths on their edges. When formed in urine they are almost always striated on the surface. If the urine is clear, the crystals are colorless and transparent; if it be colored, the crystals will retain some of the coloring matter.—*Phosphate of Soda and Ammonia, Ammonio-sodic Phosphate*, is frequently found in putrid urine containing the triple phosphates; the prismatic crystals of this salt, although strongly resembling those of the ammoniomagnesian phosphates, are quadrangular, or of some form derived from it. This salt has also been called *microcosmic* and *fusible salt*. It is only found in urine commencing to decompose, or that has become putrefied. Its chemical reactions are similar to those of the ammonio-magnesian phosphates.—For volumetric and other solutions of soda, see *Soda, Phosphate of*.

Phosphates in General. Phosphates that may be met with in urinary deposits, have heretofore been treated upon. (See *Phosphates, Phosphate of Lime, Ammonio-magnesian Phosphate*.) Under the present caption will be given processes of detection, separation, and estimation, applicable to phosphates in general.—Urine contains alkaline and earthy phosphates; the *alkaline* are the phosphates of soda, either acid, neutral, or basic. *Phosphate of potassa* does not normally exist in the urine; introduced into the economy it is decomposed in the presence of carbonate of soda, giving phosphate of soda and carbonate of potassa. The *earthy* phosphates are those of lime and magnesia. The latter readily unites with ammonia forming the ammonio-magnesian phosphate of urinary deposits.—The presence of phosphates may be recognized by adding nitric acid and a little molybdate of ammonia to the deposit diluted with distilled water; on heating the mixture a characteristic yellow color is produced.

As the physician is not, and can not be by profession, a chemist, it must not be expected that he should enter into the more difficult and complicated chemical processes for detecting and estimating the various sediments that may occur in urine. But he must have reliable agents to prevent the least error, and, in the present instance, those, the use of which is quite easy, and at the same time quite exact, have been selected, and which may be prepared by any good pharmacist.

Detection and Separation of Phosphates.

Two cases may present themselves to the physician:—1st. The urine is recently passed, and is clear and limpid. See I. *Detection*.

2nd. The urine is turbid or sedimentary. In this case filter or decant, and test the clear liquid according to I. *Detection*. Examine the sediment under the microscope, and if it be desired to analyse it, see II. *Separation of Phosphates, etc.*

‡ I. Detection.

Add an excess of ammonia to the urine under investigation, agitate, and then allow it to rest.	*Earthy phosphates* are precipitated. See II, A. *Alkaline phosphates* remain in solution. See II, B.

Upon examining the precipitate obtained, as above, under the microscope, it will be found to consist of triple phosphates in fern-like, foliaceous, or stellar crystals (see *Fig.* 24), and some very pale amorphous phosphate of lime. (See *Fig.* 24.) If it be desired to precipitate only the phosphate of lime, add to the urine, instead of ammonia, a solution at the ⅛th of carbonate of soda.

X. Table.

II. Separation of Phosphates, or of their Bases.

A. Earthy Phosphates.

1. Throw upon the precipitate *acetic acid*, q. s., until it is completely dissolved.	Neutralize with a few drops of ammonia, and then add a little hydrochlorate of ammonia.	Then add slowly, and in excess, *oxalate of ammonia*.	A precipitate. Lime, in the state of oxalate; ascertain its characters under the microscope. *Ammonio-magnesian phosphate* remains in solution; decant or filter. See 2.
2. To the filtered or decanted liquor add an excess of *ammonia*.	A precipitate.		Ammonio-magnesian phosphate; recognize its characters under the microscope.
	No precipitate.		Place a little of the liquor in a sample tube and close it; examine it in 24 hours afterwards; if there is no precipitate, it is because there is no phosphate of magnesia in the urine.

1.—If, as frequently occurs, too much ammonia has been added, so as to render the urine turbid, a drop or two of acetic acid will cause the turbidity to disappear, and we will continue adding the hydrochlorate of ammonia to it, the presence of which prevents the precipitation of the magnesia by oxalate of ammonia. This process detects very slight traces of lime.—The oxalate of lime, thus precipitated, appears under the microscope in the form of very black crystalline points, united in groups, sometimes in horse-shoe form, or in rosette. After a longer time the octohedral forms appear. Should there be any doubt concerning the formation of a precipitate, a little of the fluid should be set aside, to be examined 24 hours subsequently. See *Fig* 23.

2.—When there is a small quantity of magnesia present, this precipitate always forms slowly in stellar or foliaceous crystals.

XI. Table.
‡ II. B. Alkaline Phosphates.

| Divide the urine with ammonia (I. Detection), into two unequal parts. | One-third (a) will serve for the detection of *phosphoric acid*. The remaining two-thirds (b) will serve for the separation of the bases of the phosphates, the phosphoric acid having been recognized. |

a. Acidulate with nitric acid.	Add at least an equal volume of *molybdate of ammonia*.	A yellow precipitate.	*Phosphoric acid*, in the state of phospho-molybdate of ammonia, insoluble in acids, soluble in alkalies.
		No immediate precipitate.	There are only traces of phosphoric acid. Heat the mixture to about 104° F.
b. Drive off the ammonia by boiling and then separate the liquor into two parts.	1. Add *alcoholic solution of bichloride of platinum*.	A precipitate.	*Potassa*, in the state of chloroplatinate, in beautiful yellow octohedral crystals.
	2. Add *solution of bimeta-antimoniate of potassa*.	A granular precipitate.	*Soda*.

It is necessary to previously recognize the presence of phosphoric acid, because potassa and soda exist in the urine in the state of urates and sulphates. The presence of phosphate of potassa in the urine is doubtful.

a. The yellow precipitate becomes attached to the sides of the glass tube if there is a little phosphoric acid; under the microscope octohedral crystals, with very dark outlines, and appearing rounded, often united in masses analogous to frog's eggs.

b. 1, 2. These two precipitates equally attach themselves to the walls of the tube; they do not form immediately, but in 12 or 24 hours.

Estimation of Phosphates. To estimate the variations of the phosphates in the urine, the general method is to determine the amount of phosphoric acid present. But the methods for this purpose are entirely too lengthy and too complicated to be performed by the physician, who must, therefore, be satisfied by observing the daily variations of the earthy phosphates, precipitating them each time from a given volume of urine.—For instance, 10 c. c. of fresh or filtered urine are placed into a tube or glass jar graduated into fifths of a cubic centimetre, upon which 4 or 5 c. c. of ammonia are to be poured; then agitate thoroughly, and allow the mixture to rest. On the next day the height of the precipitate in the tube may be read off. This process gives accurately comparable results. By repeating this very simple process every day at the bed-side of the patient, a curve may be traced representing the daily variations of the terreous phosphates during the disease.— It has been ascertained that 1 c. c. of this precipitate of 24 hours,

Fig. 25.

Jar Graduated into 1-5 c. c.

is equivalent on an average to 0 grm .02 of earthy phosphates; with this data, and knowing the volume of the urine from which this precipitate has been obtained, daily, the quantity of earthy phosphates passed per day in each litre of urine, can be readily calculated.

Phosphoric Acid. About 2.6 to 3.5 grammes of phosphoric acid, not free, but united to alkalies and earths, is normally passed during each 24 hours. It may sometimes be desirable to estimate the quantity of phosphoric acid existing in the urine, and for which purpose several processes have been recommended. Among these, the following has been considered the best:— Into a beaker or porcelain capsule place 50 c. c. of the clear urine * to be tested, and 5 c. c. of *acetate of soda solution*, and heat on a water bath to about 200° F. While still warm add to it, drop by drop from a graduated burette, *standard solution of acetate of uranium*, until a precipitate is no longer formed, or what is still better, until a piece of *potassic ferrocyanide paper* gives a brown or reddish-brown color when a drop of the mixture in the beaker is placed in contact with it, by means of a glass rod; then the operation is terminated. Ascertain the number of cubic centimetres of uranium solution required in the process, and multiply them by .005, which will give the amount of phosphoric acid contained in the 50 c. c. of urine. As 1. c. c. of the uranium solution is equal to 0.005 gramme of phosphoric acid, the amount of this acid existing in the urine of 24 hours (when mixed together) may be readily calculated.

* Should there be a deposit of earthy phosphates in the urine, a few drops of hydrochloric acid should be added to dissolve it, and then filter to remove any mucus present. If albumen is contained in the urine, separate it by boiling and then filtering.

The above named solutions, and paper, are to be prepared as follows:— *Standard Solution of Acetate of Uranium.* (*Uranic, or Uranium Acetate.*) To 20.3 grammes of pure uranic oxide add enough strong, pure acetic acid to dissolve it, and then add distilled water in sufficient quantity to make the whole measure exactly one litre or 1,000 c. c. Each cubic centimetre of this solution will correspond to 0 grm .005 of phosphoric acid.—The accuracy of the strength of this solution may be determined, if considered necessary, by placing in a beaker 50 c. c. of standard solution of phosphate of soda with 5 c. c. of acetate of soda solution. Warm this and drop standard solution of acetate of uranium into it, exactly in the manner described above in the process for testing the urine. When the brown color has been produced, read off the amount of uranium solution used, and which corresponds to .1 gramme of phosphoric acid; if it be not dilute enough to make 1 c. c. equal to 0.005 gramme (or 5 milligrammes) of phosphoric acid, add a sufficient quantity of distilled water to make it so.—*Acetate of Soda Solution.* (*Sodic, or Sodium Acetate.*) Dissolve 50 grammes of pure acetate of soda in 50 c. c. of pure acetic acid, and add enough distilled water to make the whole measure exactly 500 c. c.—*Ferrocyanide of Potassium Paper.* (*Potassic, or Potassium Ferrocyanide.*) Saturate strips of white filtering paper with saturated solution of ferrocyanide of potassium. Dry, and keep them in well stopped bottles.— *Standard Solution of Phosphate of Soda.* Dissolve 10.085 grammes of crystals of pure phosphate of soda, that have not undergone efflorescence, in distilled water, so as to make exactly 1,000 c. c. of the solution, 50 c. c. of which contain exactly 0.1 gramme of phosphoric acid.

Phosphoruria. *Phosphuria.* Luminous or phosphorescent urine. But chiefly applied to urine in which phosphates exist in abnormal quantity.

Phosphorus. Ordinary phosphorus is sometimes employed for the purpose of suddenly or slowly destroying life. When slow poisoning is resorted to, the physician is very apt to erroneously suppose the patient to be laboring under gastritis, fatty degeneration of the liver, or other internal disease. In these cases the phosphorus having been absorbed by the digestive organs is eliminated in the urine in the state of hypophosphoric acid. This acid may easily be detected in the urine by the following process : Place 10 or 20 c. c. of the suspected urine in a platinum or porcelain capsule, and add to it some pure nitric acid (15 or 20 drops). Slowly evaporate the fluid over a spirit lamp, and as the specimen of urine approaches the condition of dryness, the mixture suddenly takes fire, burning like a bundle of matches, which indicates the presence of hypophosphoric acid due to poisoning with phosphorus. *Poulet.*

Photuria. Luminous or phosphorescent urine.

Phthisuria. Diabetic urine.

Pimeluria. *Pimeluric.* Urine containing fatty matter.

Pipette. *Drop Glass.* There are several kinds of pipettes. One, used for taking up substances and conveying a drop or more upon a slide, or into a preparation, consists of a slender glass tube, with or without a bulb, from 6 to 12 inches in length, and having an internal diameter of from $\frac{1}{10}$th to $\frac{1}{20}$th of an inch ; sometimes they are made of larger caliber, but drawn out at one extremity to the desired fineness. This is used by placing one finger upon the upper orifice to close it, and then carrying the other small end to the bottom of a vessel from

Fig 28.

Fig. 27.

Fig. 26.

Pipette.

Volume Pipettes.

Pipette graduated into 5ths.

which a sediment is to be taken ; upon raising the finger and quickly closing the orifice again with it, a portion of the sediment will pass into the pipette

and may be conveyed to any desired point. *Fig.* 26. Other pipettes are made with bulbs, and marked to carry certain quantities of fluid; thus, some carry 1 c. c.; others 5 c. c.; others again 10, or even 20 c. c., and so on. *Fig.* 27. A third kind of pipette is made to answer the purpose of a burette in volumetric analysis; it is a straight glass tube holding 50 c. c., which are graduated into 5ths or 10ths of a c. c. *Fig.* 28. These are either plain for india rubber attachment, or are made with a glass stop-cock, to admit of the flowing or dropping of the fluids employed.—The best pipette for collecting urinary sediments, is a glass tube having a caliber of about one inch in diameter, and a length of 10 or 12 inches; one extremity of this tube is tapered off to a very small orifice, and is furnished with a glass stop-cock. The deposit of a given urine is poured into this tube, and is allowed to again deposit itself. By carefully opening the stop-cock the most dense part of the precipitate may be obtained, drop by drop, as required.

Planuria. The voiding of urine through other than the natural passages.

Platinum. It must be borne in mind that platinum is injured, during chemical processes, by contact with lead, or with nitro-hydrochloric acid. This metal is also injured by the nascent silicon arising from heating silicic acid and carbon together in a vessel composed of it. Free chlorine, bromine, or iodine injure it; caustic alkalies must not be evaporated in it to the point of fusion of the residue.—*Solution of Bichloride of Platinum. Platinic* or *Platinum Bichloride.* Take of pure bichloride of platinum 1 part, alcohol of 60° 10 parts by weight; mix, dissolve, and place in a well-stoppered vial. Label, "*Alcoholic Solution of Bichloride of Platinnum. Detection of Potassa.*" This solution at the $\frac{1}{10}$th is advised, because it is fully sufficient to detect feeble traces of potassa; it keeps well. This reagent precipitates potassa in fine, yellow, often voluminous, octohedral crystals. It also precipitates ammonia, but there is hardly any necessity for error in this respect, because ammoniacal compounds being volatile, they should, if there be necessity for it, be removed by heat.

Polarizing Apparatus. A very accurate mode of determining the presence of sugar, or of albumen, in the urine, has been based upon the optical properties of such solutions, and to which attention was called by M. Biot. An apparatus designed for the detection of sugar, is termed a *saccharimeter*, and, for the detection of albumen, an *albuminimeter*. M. Soleil, of Paris, and M. Mitscherlich, of Berlin, have each constructed instruments for this purpose; the latter being more commonly employed in this country. The high price of the instrument has prevented it from coming into general use, and yet it is a most desirable instrument for one to have who devotes especial attention to diabetic affections. Mitscherlich describes the manner of using his polarizing apparatus as follows:—" Before commencing to determine the percentage of grape sugar in a liquid (urine), the operator must be satisfied that the apparatus is properly adjusted when at zero (0°). This is ascertained by placing the empty tube in its proper position, and then placing a lighted lamp an inch or two distant from the posterior opening of the tube, and in the axis of the tube; then, on looking through the anterior opening, if the instrument be properly adjusted at 0°, the posterior opening will appear perfectly dark, almost black. If the arm which moves upon the graduated circle surrounding the anterior opening be now moved from 0° to the right or to the left-hand side, the posterior opening will become more and more lighted up, until at last, when the arm is moved to 90°, an intensely bright circle will be observed. On approaching the arm from 90° to 0°, the observer will notice that the posterior aperture, although the arm be exactly at 0°, will be faintly illuminated on the two opposite margins, and that the diameter of the circle, which runs parallel with these two margins, is intensely black. This black-

Fig. 29.

Mitscherlich's Saccharimeter.

est part must exactly divide the spectrum into two halves, the arm being at 0°; the instrument is now ready for use.

(Diabetic urine is usually quite colorless, scarcely ever requiring treatment with animal charcoal, but simply filtration. If a solution of neutral acetate of lead be added to highly colored urine, nearly all the coloring matter will be precipitated.)

To use the polarizing apparatus, a colorless solution of sugar (filtered diabetic urine) is poured into the tube, which, when filled, is put into its former place; then upon looking through the anterior opening, it will be observed that the dark spectrum appears illuminated in colors, and on moving the arm towards the right, from 0° to 90°, the colors will appear in the following order, yellow, green, blue, violet, red.—The line between the blue and violet colors is the same as that which formed the darkest point of the spectrum, and on this line all the determinations are based.— A little practice will soon accustom the eye to determine the point at which the spectrum is divided into two equal halves, one of which is violet, and the other blue,—each color being of about the same intensity.—If the above named succession of colors are obtained on moving the arm towards the right, which is the case in solutions of grape sugar or saccharine urine, the expression is, "the liquid rotates towards the right," or, "turns a pencil of polarized light to the right." In the opposite case, " the liquid rotates towards the left."

The angle of rotation to which the arm is moved, is proportional to the concentration, and the length of the column of fluid through which the polarized light has to pass, that is, to the length of the tube into which the fluid is poured. The length of this tube in Mitscherlich's saccharimeter is 200 millimetres.—If a fluid of a certain saccharine strength be put into the tube, the arm being at 0°, and it should require the arm to be moved 40° to the right, in order to see the right half of the spectrum red, and the left half, blue, the same fluid, when placed in a tube of half the length, would only require a movement of the arm to 20°.—On the other hand, if a solution of 15 grammes of sugar be poured into the tube, so as to fill it, and it requires a rotation of 15° to show the two colors in their proper positions; then a solution of 30 grammes of sugar in the same quantity of water as the 15 grammes were dissolved in, will require double the rotation, 30°, for the test colors to appear.

It has been ascertained by accurate experiments that a solution of pure and dry grape sugar 15 grammes, in distilled water to make exactly 50 c. c., will, in a tube 200 millimetres in length, turn the plane of polarization 40° towards the right.—Upon the basis of this experiment, the amount of sugar in any fluid may be readily determined. Suppose the above-name. tube, 200 mm. in length, be filled with a yellowish diabetic urine, or any other saccharine solution, and that to make the line between the blue and violet fall in the center of the spectrum, requires the arm to be moved from 0° to 30°, then the calculation would be as follows: As 15 grammes of sugar in 50 c. c. of water (at 14° C., or 57° 2' F.), have an angle of rotation of 40°, it follows that the quantity of sugar present in any liquid is in proportion to this angle. Hence in the case above supposed, 40 : 30 : : 15 : 11.25—or, 11.25 grammes of sugar will be contained in 50 c. c. of a solution, having an angle of rotation of 30°. With this data, we can calculate the amount of sugar contained in any quantity of the same solution; thus, if 50 c. c. of urine contain 28.5 grammes of sugar, how much sugar will be contained in 600 c. c. of this

c c. grammes. c.c. grammes.

urine? 50 : 28.5 : : 600 : 342. A printed direction as regards the manner of using the apparatus, etc., accompanies each instrument.

Polydipsia. Diabetes; with excessive thirst.

Polyuria. Diabetes, or excessive secretion of urine; with excessive secretion of urea or other solids.

Polyuric. See *Lithuria.*

Porphyruria. *Porphyuria.* A condition of the urine tending to abnormal precipitates of its coloring matter with the precipitated solids; as, urates, etc.

Potassa. *Potash.* As the salts of potassa almost exclusively prevail in the muscles, but a very small quantity is met with either in the blood or in the urine From 1.65 to 7.13 grammes of potassa pass in the urine of 24 hours. When these salts are taken internally their presence or their influence in the urine varies, according to circumstances, not satisfactorily understood. Carbonate of potassa renders the urine alkaline in health, but in some diseased conditions it increases the acidity of this fluid. Nitrate of potassa is rapidly eliminated by the urine without affecting its acidity, although it appears after a few days to diminish the quantity of urea. Acetate of potassa lessens the amount of water, urea, extractives, and especially the earthy salts, increasing the solids and carbonates; occasionally, however, **the** water is augmented. It most generally causes alkalinity of the urine from formation of carbonate. Citrate and tartrate of potassa likewise occasion the urine to become alkaline. *Chlorate of potassa* increases the acidity as well as the urates and pigment. Sulphocyanide of potassium passes out unchanged. Ferrocyanide of potassium becomes ferricyanide. Iodide of potassium varies in its effects upon the urine, although after it is taken internally, *iodine* soon appears in this fluid. *Parkes.* Experiments upon this alkali, as to its effects upon the urine, have not certainly given any certain or positive results. When potassa exists in urine, it may be detected by *solution of bichloride of platinum.* To detect chloride of potassium, crystallize the salts contained in the fluid to analyze, separate the cubic crystals which may be chlorides of sodium and potassium, and purify them by a new crystallization. When once assured that there are no other chlorides mixed with these chlorides, redissolve them in a little distilled water, and add a solution of bichloride of platinum; if a precipitate occurs of the double chloride of platinum and potassium, it indicates the presence of chloride of potassium in the urine. See *Urate of Soda* and *Urate of Potassa, Bromide of Potassium, Carbonates, Chlorate of Potassa,* etc.

The following potassa tests are employed in urinary analysis:—*Solution of Potassa. Liquor Potassa.—a. Milder Solution.* Take of caustic potassa 1 part, distilled water 10 parts by weight; mix.—*b. Stronger Solution.* Take of liquor potassa 1 part, distilled water 2 parts; mix.—*Solution of Neutral Chromate of Potassa.* (*Neutral Potassic,* or *Potassium Chromate.*) Take of pure neutral chromate of potassa 1 part, distilled water 12 parts, by weight; mix. This is a saturated solution. Label "*Solution of Neutral Chromate of Potassa. Estimation of Chloride of Sodium.*"—*Solution of Bimeta-antimoniate of Potassa.* (*Potassic,* or *Potassium Bimeta-antimoniate. Granular Antimoniate of Potassa.*) Preserve any quantity of bimeta-antimoniate of potassa in a well closed flask, and labeled. Its solution can not be prepared in advance because it decomposes in a short time, forming a neutral ammoniate. At the moment it is desired to use it, add 1 part of this salt to cold distilled water 250 parts, and agitate it for quite a length of time, as it is not very soluble; then filter to separate the undissolved part. The filtered solution is an excellent reagent for *soda.* But its employment requires some precautions. Thus, only such solutions can be tested with this reagent as contain no other bases besides soda and potassa. All acid liquors, must, previous to using this reagent, be neutralized with a little carbonate of potassa. Finally, the crystalline precipitate of antimoniate of soda, is slowly thrown down, and is not produced at all if the liquor be very dilute, in which case it must be concentrated. When the precipitate occurs promptly, the crystals are boat shaped; when slowly, they are cubic octohedrons, or four-sided columns tapering pyramid fashion. *Standard Solution of Permanganate of Potassa.* (*Potassic,* or *Potassium Permanganate.*) *For Detection of Iron.* Dissolve pure crystallized permanganate of potassa in some distilled water, and determine the strength of a given volume of this solution as follows:—Take 10 c. c. of solution of ferrocyanide of potassium,* (containing 10 mgrms. of iron), and dilute it with about 50 c. c. of distilled water; acidify this with pure hydrochloric acid. Place the beaker containing this solution upon a sheet of white paper, and from a graduated burette let drop into it some of the dilute solution of permanganate of potash, named above, keeping the fluid constantly in rotary motion by means of a glass rod. The appearance of a yellowish-red color in the liquor indicates the termination of the process. Suppose it required 20 c. c. of the permanganate solution to produce the red color, then 1 c. c. of this permanganate solution will correspond with $\frac{0.010}{20}$=0.5 milligrammes of iron.—Solution of permanganate of potash is not permanent, and should therefore be made only as required; and, as it is acted upon by caoutchouc, being decomposed, the burette employed should have a glass stop-cock.

* *Solution of Ferrocyanide of Potassium.*—7.543 grammes of pure, dry, crystallized ferrocyanide of potassium (equal to 1 gramme of iron) are dissolved in some distilled water, and the solution diluted to 1,000 c. c. Of this solution 10 c. c. correspond exactly with 0.010 gramme of iron. Keep in a well stoppered bottle.—Neubaur and Vogel prefer to this ferrocyanide solution, the following:—Take of crystallized oxalic acid 1.125 grammes (equal to 1 gramme of iron), and dissolve in distilled water, to make exactly 1,000 c. c. of solution. 10 c. c. of this solution, corresponding with 0.010 gramme of iron, are placed in a beaker, heated to boiling, treated with a little dilute sulphuric acid, and then subjected to the action of the permanganate of potash solution until the red color appears. The quantity of this last solution required for this purpose will correspond with 0.010 gramme of iron.— *Standard Solution of Sulphate of Potassa. (Potassic, or Potassium Sulphate.)* Take of chemically pure sulphate of potassa, dried at 212° F., 21.778 grammes, dissolve in distilled water and make 1 litre of solution. One c. c. of this solution contains 10 milligrammes of sulphuric acid, and is exactly equivalent to the strong solution of the chloride of barium. Label "*Standard Solution of Sulphate of Potassa. Analysis of Sulphuric Acid.*"

Purpurate of Ammonia. (*Prout.*) *Murexide.* (*Liebig.*) These names are given to the rich purple tint produced, when to a small quantity of urine is added a drop or two of strong nitric acid, upon a porcelain plate, then evaporated to dryness over a spirit lamp, and the yellowish-red residue (alloxantin) exposed while warm to ammoniacal vapors, or to a drop or two of ammonia. It is employed as a test for uric acid. If traces of uric acid are sought, an excess of ammonia must be avoided; a glass rod should be moistened with this alkali, blowing the vapors from this upon the residue, while the rod is held near it.—According to Hardy, the characteristic coloration of the uric acid is principally due to its modified anhydrous alloxan, then, after the addition of the ammonia, to the isoalloxalate of ammonium. The same reaction is said to occur with caffein, hypoxanthin, tyrosin, and xanthoglobulin; Schiff consequently advises *Carbonate of Silver*, which see; also *Uramile.*

Purpuric Acid. If hydrochloric acid be added to purpurate of ammonia, murexan is separated, and the solution contains alloxan, alloxantin, urea, and ammonia. We have various formula given by different chemists for these two substances; thus, *Murexide*, $C_{16} H_4 (N H_4) N_5 O_{12} + 2$ Aq., and *Murexan*, $C_{16} H_5 N_5 O_{12}$, consequently there is no certainty as to their composition. Purpuric acid in a pure state is unknown.

Purpurin. *Madder Purple.* A red coloring principle obtained from madder root. The name is sometimes erroneously applied to the murexide formation in urine.

Pus Corpuscles. It is somewhat difficult to define pus. Its detection in urine absolutely necessitates the employment of the microscope. The old chemical processes are justly abandoned, because they do not give characteristic reactions. Pus may be described as a pathological, sero-albuminous fluid, holding in suspension anatomical elements called leucocytes (white globules of pus), giving to the liquid containing them an opaque milky aspect, and a more or less creamy consistence. Indeed, there appears to be a direct relation between pus, properly so called, and the blood of leucocythemia. The *leucocytes* may be discovered in urine by means of a microscope having a magnifying power of 300 or 400 diameters; and whenever they are found in this fluid in considerable quantity, it may be affirmed that pus is present. But should these globules not be found, it must not be concluded that pus is absent, because they are rapidly destroyed under the influence of the ammoniacal fermentation of the urine. It must be stated that the most eminent histologists are not in accord concerning the question of the genesis of pus globules.

As it is not designed to enter into a consideration of the several conditions and circumstances tending to the development of pus, but simply to its recognition in the urine, we will, therefore, proceed at once toward this object. In this investigation two conditions may be present:—1st. *The urine may be acid or neutral.* If it contains pus, it will be turbid, and there will be rapidly deposited an opaque white sediment, which, examined under the microscope, will show the pus corpuscles. If these be separated by filtration, the presence of *albumen* in the filtered liquid may be determined by the means heretofore advised —When only a very small quantity of pus is present, and when it has been suddenly conveyed by the passage of the urine through the urinary canals, it appears in this fluid in the form of more or less numerous filaments. This fact is observed with blennorrhagic patients, and particularly with leucorrheic women, likewise with persons after their 40th or 45th year, in whose urine a few pus corpuscles are often found; in these cases, albumen is never found in the urine, and which is accounted for by the small quantity of pus contained in it.

2d. *The urine is alkaline and strongly ammoniacal.* In this case the pus corpuscles may not be detected, their morphological identity having been destroyed by the carbonate of ammonia, and converted into a greyish mucogelatinous, closely adherent mass; the presence of pus may, therefore, be

11

suspected, but not demonstrated. It is not rare, however, to find in the mass, or at the upper portion of the deposit, a certain number of still recognizable leucocytes. They are always accompanied with crystals of ammonio-magnesian phosphate, vibriones, and filaments belonging to fungi on their way of development. In the filtered urine, *albumen* may be determined by the means heretofore indicated.—As this rapid decomposition of the urine is frequent in cases of chronic maladies of the bladder, the investigation for pus should be made immediately after micturition. Sometimes, however, the decomposition is completely effected in the bladder before urination; in such case, we may be certain the urine contains pus.

Chemical Characters. Acetic acid exerts, what is considered, a characteristic action upon pus. A drop of this acid added to a drop of pus upon the glass slide, determines in the interior of the leucocytes the formation of from one to four, generally three, distinct, very brilliant, nuclei, small, roundish, or curved in horse-shoe form, while the rest of the globule is so pale that its outline becomes nearly, if not quite, invisible.—Ammonia very rapidly dissolves them, while it merely renders epithelial cells paler.—*Leucocytes* are sensible to coloring reagents; they color much better when they are fresh. In the application of this property they may be readily rendered conspicuous in the midst of other deposits accompanying them, as, crystals, and especially spores, which do not retain the coloring matter.—Liquor potassa added to pus, as well as the carbonate of ammonia in the urine, converts it into a viscid, gelatinous mass, so ropy and tenaciously adherent to the walls of its containing vessel, as to render it difficult, if not impossible, to pour it from one vessel to another. Mucus is rendered more fluid and limpid by the addition of caustic alkali, while acetic acid develops a filamentous appearance. The glairy mass occasioned by the action of an alkali upon pus, contains triple phosphates, and the filtered urine, albumen, which is seldom present in pure mucus deposits.

· *Microscopic Examination.*—This reveals the presence of leucocytes; if the urine is fresh, and not alkaline, they undergo no special change; some of them may contain one or two nuclei, determined probably by the action of the urinary fluid; the others present the granular or mamellonated aspect common to them. The action of acetic acid has already been referred to. The diameter of these leucocytes varies from .008 to .011 of a millimetre, being larger than blood globules. They are found isolated, or united in a group by mucus; then they undergo deformations, becoming polygonal, or elongated in stick forms. In their interior, more or less voluminous and refracting fatty droplets may often be observed, and which is an indication of oldness; at a more advanced stage they become disaggregated, and then the fatty matter becomes freely diffused throughout the fluid. Or else the fatty globules unite in blackish rounded groups, termed *granulous bodies of Gluge.*— Water added to the leucocytes causes them to swell, and one or more nuclei to appear within them. Even in feebly ammoniacal urine, the leucocytes are swollen, very pale, and show one or two nuclei. Their dimensions may be such that they double in size and become vesicular, then they terminate by rupturing and becoming destroyed.

If, when liquor potassa is added to urine containing a deposit, no change is effected, the deposit consists of phosphates; if the urine is rendered clearer, but not viscid, urate of soda is indicated; if it becomes transparent, viscid, or thread-like, pus is present; if it becomes gelatiniform without clearing up, pus, as well as phosphates, are present. Purulent urine becomes clear and transparent on standing; if phosphate be present also, the deposit presents two layers, a lower one, greyish, and formed of the phos-

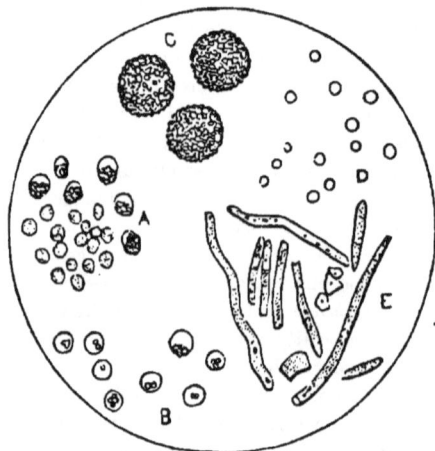

Fig. 30.

A. Pus globules, as they appear in urine, 400 diameters.
B. Pus globules acted upon by acetic acid.
C. Large organic globules, 400 diameters.
D. Small organic globules.
E. Coagulated albumen, in Bright's disease.

phates, and an upper one, very fluid, of a dull opaline blue color, and formed of the pus corpuscles.

Clinical Import.—Whatever may be stated to the contrary by others, the presence of a rather considerable quantity of pus in the urine is always an inauspicious symptom. It is only with women that its constant presence may be observed without leading to the conclusion that there is a catarrhal inflammation of any portion of the urinary apparatus. In hospitals, the origin of this pus in the urine may always be discovered, and, in most instances, will be found to proceed from the vagina or uterus. This origin may, however, be excluded, by obtaining the urine by catheterism.

The chief difficulty is here:—Pus is found in the urine; where does it come from? It does not enter into the province of this work to dilate upon this point. It will merely be stated that pus, in small amount, associated with some mucus and voided with the first jet of urine, in the form of filaments, certainly indicates a chronic inflammation of the urethral mucus membrane, and a stricture in the way of formation. In prostatic disease, shreds or plugs of mucus are apt to be discharged, which are found to contain aggregated pus corpuscles and prostatic epithelia.

The nature and appearance of the pus corpuscles may also furnish some indications. When they are entirely normal, sensitive to the action of acetic acid and of coloring agents, it may be concluded that it is laudable pus, of good character, furnished by a simple inflammation of the urinary mucous membrane.—If, on the contrary, the leucocytes are found more or less altered, accompanied with granular bodies, and cells of various irregular forms, partially destroyed or in fatty degeneration, the existence of a tuberculous or cancerous affection may be suspected. However, the importance of these remarks must not be exaggerated.—Pus accompanied with a large number of renal casts, especially when several of these casts contain within their interior some pus corpuscles, very probably comes from the kidney itself (*pyelitis*). Undeniably, the most frequent cause of purulent urine is vesical catarrh.

Pycnometer. *Picnometer.* This is a specific gravity bottle, containing a small thermometer in its tube for ascertaining the temperature of the fluid contained in it. This bottle is made to hold exactly 50 c. c. or 100 c. c. of distilled water at 60° F., and is provided with a glass stopper having, at its upper and external part, a long tube with a fine capillary canal running through it, which permits any air in the fluid to escape. In order to ascertain the specific gravity of urine, this vessel, well cleansed, is filled with the urine, the stopper properly placed in, the bottle carefully dried externally, and then weighed with the fluid at 60° F. By subtracting from the gross weight thus obtained, the weight of the empty bottle, we obtain the weight of a volume of urine equal to a similar volume of distilled water. By dividing this weight by the known weight of distilled water, the quotient gives the sp. gr. of the urine. Thus, the bottle filled with distilled water weighs 80 grammes, but when empty weighs 30 grammes; consequently the water weighs 50 grammes. Now, when filled with urine the bottle weighs 81.2 grammes, from which subtracting the weight of the empty bottle 30 grammes, there is 51.2 grammes left for the weight of the urine. Then, as the weight of the water is to the weight of the urine, so is the sp. gr. of water 1.000 to the sp. gr. of the urine; thus, $50 : 51.2 :: 1.000 : 1.024$ sp. gr. of the urine.

Pyin. A peculiar albuminous substance detected in pus by Gueterbock, which approaches in character nearer to casein than to fibrin; it is precipitated by corrosive sublimate.

Pyoid Corpuscles. Certain globules observed in pus which give no nuclei when treated by acetic acid, but simply granulations. *Lebert.*

Pyoturia. *Pyuria.* Terms applied to urine containing pus, of which there are several varieties, to distinguish the origin of the pus, as, pyuria vesicalis, urethralis, or renalis.

Q.

Quinia. When this substance is administered internally, in rather large doses, from 15 to 50 per cent. of it passes into the urine, in the course of from two to eight hours. It appears to lessen the formation of uric acid. It may be detected in urine as follows:—1. To 10 c. c. of the clear urine add 5 c. c. of chlorine water, and add to this one drop of ammonia. The formation of a green zone indicates the presence of quinia; of a dark brown, morphia. This will detect $\frac{1}{50000}$th of quinia, and $\frac{1}{7000}$th of morphia. If morphia be likewise present with the quinia, the addition of one drop of nitric acid will convert the green into an orange red, yellow, or brown color.—2. To 50 c. c. of urine add solution of pure tannin until a precipitate no longer forms; separate this by filtration, add milk of lime to it, allow it to stand for a time, then separate the precipitate formed, wash it with distilled water, and then exhaust it with ethereal alcohol. Evaporate the solution, and to the residue add chlorine water, and then ammonia, when the green color characteristic of quinia, will appear.—3. Render the urine alkaline by the addition of liquor potassa, and then add ether and shake thoroughly together; the ether takes up the quinia. Separate the ether, and evaporate it. Now place one drop of test fluid on a glass slide, and add to it one drop, or a little, of the ethereal residue, obtained as above; allow time for it to be dissolved. Then add a very minute drop of an alcoholic solution of iodine, by means of a delicate capillary tube. A yellow or cinnamon colored compound of iodine and quinia is formed at first, and finally the beautiful rosettes or crystals of sulphate of iodo-quinia are formed; no heat is required. Under the microscope, with the selenite plate, and a Nicol's prism beneath, these crystals assume the two complementary colors of the stage,—red and green, if the pink stage is employed,—blue and yellow, if the blue selenite stage is used. See *Vitalis' Method. Test Fluid.*

R.

Rabuteau's Method. This is for the determination of the amount of ammonia in urine, and is based upon the fact that the salts of ammonia decompose with great facility under the influence of the hypochlorites, and that all their nitrogen is set free. As urine contains urea which is also decomposed by hypochlorite of soda and gives off nitrogen, two processes are required in order to estimate the amount of ammonia which this fluid may contain in combination, or in a free state.—The hypochlorite of soda is prepared by exhausting 100 grammes of powdered chloride of lime with water recently boiled, and cooled; then dissolving in this liquid, filtered, 200 grammes of crystallized carbonate of soda, finely powdered. Filter and wash the precipitated carbonate of lime, add the liquors together and obtain 2 litres, and preserve this solution in a well-closed vessel. Squibb's solution of hypochlorite of soda may be substituted for this.

a. Ten grammes of the urine are introduced into a small balloon of 200 c. c. capacity; then the vessel is filled with the solution of hypochlorite of soda and at once closed with a cork furnished with an abductor tube, the extremity of which enters within a graduated tube filled with water. Heat the balloon until there is no longer any disengagement of gas and then divide the volume of nitrogen which occupies the graduated tube, by 34. Let V represent the volume of nitrogen obtained.

b. Lastly, boil 10 grammes of the urine with 1 gramme of carbonate of soda; at the end of 5 minutes of ebullition, there will be no longer any ammoniacal compound. The filtered and cooled liquids are now treated with hypochlorite of soda, the same as named above in *a.* Let V^1 represent the volume of nitrogen obtained this time. The difference $V - v^1$ will represent the volume of nitrogen proceeding from the ammoniacal compounds that were present in the urine. Now, one volume of the nitrogen obtained corresponds to 2 volumes of ammonia; it suffices then to multiply the volume of nitrogen obtained by 2, to have that of the ammonia. If we have $V - v^1 = 0$, the urine contained no ammoniacal compound.

Ratesi's Test for Sugar. This test gives a very sensible reaction with a mixture containing 5 parts of sugar to 1,000 of liquid. Take of concentrated liquor of silicate of potassa 60 grammes, bichromate of potassa 2 grammes, caustic potassa 2 grms .50; dissolve without heat. Upon one of the extremities of several strips of tin, let fall one or two drops of the above reagent, and dry them with heat; then let one or two drops more fall upon the spot, and again dry with heat; repeat this the third time and the apparatus is ready for use. Upon this dried spot let fall a drop or two of urine, or of the fluid suspected to hold sugar, and gently heat it. If sugar be present, the yellow color of the spot assumes a more or less beautiful green color of the oxide of chromium.—This reagent can readily be employed in city or country practice; the small strips of tin thus prepared can be carried in the physician's case, or in a portable sheath, and their use is nearly as easy and expeditious as that of test paper. They may be preserved for several months. M. Ratesi advises to examine the urine, before submitting it to a chemical examination, with the areometer, because if the density marks 0, or — 1, we may be certain it contains no sugar.

Renal Tube Casts. *Renal Casts. Urinary Casts or Cylinders.* It is often the case that peculiar tubular or cylindrical bodies are met with in the urine, especially in those renal affections that have been termed Bright's disease. As it is of considerable importance that these should be detected when present, and especially in an albuminous urine, this fluid should be placed in a narrow conical glass, and allowed to stand for a sufficient length of time, when a portion of the sediment may be taken up with a pipette, or, what is still more preferable, the urine may be filtered through fine cambric linen, from which the deposit may be removed by scraping it off from the linen. In whichever way they may be separated, they should be colored in order to render them more distinct under the microscope. If but few casts are present in the urine other means may be adopted for procuring them in order to examine; thus, the urine may be acidulated with a little acetic acid, which will throw the casts down with the uric acid precipitated,—or, if the specific gravity of the urine be high, the urine may be diluted with distilled water, set aside for an hour, and the deposit be then examined.—Sufficiently firm pressure made with a needle in its handle upon the thin glass cover, just over the body under examination, will crush it if it be a renal cast, while the thin glass may be broken by the pressure, or merely flatten the object should it be a fiber of wood, cotton, flax, wool, or silk, a hair, or other foreign body.

The casts are probably chiefly formed in the straight uriniferous tubules; they are formed by the escape of blood into the renal tubules, from capillary rupture or otherwise, where, the fibrin coagulating, a mould is formed of the shape of the tubule into which the blood has been extravasated, and which subsequently passes out into the pelvis of the kidney, through the ureters into the bladder, and from, thence is discharged with the urine. No doubt many of the hyaline casts are formed in this manner. But the evidence is by no means wanting to favor the view that the epithelial and granular casts are the result of a degeneration and desquamation of the renal epithelium. The investigation of renal casts should, at first, be made with a weak magnifying power, say of 100 or 150 diameters, which will enable us to at once examine a greater quantity of urine; and if the cylinders be observed, they can subsequently be examined under a much higher magnifying power. With a little experience, the physician will soon become familiar with the appearance of these casts, and at once determine them from foreign bodies in the urine, if he has made himself acquainted with the characters of these. A small drop of the urinary sediment is taken up with a pipette, and spread upon the central surface of a glass slide; this is examined under the microscope *without a thin glass cover being placed over it*, the pressure of which causes the cylinders to glide outside from beneath it. The breadth of these casts equals the diameter of from 2 to 6 blood corpuscles [$\frac{1}{1500}$th to $\frac{1}{500}$th of an inch], their length varying from $\frac{1}{150}$th to $\frac{1}{50}$th of an inch (0.1693 to .5079 of a millimetre.

Renal tube casts never become twisted on themselves, as is the case with cotton fibres. They may be rendered more distinct by coloring them with *solution of iodine, solution of carmine*, or with a drop of *solution of fuchsin*.

In the examination of casts, the action of acids upon them, and upon their contents, must be especially noticed. When they resist the solvent action of hydrochloric acid, it has been supposed that the renal inflammation is correspondingly intense. If the granules of the cast are formed of protein, acetic acid will cause them to disappear; if of olein, they will become more distinct. The width of the cylinders is of some importance, as it is supposed that very broad casts are formed in tubules completely deprived of their epithelium, and that the prognosis is more serious when these wide casts show no nuclei on their sides, or an attempt at a reformation of epithelium. However, from recent observations, the importance of the breadth of the cast becomes less. Two or three slides should always be examined before the examiner decides upon the character and significance of the casts. The following four varieties of casts have been admitted:

‡ Renal Tube Casts.

Very pale or transparent amorphous cylinders.	With badly defined margins, often twisted or varicose.	*Mucous casts. a.*
	With very clear margins, sometimes interrupted by fractures.	*Hyaline casts. b.*
More or less dark epithelial or granular casts.	No line of contour, epithelial cellules united into a cylinder, proteinous or fatty granulations.	*Epithelial casts. c.*
	A more or less distinct line of contour. Fundamental substance finely granular studded with blood globules.	*Fibrinous casts* with red blood globules. *d.*

a. Mucous Casts. In normal urine very pale cylinders are frequently met with, formed of a finely granular matter, and which have badly defined margins; they often hold renal cellules, or leucocytes, in suspension. Funcke considers them to be formed of mucin. They are often very abundant in the deposits of albuminous urine; they are not colored by carmine, but acetic acid renders them fibrillary and punctated. As they have no signification, care must be taken not to confound them with the hyaline casts.

b. Hyaline or Transparent Casts. These are very rare in normal urine, and the presence of a large quantity of them in the sediment, is a certain indication of albuminous nephritis. They are distinguished from the preceding by having well defined margins, delineated by a distinct line. They are straight, or spirally twisted, with parallel borders, and are formed of a homogeneous transparent matter of proteinous nature. Their extremities are marked by a clear and distinct fracture, and they often present transverse slits at points along their longitudinal surfaces. Acetic acid difficultly attacks them. Iodine solution, or solution of carmine colors them, but a great deal better if the disease is of long standing; in this case, they have naturally a yellowish reflection. Finally, they often have some epithelial cells upon their surface, from the renal tubules, generally granular, and which sometimes form a complete cortex to the cast; this is especially observed in an advanced stage of albuminous nephritis. We may have all the varieties in one sediment.

A hyaline cast, of more solid aspect, is termed a *waxy cast,* and as these cylinders have none of the physical or chemical characters of fibrin, it is an error to call them *fibrinous.* By the term *hyaline,* nothing concerning their composition is prejudged, while it recalls to mind an important physical character.

c. Epithelial or Granular Casts. These are the internal casts, derived from the renal tubules, the epithelium of which is often altered, granular, and infiltrated with proteinous granulations, or with fatty droplets. These casts are usually wide, never very narrow. When the granular aspect of a cast is of a dark, somewhat solid character, it is termed a "granular cast." The "fatty, or oil cast" is a variety of the granular, produced by the granules of olein running together into globules of fat. Epithelial cells, blood corpuscles, leucocytes, pus corpuscles, urates, uric acid and especially oxalate of lime crystals are often observed in these casts. When blood corpuscles exist in these cylinders they are termed "blood casts."

d. *Fibrinous Casts.* These are casts composed of a finely granular matter which swell and become clearer under the action of acetic acid, and which contain blood corpuscles in their interior. They indicate a hemorrhage, the rupture of some capillary vessels, or an effusion of blood in the interior of the uriniferous tubules.

All these casts may be colored by the bile when there is an icteric complication, or they may be interspersed with granulations of urates, or with crystals, which may be cleared away by a drop of acetic acid.—With the exception of the hyaline, these casts have often been observed in sediments of non-albuminous urine, as, in certain forms of purulent infection, in icterus consecutive to an attempt at asphyxia by charcoal, and which appear to be owing to a state of renal congestion.

Clinical Import. See *Albumen.* "The presence of casts in the urine is a sure sign of disease of the kidney, but not, however, necessarily of a *permanent* disease. They are present in many acute diseases accompanied by more or less albumen in the urine, as, in bronchitis, pneumonia, convalescence from scarlatina, etc. But if they are found for several weeks together, after all pyrexia has subsided, permanent disease of the kidney may be inferred. Casts are constantly present in the urine in all cases of renal congestion, and of acute or chronic Bright's disease. But no certain information as to the nature of the renal disease, that is, whether lardaceous or fatty, can be obtained from the characters of the casts, since all forms of Bright's disease terminate in fatty changes. Some assistance, may, however, be derived from the appearance of the casts in forming a judgment of the acute or chronic character, or, a prognosis, of the disease. If, for example, there be found in the urine any epithelial casts which have undergone little, or no, granular change, and also casts studded with red-blood corpuscles, together with a large quantity of epithelium from the renal tubules, having a natural or only slightly cloudy appearance, there can be little doubt that the patient is suffering from an acute attack of Bright's disease; while if the casts be chiefly fatty, or intensely granular, and the epithelium be small in amount, and the cells withered and contracted, or containing globules of olein, it will be more than probable that the case is one of chronic Bright's disease.—Since little reliance can be placed on the characters of the casts as an aid to special diagnosis, some of the renal derivatives in the chief forms of kidney affection have been subjoined.

"*Congestion of the Kidney.* The casts are chiefly hyaline, seldom showing any marks of fatty change. Very rarely are blood or epithelial casts discovered.—*Acute Bright's Disease.* At the commencement, the urine deposits a sediment which consists of blood corpuscles, narrow hyaline casts, and casts covered with blood corpuscles, the 'blood casts' of some authors. In the next stage, the amount of blood present is not so great, but a considerable desquamation of the renal tubules taking place, renal epithelium and epithe-lial casts are found in great numbers; the epithelium has undergone very little, if any, granular change; hyaline casts are observed together with epi-thelial. In the next stage, the changes in the epithelium may be almost daily observed; at first, they become granular, cloudy in appearance, which alteration (the sequel of the catarrh) often proceeds to fatty degeneration, and the epithelial cells then contain large fat drops, while the epithelial casts undergo similar change, and become distinctly granular and even fatty. If the patient recover, the casts and epithelium gradually disappear from the urine; but if the case becomes chronic, the renal derivatives show the characters described in the next paragraph.

" *Chronic Bright's Disease*. Numerous forms of casts are met with; the hyaline, both narrow and wide forms; the larger are often beset with gran-ules which dissolve on addition of acetic acid; the granular, whose surface is often covered with fatty or shriveled up epithelial cells; fat drops may stud the cylinder. Epithelial casts are rare, except in febrile exacerbations, when the renal derivatives found in acute Bright's disease are present, together with granular and fatty casts, evidence of the previous alteration of the kidney.—*Lardaceous or Albuminoid Kidney*. The urinary deposit contains hyaline casts, which are often accompanied by pus corpuscles. Atrophied epithelial cells, becoming fatty in the later stages of the disease, are almost invariably present." *W. Legg.*

Resins. Resins administered internally affect the urine variously, while some have no influence whatever, others develop peculiar odors, while a third portion affects the color of this fluid. These investigations are, however, of an imperfect and limited character. The greyish-yellow color imparted to urine by resin, is converted into blue by the action of perchloride of iron. Gamboge, taken in large quantity, colors the urine yellow. Aloes imparts a deep-red color, etc. In many instances the odor or the color does not depend upon the resin itself, but upon a peculiar acid or other principle associated with it; thus, it appears to be the resinous acid of copaiba that influences the urine. See *Color of Urine. Odor of Urine.*

Rhubarb. According to the degree of acidity or alkalinity of urine, its color assumes a greenish, or more or less deep yellow color, when rhubarb has been taken internally; the addition of ammonia or potassa to the fluid, changes the color to a fine blood red.

Rosacic Acid, A name that has been applied to neutral urate of soda, likewise, to the coloring matter found in the roseate sediments of urine. As these sediments contained a peculiar acid, Proust termed it rosacic acid, but subsequently found the acid to be merely uric acid and a red coloring sub-stance. This coloring substance is now considered to be one of the modifica-tions of urohematin.

S.

Sabulous. Like sand, as, sabulous urine, gritty or sandy urine.

Saccharomyces Cerevisiæ. A fungus growth supposed to exist only in saccharine fluids, and to be distinct from *torula cerevisiæ*, with which it has generally been associated. It consists of numerous round or ovoid, granular, brownish bodies or cells, presenting one or several distinct, clear, granulated, or dark, nuclei. These cells, during the process of budding, will present an appearance, as if two, three, or more, were united in irregular forms. If an aqueous solution of a small quantity of honey and white of egg, be kept for 12 or 15 hours at a temperature of from 96° to 105° F., these bodies appear, and may be studied under the microscope, in their various stages.

Salicin. According to Landerer and Ranke, this agent, when taken in considerable doses, partly passes away in the urine unchanged; the remainder undergoes changes similar to those effected artificially by oxidizing agents. The changes are supposed to be from salicin to saligenin, saliretin, salicylous acid, salicylic acid, and carbolic acid; the first two have not been detected in the urine, but the remainder have. Bertagnini found salicyluric acid in the urine, supposed to be formed by the combination of salicylic acid and glycocoll, somewhat similar to the manner in which hippuric acid is formed by the combination of benzoic acid and glycocoll.—As with salicylic acid, salicin, when passed in the urine (salicyl hydride), forms a purple-red color with perchloride of iron.

Salicylic Acid. This agent when taken internally is rapidly eliminated by urine; on this account it should be administered in small doses frequently repeated. A few drops of urine, from a person who has taken salicylic acid, added to a small portion of a very dilute solution of perchloride of iron, will occasion a precipitate of a beautiful intense violent color, indicative of the presence of this acid. *Prof. Seé.* If the urine be high colored, previous to applying the above test, an excess of acetate of lead should be added to throw down the coloring matters; then filter, and to the filtered liquid carefully add an excess of sulphuric acid, which will precipitate the lead. Filter again; and to the resulting clear liquid, add a few drops of the test, as above.

Sand. See *Gravel.*

Santonin. *Santonic Acid.* Passes quickly in the urine, to which it imparts an orange-yellow color. This yellow color is converted into a deep red when an alkali is added to the urine (or when the urine itself is alkaline); liquor potassa giving the best results. Rhubarb produces similar results, but crystals of oxalate of lime will likewise be found in the urine, which is not the case when santonin is given, unless these crystals previously existed, pathologically, in this liquid.

Sarcinæ. Sarcina ventriculi is a fungus discovered by Goodsir, in the matters ejected from the stomach, under certain conditions. A number of minute greenish cells, roundish or square, are aggregated together in series of 4, 8, 16, 32, 64, etc., forming a large cube. A similar fungus has been found in the pelvis of the kidney, by Hepwood; and Beale, Begbie, Johnson, Heller, Munck, Welcker, and others, have found it occasionally in the urine. The cells of the urinary sarcina are smaller, and less regularly arranged than those from the stomach and lung, and frequently have a darkly punctated center. Their clinical import is unknown; many regard them as mere accidental formations. They are found in acid, neutral, or alkaline urine, more commonly the last, as, in vesical catarrh, painful urination, renal pains, also, during the presence of dyspepsia, hypochondria, etc. The urinary fungus has been termed merismopœdia punctata (*Meyen*), and Sarcina Welckeri (*Rossmann*). See *Fungi*. *Vegetable Organisms*.

Sarkine. *Sarcine*. See *Hypoxanthin*.

Sediments. See *Urinary Sediments*.

Senna. Senna taken internally imparts a brownish color to the urine, or, according to Gubler, an intense yellow with greenish reflection, similar to icteric urine; but the characteristic reaction of the coloring matters of bile by nitric acid, does not exist. Moreover, if a fragment of caustic potassa, or a drop or two of ammonia, be allowed to fall to the bottom of a tube containing senna-colored urine, a beautiful purple color is formed, which is not the case when the color of the urine is due to bile matter. Prof. Hirtz states that the biliform color produced by senna only occurs in neutral urine, and when the urine becomes alkaline the red color is established; if a drop of nitric acid be added to a small quantity of the urine, the red color instantly disappears. Gubler intimates that these reactions are owing to chrysophanic acid in the senna, as, similar, but less marked, reactions occur with rhubarb.

Silica. Si O_3. *Silicic Acid*. *Oxide of Silicon*. This substance is found in various parts of the body, as, in the blood, muscles, hair, bile, etc. It is also met with in minute quantity in the urine, and has formed a small element in the formation of biliary and urinary calculi. It is chiefly derived from plants belonging to the ceralia, as, wheat, rye, etc., which are rich with silica. It has never been found as a deposit in urine, except when placed there for purposes of deception. It may be obtained by evaporating a large quantity of the urine containing it, to dryness, and reducing the residue to an ash by calcination. The silica will be found in the ash, if present at all; it is insoluble in water and acids, but entirely soluble in boiling liquor potassa, or soda, and in hydrofluoric acid.

Silk. Fibres of silk are sometimes observed in urine; they are of small diameter, of regular size, have a clear outline, and a smooth, shining aspect, and bear no resemblance to renal casts. They swell in nitric acid, and are readily soluble in liquor potassa or soda; in solution of neutral chloride of zinc.

Soda, Carbonate of. See *Carbonate of Soda*.

Soda, Phosphate of. See *Phosphate of Soda.—Phosphate of Soda, Ammoniacal Solution of.* Take of pure phosphate of soda 1, distilled water 10; mix, and when dissolved, add ammonia 2. Label, "*Ammonial Solution of Phosphate of Soda. Detection of Magnesia.*"

1. **Soda, Solution of.** Take of caustic soda 1, distilled water 10; mix. Label, "*Solution of Soda at the $\frac{1}{16}$th."—Solution of Caustic Soda*, sp. gr. 1.12, may be procured by treating the caustic soda ley (of soap boilers) of 36° Baumé, (1.334 with the densimetre), 240 grammes, with distilled water 360 grammes. A total volume of 600 grammes of sp. gr. 1.12, is obtained, containing about 80 grammes of soda in the solution.—2. *Standard Solution of Soda. For Analysis of Ammonia.* Into a small beaker glass place 10 c. c. of the *standard solution of sulphuric acid* (for analysis of ammonia), and add to it a few drops of tincture of litmus until the solution is rendered slightly red. To this add, from a graduated burette, a carefully prepared dilute alcoholic solution of freshly made soda, free from carbonic acid, until the blue color of the litmus is restored. Then read off upon the burette the amount of this caustic soda solution required to neutralize the 10 c. c. of the solution of sulphuric acid. Suppose 30 c. c. of the soda solution have been required, then we know that every cubic centimetre corresponds with .00715 gramme of ammonia, as the 10 c. c. of standard solution of sulphuric acid, neutralized by 30 c. c. of the soda solution, correspond to 0.2146 gramme of ammonia, $\frac{0.2146}{30} = .00715$ gramme.—3. *Standard Solution of Soda. For Analysis of Lime.* Place 10 c. c. of the *standard solution of hydrochloric acid* (for analysis of lime) into a beaker glass, and add a few drops of tincture of litmus until it is slightly reddened. To this add, from a graduated burette, a solution of freshly made soda, perfectly free from carbonic acid, until the blue color of the litmus is restored. Then read off upon the burette the amount of this soda solution required to neutralize the 10 c. c. of hydrochloric acid solution. The soda solution must be of a strength that 10 c. c. of it will accurately neutralize 10 c. c. of the standard solution of hydrochloric acid. Suppose, in the process just given, 8 c. c. of the soda solution were required to neutralize the acid solution, we then measure off 800 c. c. of the soda solution, and add enough distilled water (200 c. c.) to make exactly one litre. (Or, if 6 c. c. of the soda solution were required, we measure off 600 c. c. of this solution, and dilute to a litre with 400 c. c. of water.) Equal volumes of the hydrochloric acid, and the soda solution, will now exactly neutralize each other.

Soda, Urate of. See *Urate of Soda.*

Solids in Urine. The solids in urine comprise the brown, strongly odorous, and bitter residue remaining after the fluid portion of the urine has been removed by evaporation, and which is composed of organic and inorganic acids and bases. See *Fig.* 34. The amount of solids passed in the urine of 24 hours by a healthy person varies from 40 to 66 grammes, depending upon certain physiological (as well as pathological) conditions, and, owing to which, the urine, even of the same person, may contain at different periods, different quantities of solids. Diet, exerts an influence in this respect, the solids being augmented by an animal diet, lessened by a vegetable, and still further diminished by a non-nitrogenized diet; while a mixed dict gives a medium amount. When large amounts of water are taken into the stomach, the solids are usually increased in the urine of 24 hours. Males excrete more solids than females. It will likewise be found that the solids are increased in certain diseases, and reduced in others; also, that the daily amount of solids is influenced by the internal use of fermented or alcoholic drinks, as well as of certain medicines, and by sedentary or active habits of mind and body. A persistent increase in the urinary solids is an unfavorable symptom, indicating an excessive destruction of the constituents of the body, tending to exhaustion and ultimate wasting of the tissues; while a persistent decrease, indicates a diminution of vitality tending to fatality. However, it would be extremely improper to form a conclusion from the amount of solids only; the quantity of fluid, the sp. gr. of the urine, and the nature of the solids must all be taken into account. When urea, the principal solid constituent of urine is retained, the urine does not increase in quantity, though it may diminish, and its sp. gr. is low, as, in uremia. In most chronic diseases, except diabetes, in which the solids are reduced, we may augur favorably when they become increased; while, in acute diseases, an increase of the urinary solids, is an unfavorable indication. See *Agents, etc.; Specific Gravity; Urinometer.*

‡ An approximate determination of the amount of solids existing in the mixed urine of 24 hours may be made in a few minutes, and with sufficient accuracy for ordinary clinical purposes by using Trapp's formula 2, or Hæser's (diabetic urine) 2.33. Thus, having determined the *correct* sp. gr. of the urine, multiply the last two figures of the number expressing this sp. gr. by 2.33, and the result gives the amount of solids, in grammes, existing in 1,000 c. c. of the urine. For instance, a person passes 1,250 c. c. of urine in 24 hours of sp. gr. 1.020. The last two figures (20) of this sp. gr., being multiplied by 2.33, gives 46.6, which is the amount in grammes of solid matters contained in 1,000 c. c. of the urine. But 1,250 c. c., were passed in the 24 hours, and a little calculation will be required to determine the solids in this quantity, which consists in simply multiplying the whole amount of urine passed in 24 hours (1,250) by the amount of solids in 1,000 c. c. of the urine (46.6), and then dividing this by 1,000; thus, $\dfrac{1,250 \times 46.6}{1,000} = 58.25$ grammes.

And so with any amount of urine in 24 hours, whether it exceeds or falls short of 1,000 c. c. Thudicum believes Trapp's formula, which is used in the same manner as above described, to be best suited for urine of low sp. gr., or below 1.018; and Hæser's for urine above 1.018, or of high sp. gr.

The above method, however useful it may be for ordinary clinical purposes, is by no means accurate enough for scientific purposes, which require another process involving more or less difficulty, great care, and expenditure of time; and, from the disgusting odor emitted during the process, it must be conducted in some out of the way place, or in a properly arranged and well-ventilated laboratory. Among the several methods that have been pursued, that advised by Rose, is probably the quickest and best: Into a clean porcelain or platinum crucible of known weight, measure exactly 20 c. c. of the urine, and gradually evaporate it until it is nearly dry. As heat decomposes a certain amount of urea, this may be limited to a minimum by keeping the urine constantly acid. Into the nearly dry mass, place some finely ground platinum sponge, of known weight (1 or 2 grms.), and stir it around by means of a platinum wire, and then continue the evaporation to full dryness. Let the crucible remain for some time in an air bath, or under an air pump, and then weigh it. From the gross weight, deduct the weight of the crucible and of the platinum sponge, and the remainder will be the total amount of solids in the 20 c. c. of urine, from which can be calculated the amount in any given quantity of the same urine. To determine the fixed inorganic salts, the crucible should be placed over a spirit lamp, igniting the mass in it, and continuing the heat until all the carbon is consumed, and a perfectly white ash remains. When the crucible is cool, again weigh it, and from the gross weight deduct the weight of the crucible and of the platinum sponge, and the amount of fixed salts in the 20 c. of urine is obtained.

Solidarity. The mutual responsibility existing between two or more persons, or parts; consolidation; joint interest; a kind of mutual dependence.

Sorrel. This plant when taken internally gives rise to considerable oxaalate of lime in the urine. Gallois states, however, that if the sorrel be used for a long time continuously, this effect diminishes, or ceases entirely.

Specific Gravity. The specific gravity or density of urine is more generally determined by a small instrument called *urinometer*. A portion of the mixed urine of 24 hours is placed in a small cylindrical glass vessel, into which fluid the urinometer is introduced, and as soon as this becomes stationary, the specific gravity is read off. The vessel should be of sufficient length so that the urinometer when suspended in the fluid will not touch the bottom; and its diameter should be large enough to prevent the urinometer from touching the sides of the vessel, which would interfere with the correctness of the result. The above-named method only gives approximative results, but which are considered sufficiently accurate for clinical purposes. See *Urinometer*. But when great accuracy is desired for scientific research, another course is pursued, requiring more time and careful attention. A bottle accurately adjusted so as to hold exactly 20 c. c. of distilled water, and furnished with an elongated ground glass stopper, within which a thermometer is fitted (termed a *pycnometer*), is employed for this purpose. The weight of the pycnometer must be known, as well as the weight of the distilled water which will fill it. The weight of the urine is determined by carefully filling the instrument with this fluid, weighing it, and from the gross weight deducting the weight of the empty pycnometer. Then the specific gravity is obtained by calculation: As the weight of water is to the weight of urine, so is the specific gravity of water to the specific gravity of urine. The sp. gr. of water is 1. See *Pycnometer.*

The indications furnished by the sp. gr. of the urine enables one to calculate the quantity of solid materials (urea chiefly) contained in this fluid; and the daily variations in the excretions of these matters may thus be observed and recorded. These solid matters in the urine are daily modified by the food ingested; thus, soup greatly increases the sp. gr. of the urine. A litre of soup will add nearly 15 grammes of solid matter to the urine; and if tannin be added, an abundant precipitate of gelatin, of modified albumen, and other organic substances, is obtained. In order that the physician may derive satisfactory information concerning the metamorphoses of the organism, from the sp. gr. of the urine, it is highly necessary that all these alimentary influences be well determined; and these physiological variations can be studied without much trouble. The corrections for temperature must not be neglected, or errors may be made amounting to 4 or 6 grammes of residue per litre.—A considerable and permanent increase of the sp. gr. of the urine, should attract attention; it may be due to an increased amount of *urea,* or to an abnormal secretion of sugar.—A diminution of the sp. gr., coinciding with an increase of the quantity of urine eliminated in 24 hours, is met with in certain cases of dropsy, etc.—A diminution of the sp. gr., with a decrease in the amount of urine, indicates an obstacle to the secretion of urea, and frequently an alteration of the kidneys, as observed in the chronic form of Bright's disease.—If the sp. gr. in polyuria be high, it reveals a true diabetes; if, on the contrary, it is low, it represents a false diabetes. See *Solids in the Urine.*

The average sp. gr. of healthy urine of 24 hours is from 1.015 to 1.0120. That which is passed after drinking much water or fluid, from 1.003 to 1.009, is generally pale, and is termed *urina potus;* that which is passed after the digestion of a full meal, *urina chyli,* has a sp. gr. of 1.030; and that which is passed after a night's rest, *urina sanguinis,* furnishes the best specimen of the average density of the whole urine, varying from 1.015 to 1.025. However, as heretofore remarked, various circumstances may cause the sp. gr. to vary. Although the sp. gr. does not yield direct indications of disease, still it furnishes important information; thus, a persistently concentrated urine of sp. gr., below 1.015, would lead to an examination for albumen,—if still lower, 1.005 to 1 008, there may have been an hysterical attack, or diabetes insipidus may exist. A pale, *apparently* dilute urine, of sp. gr. 1.025, may be due to sugar, most certainly so if the sp. gr. ranges between 1.035 and 1.065.

Prof. Haughton states that with urine of persons in health or in disease, but containing neither sugar nor albumen, the quantity of *urea* may be determined by simply subtracting 1,000 from the sp. gr., and then multiplying the remainder by 10. The product will be the number of grains of urea in a pint of the urine. In 19 cases, the averages have varied only from 149 to 156 grains, which, in practice, is of slight importance.—As simple an operation is used for determining the amount of *sugar;* instead of multiplying the remainder by 10, as in the preceding case, multiply by 20, and the solution is given. In 16 experiments the averages did not differ 2 grains.

The following Table has been given for the determination of the proportion of solid extract in any quantity of diabetic urine, by its specific gravity, this sp. gr. being compared with 1,000 parts of water at 60° F.

XII. Table.

Specific Gravity.	Grains of Solid Extract in a *Wine Pint* of Urine	Specific Gravity.	Grains of Solid Extract in a *Wine Pint* of Urine.
1.020	382.4	1.036	689.6
1.021	401.6	1.037	708.8
1.022	420.8	1.038	728.0
1.023	440.0	1.039	747.2
1.024	459.2	1.040	766.4
1.025	478.4	1.041	785.6
1.026	497.6	1.042	804 8
1.027	516.8	1.043	824.0
1.028	536.0	1.044	843.2
1.029	555.2	1.045	862.4
1.030	574.4	1 046	881.6
1.031	593.6	1.047	900.8
1.032	612.8	1.048	920.0
1.033	632.0	1.049	939.2
1.034	651.2	1.050	958.4
1.035	670.4		

(See *Table of Corrections for Saccharine Urine in Appendix.*)

Spermatozoids. *Spermatozoa. Zoosperms. Seminal Filaments. Spermzoons. Seminal Animalcules.* See *Fig.* 31. Spermatozoids can be detected in the urine only by means of a microscope; using a low power at first, to rapidly pass the suspected deposit under review, should only a few of these be present; and subsequently, a higher power of 350 or 400 diameters, should filaments resembling spermatozoids have been discovered in the fluid by the low power. A power of 500 diameters shows them well. They consist of an anterior, flattened pyriform enlargement (the head), and a finely tapering filiform appendage (the tail); all these parts appearing completely homogeneous and brilliant. Their entire length is about 0 mm .050, the head alone measuring about 0 mm .005 in length. At its orgin, the tail measures about 0 mm .001 in diameter, from which point it gradually diminishes in diameter up to its extremity. In semen these filaments move rapidly through this fluid, and in a peculiar, undulating manner. The rapidity of movement has been estimated at 0 mm .060 per second, that is, they move in a second through the space of a linear quantity nearly equal to their own length. The force of projection developed by this undulatory motion is sufficiently powerful, not only to displace the filament in the midst of the medium in which it moves, but even to push and displace many epithelial

Fig. 31.

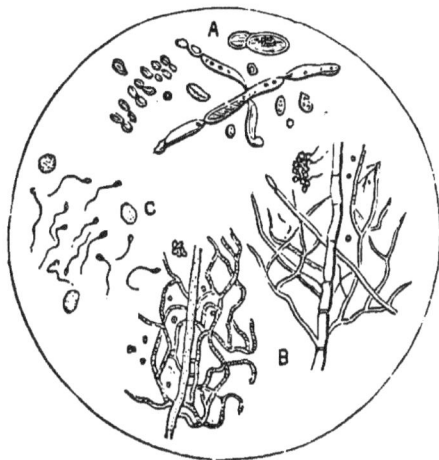

A. Torula cerevisiæ, found in urine after it has undergone saccharine fermentation.
B. Penicilium glaucum, a fungus growth in acid urine
containing albumen, when exposed to the air.
C. Seminal animalcules or spermatozoa, and seminal
granules.

cells and crystals larger than it, which have formed from the evaporation of the fluid on the glass slide, and against which it strikes in the course of its movements. As the liquid in which they move becomes condensed by evaporation, phosphatic crystals form in the field of the microscope, the motions of the spermatozoa become less and less active, and when motion is about to cease, they fold themselves so as to form a loop, or kind of ring; but they are not dead, as an alkaline solution will, even after an hour or two, restore their movements.—In seminal urine may also be detected a few minute granular corpuscles, of a round or oval form, and rather larger than the bodies of the spermzoons, termed "sperma'ic granules." Traces of albumen may also be quickly detected in urine, by the application of heat and nitric acid; and large octohedra of oxalate of lime are of common occurrence.

Water or acid fluids added to semen arrests the movements of the spermatozoids, and hence, when observed in urine, they are always motionless, unless a considerable amount of pus be likewise present. These anatomical elements are very easy to recognize, and can not be confounded with others. In sedimentary urine, they are found mixed with the deposit, and are very often held in some filaments of mucus. In urine, *post coitum*, they are thus presented. A remarkable character is their decided resistance to nearly all causes of destruction. They have been found in putrefied urine at the termination of three months. Mineral acids and caustic alkalies attack them only when heated. Acetic acid effects no change in them, except to render them more conspicuous. After drying and softening in water, they preserve a recognizable and perfectly characteristic form, which is of great importance in medico-legal examinations. According to Valentin, calcination itself leaves their form intact. Ammonia, however, rapidly acts upon them, even when cold. Solution of carmine, iodine water, etc., which color leucocytes and epithelia, have no action on spermatozoa. In cases where they are mixed with certain deposits, so as to interfere with their detection, these deposits may be removed and the field of the microscope be cleared up, by the addition of a little acetic acid, if phosphates be present; and, if urates, by a solution of soda, or of potassa, at the $\frac{1}{10}$th.

Clinical Import. The presence of spermatozoids in the vaginal mucus, or upon the linen of a woman, is considered as being proof of coition, or of an attempt at it; and their detection, therefore, is of immense importance in cases of suspected rape. By soaking a piece of muslin or linen, which has a stain upon it, in some water for an hour or so, and then examining the sediment in the water, these filaments, when present, can readily be detected. Spermatozoids are frequently found in the urine of males who are in a state of health. But when they are observed more or less constantly in the urine, accompanied with other more important symptoms, they demand interference. Frequently, their presence signifies masturbation, and thus may be revealed to the physician unacknowledged habits which he can only suspect. They are met with in urine after coition, as well as after nocturnal pollutions; they have been observed in the urine of patients laboring under typhus, and under the badly defined disease, known as involuntary losses, or spermatorrhea; in severe cases of this last named malady, they may be broken, imperfect, and deformed. A few spermatozoids in the urine, requiring great care and delicacy to collect for examination, are of no clinical importance whatever.

Spirillum. Small, short filaments of corkscrew form, moving rapidly by a very remarkable spiral or gimlet-like motion. They are insoluble in boiling potassa. They vary in size from $\frac{1}{1400}$th to $\frac{1}{800}$th of an inch in length, and from $\frac{1}{20000}$th to $\frac{1}{12000}$th of an inch in diameter. See *Vegetable Organisms.*

Stains, to Remove. *Nitrate of Silver.* Take of cyanide of potassium 10 grammes, distilled water 125 grammes; mix. At the time of using this, add tincture of iodine 20 drops, or, in proportion to the amount of solution employed. Place a few drops of this mixture on the spot, and gently rub the moist spot with the fingers; then wash with soft water. The operation should be performed in a darkened place, as a bright light destroys one of the principal agents, the iodide of cyanogen. It must be remembered that in presence of acids the cyanide of potassium disengages prussic acid, and that any reaction of this kind must be avoided, or serious results may occur.—2. Dissolve half a grain or a grain of iodine in 10 or 15 drops of ammonia, then, by means of a brush, apply some of the solution to the stains; after the stains disappear, wash with water. Throw away the solution when done with it, because when dry it forms the explosive iodide of nitrogen. This is a prompt procedure, free from the danger attending the use of cyanide of potassium. *Nitric Acid.* When the spots are fresh, wash them repeatedly with a concentrated solution of permanganate of potassa, and then wash with water. Remove the brown stain produced by the permanganate, by an aqueous solution of sulphurous acid. Old stains, from nitric acid, will only disappear with the epidermis.—*Picric Acid,* likewise stains the skin yellow, and which may be removed by repeated washings.

Stanhope Lens. A kind of toy, consisting of a thick double-convex lens, one surface of which is of greater convexity than the other. The object is placed upon the least convex surface of the lens, while the eye is applied to the greater convex surface. When the least convex surface is applied to the eye, the lens forms a simple microscope having a focus of one-fourth to one-eighth of an inch. See *Coddington Lens.*

Starch Granules. These are frequently found in urinary sediments, having entered them accidentally, or intentionally for purposes of deception. They may be detected by boiling them in a test tube with water, when they swell and become changed into a jelly-like mass. Tincture of iodine, or solution of iodine causes them to assume a blue color. Under the microscope they present the following appearances :—*Potato starch*, the granules are ovate or egg-shaped, and the hilum or point around which a number of concentric lines are arranged, is situated near one extremity. *Wheat starch*, the grains are circular, with a central hilum which is seldom visible. *Rice starch*, granules very minute and of irregular form.—In bread crumbs the starch granules, although preserving their general form, are much swollen, transparent, and sometimes appearing as if cracked.

Strangury. Difficult urination, attended with tenesmus, pain, and a sensation of scalding, as the urine passes drop by drop.

Strychnia. When taken internally, strychnia passes off by the urine, but whether all that is swallowed is eliminated through this fluid is not positively known. Strychnia crystallizes in octohedrons with rectangular base, or in quadrilateral prisms, terminated by pyramids with four faces ; it is soluble in alcohol and benzine, insoluble in ether, fat oils, and liquor potassa. Its alcoholic solution turns the plane of polarization to the left. Chloride of gold precipitates it from its solutions in fine crystals ; if to the isolated crystalline precipitate, a few drops of concentrated sulphuric acid be added, or enough to dissolve it, and then a very dilute solution of chromic acid be gradually added by drops, a beautiful blue color will be produced. If too much chromic acid be added the color will disappear. Alcohol must be absent from the solution tested. Strychnia may be detected in the urine by one of the following methods :—1. See *Morphia*, Erdmann's process, No. 5. By this process, if strychnia only be present in the urine, no coloration will be produced, after the addition of the mixture of sulphuric and nitric acids and the 2 or 3 drops of water. But after a few fragments of binoxide of manganese, have been added, a purple-violet tint changing to deep onion red, will indicate the presence of strychnia; and then, if the distilled water and ammonia be added, the violet purple changes to a yellowish green and yellow.—2. Remove any albumen that may be present; add subacetate of lead, filter, and treat the filtered urine by sulphureted hydrogen to remove the lead. Again filter, and evaporate the remaining liquid to dryness; place the residue in contact with ammonia, for 24 hours, then agitate with double its weight of chloroform, and again evaporate. Dissolve the residue in 2 c. c. of water acidulated with pure nitric acid; filter, place the liquid in a watch glass, and add a drop or two of solution of bichromate of potassa. After several days crystals of chromate of strychnia will be deposited, visible to the naked eye, from which the chemical characters of strychnia may be recognized. By this process the $\frac{1}{2110000}$th part of strychnia has been detected. *Cloetta.*—3. Saturate the suspected urine with ammonia, and allow it to evaporate spontaneously; heat the residue with a little amylic alcohol, and then add a few drops of this solution to sulphuric acid and bichromate of potassa; if strychnia be present, a blue color will be produced. *Schachtrupp.*—4. Acidulate the urine with concentrated sulphuric acid (after having removed albumen and precipitates), and then add a small quantity of sesquioxide of cerium; on agitating, a beautiful blue color will be produced if strychnia be present, which gradually changes to a cherry red. If *brucia* be present, the color will be orange and then yellow; *morphia*, an olive brown, becoming brown; if *narcotine*, a red brown, becoming a cherry red; if *codeia*, an olive green and then brown; *quinia*, a pale yellow. *Sonneschein.*—5. Dr. R. Southey considers a saturated solution of iodic acid as a very delicate and exceedingly sensitive test for detecting strychnia; he adds some of this reagent to the final chloroform extract obtained by Sta's method (which it is not necessary to give here); if strychnia be present, a pink-rose color is produced, which, after some time, fades gradually to a fawn color. In Cloetta's method (2) above named, iodic acid may be used instead of the solution of bichromate of potassa.

Succinic Acid. There is uncertainty with regard to this acid in the urine. Kletinsky states that when he took a succinnate he found this acid in the urine; Buckheim, and Hallwachs, could not detect it in the urine, even when large quantities of a succinate were taken internally. Wöhler, Meissner, Koch, and Salkowski have found this acid in dog's urine; the first named, also in that of the wolf.

Sugar. *Grape Sugar. Glucose.* This is an abnormal ingredient of urine. When present, the urine is apt to be paler than natural, clear, and having a greenish tint, unless urates or phosphates are present. Its taste is sweet, and its odor faintly agreeable, like that of new-mown hay, and it readily ferments. On pouring the urine from one glass into another, it will froth readily. Its sp. gr. varies from 1.025 to 1.055, depending upon the conditions present; thus, an excess of urea may increase its sp. gr., while the presence of albumen may greatly diminish it. After testing the urine for albumen, the next step should always be to determine whether sugar be present, more especially in cases where the symptoms would lead to a suspicion of its presence. Before undertaking the analysis, any albumen in the urine must be precipitated by heat and acetic acid, and then separated by filtration. The presence of albumen interferes with the proper action of the reagents employed. If the urine be high colored, from bile pigment or otherwise, it may be rendered colorless by placing an ounce or two of it in a half-pint bottle, together with a tablespoonful of animal charcoal, and 15 or 20 grains of carbonate of soda, and then shaking it well for 5 or 10 minutes; on filtering, the urine will be clear. Another method, is, to add a solution of sugar (*neutral acetate*) of lead, which precipitates the coloring matters, tannin present, and also a small amount of sugar; this does not interfere with the qualitative analysis. (See *Tables of Correction of Saccharine Urine in Appendix.*)

Qualitative Analysis.

The detection of glucose in the urine by Fehling's solution, or any of its modifications, rests upon the fact that the salts of copper (the acetate, sulphate, and tartrate, but not the nitrate) are reduced by glucose in presence of a fixed alkali, and which reduction is hastened by the action of heat; a red oxide of copper being precipitated. Under certain circumstances, this action may be prevented, as, by the presence of ammoniacal salts, or albuminoid matters; or it (the action) may occur when sugar is absent, because various organic substances determine it, as, allantoine, chloroform (Beale), cellulose, creatine, creatinine, leucin, tannin,—uric acid? (Berlin). There is still a minor cause of error, the precipitate of the earthy phosphates of the urine under the influence of heat and Fehling's solution, which is alkaline. This precipitate is flocculent and greenish, and but little experience is required to enable one not to confound it with the reduction of the copper salt; however, it should be avoided if possible.

We may, at the start, avoid these causes of error, as, heretofore stated, by removing the ammoniacal salts and albuminoid matters; the organic substances, referred to, have a much less energetic reducing power than glucose, and require a prolonged boiling, which it is easy to avoid,—finally, the precipitate of the phosphates is troublesome only when the urine and the alkaline liquor are thoroughly mixed, and such mixture can be abstained from. [See other causes of error, under *Copper, Fehling's Solution, Trommer's Test.*] The following course may be pursued:

‡ Preliminary Steps.

1. The cold urine, filtered or decanted is not albuminous. { See 3.
2. The cold urine contains albumen....... .... { See 5.

3. The urine has an acid or neutral reaction (diabetic { urine is almost always acid)............................. .... { See *Detection.*

4. The urine has an al- { Boil it in a tube with a small { kaline reaction due to am- { piece of caustic soda; filter or { moniacal salts. { decant, and pass to...... .... { See *Detection.*

5. Add a few drops of acetic acid to the urine, coagulate { the albumen by heat and filter. It is useful to neutralize { the filtered urine with a little carbonate of soda, and then { test as per.. .... { See *Detection.*

Remarks.—3. The alkalinity of the urine should be attended to only in case it is ammoniacal, whether this occurs from decomposition after its emission, or, is passed by a patient laboring under vesical catarrh. Upon boiling it for two or three minutes, a strong ammoniacal odor is given off, and at the same time the phosphates are precipitated. When the odor, just referred to, ceases to be evolved, the phosphates should be allowed to precipitate, or, the urine may be filtered.—After a great number of trials, this is the process upon which we have determined. The presence of the ammoniacal salts in a saccharine urine is a very common cause of error, and one quite difficult to overcome. It simply suffices to add a few drops of solution of hydrochlorate of ammonia to a strongly saccharine urine, to prevent any reaction from occurring, even on prolonging the boiling.

Now the result of my experience is as follows: If *normal urine* be boiled with a piece of caustic soda or potassa, a flocculent precipitate of phosphates will be obtained, and very frequently the liquid will also become darker, assuming a more or less brown-amber color. This fact has long been known, but it should be constantly present to the mind, otherwise the presence of sugar may be erroneously diagnosed. Now, if the liquid be filtered, and the clear urine be poured into a tube containing Fehling's blue liquor, a bottle-green tint will take the place of the blue. On heating the fluid to boiling the tint will become darker, but *no remarkable turbidity will appear in the mixture.*— If an *ammoniacal saccharine urine* be treated in the same manner, being careful to prolong the boiling, and to agitate the tube from time to time until an ammoniacal odor can no longer be distinguished, and until a piece of reddened litmus paper placed at the orifice of the tube does not turn blue,—then, if the filtered and clear fluid be mixed with Fehling's solution and heated to boiling, *a precipitate will take place, of a milk and coffee, or chocolate color*, the abundance of which will be in proportion to the amount of sugar present.— We are far from guaranteeing the infallibility of this method, which is the result of comparative experiments made with urine artificially rendered saccharine. It may be observed, however, that this method has enabled us to detect the sugar after having previously and fruitlessly attempted to do so with the cupric solutions.

The coloring assumed by urine upon boiling it with caustic soda, when it is very deep, is an almost certain sign of the presence of sugar. The ulterior precipitate of the cupric oxide has presented very variable colorings, which we can not explain. The more common colors are those named above, and chestnut brown.—In doubtful cases, a counter experiment may be made, by adding the piece of caustic soda to the blue liquor of Fehling, then pouring the urine into this, and boiling the whole together. The reduction is effected with disengagement of ammonia; if this does not occur, which is often the case, add new quantities of the alkali. It frequently happens, that following one of these successive additions, the reduction becomes suddenly effected with the variable tints heretofore mentioned.—However, it is unnecessary to give too much attention to these difficulties, as it is very seldom that an ammoniacal urine will require to be examined for the presence of sugar.

5.—A small amount of albumen does not prevent the detection of sugar; and to facilitate the separation of the albumen from the urine, in addition to the acetic acid, a few crystals of sulphate of soda may also be added to it.

‡ Detection.

(Before proceeding with this, see *Preliminary Steps.*)

1. Place 5 c. c. of *Fehling's* or *Pavy's Solution* in a beaker glass, and boil it.
> *a.* The liquor is turbid; it has become changed and is not proper for the test. *Reject it.*
> *b.* The liquor remains clear. Add the urine by pouring it along the tube so inclined that the two liquids do not mix together, and the urine forms a layer 1 or 2 centimetres in height above the blue liquor, and then proceed as in 2, below.

2. Heat the surface of contact of the two fluids, gently turning the test tube between the fingers.
> Production of an ochreous yellow ring.........*Sugar.*

Observations. This very practical process is quite sensitive and exact. Mehu lauds it highly, and we have become satisfied that it detects very slight traces of sugar—1. It must be understood that it is useless to employ a larger quantity of the reagent. Boiling for one minute is sufficient.—2. When there is much sugar, the ochreous ring of cupric oxide appears before the liquid commences to boil; in any case, the boiling having once occurred, must not be continued for too long a time in order to avoid the reducing action of the organic substances, heretofore referred to. Besides, the reaction continues during the cooling, and the coloring becomes more marked.—If the yellow ring of cupric oxide be not obtained, it may be affirmed that the urine tested contains no glucose, for if by fraud or otherwise, it was sweetened with ordinary sugar, no reaction would occur.

Other Methods.—1. *By Potassa.* (Moore's Test.) Add to the urine an equal volume of freshly made solution of potassa, sp. gr. 1.060; agitate so as to mix the two, and then carefully heat to boiling the superior portion of the liquid. If sugar be present, the heated portion will become colored yellow, then reddish-brown, or even dark purple when the sugar is in excess, while the inferior portion will preserve its original color. The presence of albumen in the urine does not interfere with this test; in the others it is necessary to remove the albumen at first. The same result is obtained by a solution of caustic soda. Neubauer recommends this test as a confirmative experiment. The coloration is due to the production of glucic, and ultimately of melassic, acid, which are held in solution. (Add a few drops of nitric acid to this brown fluid, the color passes off, and an odor is given off, somewhat resembling that of burnt molasses. (*Heller.*) *Causes of Error.* Several specimens of non-saccharine urine have been met with, which, boiled with caustic potassa, acquired quite a dark Madeira-wine color. Besides, the potassa and soda of commerce, contain foreign substances that may disturb the reaction, as, a contamination with lead, from being kept in flint, and not green-glass bottles, or from being prepared and evaporated in glazed earthen-ware vessels, etc.—2 *By Fermentation.* Procure a large test tube, 6 or 8 inches in length, and from ¾ to 1 inch in diameter, and adapt a cork to its open extremity.—Also, bend a piece of ordinary glass tubing, syphon-like, so that one arm shall be 8 or 10 inches long, and the other 4 or 5 inches.—Through the center of the well-fitted cork, above referred to, pass the long arm of the bent tube until it nearly touches the bottom or closed extremity of the test tube; the short arm remains on the outside of this tube. The instrument is now ready for use. Place a small quantity of baker's or brewer's yeast, 3 or 4 c. c., into the test tube, which is then to be filled brimful with the urine. Fit the cork tightly into the test tube, with the syphon tube attached to it, so that no air remains within the test tube. Place the whole in a warm situation, or, the test tube may be introduced into a vessel of warm water. If sugar be present, fermentation will ensue with generation of carbonic acid, which will drive the fluid through the syphon so that it escapes from the orifice of the short arm, beneath which a glass may be placed to receive it. This test may be relied upon if the process be properly managed, though it does not indicate the kind of sugar.—3. Take a compress of thread or of cotton (an old cotton handkerchief will answer), and allow a drop of the urine to fall upon it, which immediately spreads out. If this part of the cotton be held over some hot coals, or near a hot fire, a very distinct chestnut-colored spot is produced, darker at the circumference than at the center, the darkness being deeper as the amount of sugar is greater, beside which an odor like that of burnt sugar is evolved from the darkened spot. This is an easy, sensitive, and readily applied process. *Goudouin.*—4. Place two fluidrachms of non-albuminous urine, and one fluidrachm of liquor potassa (or soda), into a test tube, to which add a few grains of ordinary subnitrate of bismuth; agitate, and boil for a minute or two. If sugar be present, a greyish or black precipitate of metallic bismuth will be precipitated on the walls of the tube. *Böetger.* However, this is not a very reliable process, as the reaction sometimes fails with saccharated urine; while, on the other hand, a black precipitate has occurred where no sugar existed in the urine.—O. Maschke has recently proposed a modification of the above test, which he considers positively reliable. Every trace of albumen, or other protein bodies are removed from the urine, by adding to a sample of it, $\frac{1}{3}$ or $\frac{1}{4}$ its volume of solution of tungstate of soda strongly acidified with acetic acid. When precipitation has ceased, filter, and to the clear filtrate add an equal volume of solution of carbonate of soda, and 3 or 4 grains of subnitrate of bismuth. Regardless of any coloration produced, the mixture is to be thoroughly agitated, and the bismuth be then allowed to deposit; if the subnitrate has now a grey, brown, or black color, alkaline sulphides are present, and a new sample of urine must be used. This second specimen of urine must be lightly acidulated with acetic acid, a small amount of the subnitrate of bismuth be then added, and the whole be well shaken. Filter; then remove albumen from the filtered urine by the process above described, add the soda solution and the bismuth, and boil. If the urine contains glucose, it will assume a brown color, and the subnitrate will be reduced to the greyish or blackish metallic bismuth, with, very likely, some undecomposed oxide of this metal. Francqui and Vyvere precipitate nitrate of bismuth by excess of potassa; then heat, and add tartaric acid until the precipitate formed is dissolved. A few drops of this fluid boiled with diabetic urine gives a blackish precipitate of metallic bismuth.—For further qualitative tests, see *Kletinsky's Cupro-potassic Test; Knapp's Test; Maumene's Reagent; Mulder's Test; Lime; Ratesi's Test; Specific Gravity*, though this is unreliable, and should not be trusted to alone in determining the amount of sugar present; *Trommer's Test.*

Quantitative Analysis.

The means employed for quantitative analysis may likewise, in most instances, be successfully used for the detection or qualitative analysis of sugar in the urine. Among these may be named *Pavy's test*, the best one for the physician; *Fehling's test*, equally as accurate; *Mitscherlich's polarizer*, very exact, but an expensive instrument; and *Wayne's analysis*. Pavy's and Fehling's tests are used similarly in quantitative analysis, the former being worked in the English system of minims and grains, the latter in the French metric system. With Fehling's titrated solution, 20 c. c. are completely decolorized by 1 decigramme (100 milligrammes) of glucose. To determine the saccharine richness of a diabetic urine, it must be ascertained what volume of this urine will decolorize 20 c. c. of Fehling's solution, or, which amounts to the same, what volume of urine contains 1 decigramme of glucose.

At first, it is necessary to be certain that the fluid reagent is exactly titrated. To this end, some pure glucose is to be dried at 212° F., of which 1 gramme is to be dissolved in 200 c. c. of distilled water; of this solution 20 grammes contain 1 decigramme of glucose, and should consequently decolorize exactly 20 c. c. of the blue test fluid. This done, a burette graduated into tenths of cubic centimetres, is to be filled with the saccharine solution up to zero. By means of a pipette, 20 c. c. of the test fluid (Fehling's), is placed into a beaker or other glass vessel, capable of holding at least 150 grammes, and is then diluted with distilled water to the volume of about 100 c. c. Heat the vessel over a spirit lamp to the boiling point, and then allow the saccharine solution in the burette to flow, drop by drop, into the heated blue test fluid. A turbidity occurs; the reduced oxide imparts a violaceous color to the liquor, which becomes more and more decolorized as the saccharine fluid falls into it. By adding to the blue liquor a few grammes of concentrated solution of caustic soda, its density is increased, and the precipitation of the cupric oxide is facilitated.—Towards the close of the operation, a few seconds must be allowed to pass after each drop has fallen into the blue liquor, in order to observe whether the decolorization is complete. The better to seize upon the exact instant of this complete decolorization, the burette or glass vessel should be held between the eye and the window in such a manner that the light passes horizontally through it. If the fluid still retains a blue tint, it must be heated anew, and another drop of the saccharine solution be allowed to fall into it, and so on until the desired effect is produced. With a little practice, the operator will be enabled to exactly decolorize the test liquor. If the amount of saccharine fluid necessary to exact decolorization has been exceeded, the liquor, above the reduced oxide, assumes a yellow tint, due to the action of the alkali upon the glucose in excess,—in which case, the operation must be repeated.—If the experiment is exact, the decolorized fluid, *filtered at boiling temperature*, will answer to the following conditions:—1. It will give no red precipitate of cupric oxide, when heated with a few drops of the saccharine solution; thus proving that the decolorization is complete, and that no reducible oxide of copper remains in solution;—2. It will give no red precipitate of cupric oxide, when heated with a few drops of the Fehling's liquor, thus showing that an excess of the saccharine solution has been added.—If the test solution of Fehling, and the solution of glucose, have been properly prepared, 20 c. c. of the latter solution (200 divisions of the burette) will decolorize 20 c. c. of the former; but, if it has required 224 divisions (of the burette) of the glucosic solution to decolorize these 20 c. c., it is because this quantity of the blue liquor corresponds to 0 grm .112 of sugar. Then, in the investigations of diabetic urine, we must ascertain what volume of this urine is required to decolorize 20 c. c. of Fehling's liquor, and the volume found will contain 0 grm .112 of glucose.—See *Copper, Fehling's Test.*

Process of Analysis. The standard of the test liquor having been exactly determined by the preceding operations, the following course is to be pursued, in order to determine the amount of sugar contained in a given volume of urine. Having rendered the urine clear by filtration, 10 c. c. of it are to be measured into a graduated test glass, and, according as it is more or less rich in sugar, its volume is increased to 100 or 200 c. c., by the addition of distilled water. (For the purpose of arriving at a more perfect titrage, the column of liquid to be operated upon should be rather long; to this end, it should, if possible, be so arranged that 100 parts of the fluid contain only 1 part of sugar. If a urine contains 105 grms. of sugar per litre, or 10 grms .5 per 100, on adding 19 times its volume of water to it, we will have a liquid at $\frac{1}{20}$th, containing $\frac{1}{2}$ of glucose in about 100, which will render the appreciation very exact. If the urine contains less than 1 of sugar to 100, it will be useless to dilute it with water.)

The urine, having been properly diluted with water, is poured up to zero, in a burette divided into tenths of cubic centimetres; into a glass beaker is also placed 20 c. c. of the titrated liquor of Fehling, to which a few c. c. of concentrated solution of caustic soda is added, and then from 80 to 100 grammes of distilled water. The fluid in this beaker, by means of a spirit lamp, is brought to the boiling point, and then

Fig. 32.

Burette Holder and Two Burettes.

the contents of the burette (urine) are passed into it, at first by cubic centimetres, and toward the latter part of the process by drops. Toward the
termination of the operation, an increased attentiveness is required, being
careful to always keep up ebullition a few seconds, before adding each fresh
drop of urine, so as to avoid the entrance of an excess of urine. The
operation should never be interrupted beyond the few seconds necessary, each
time of letting a drop fall, for the precipitation of the cupric oxide, otherwise (the liquid rapidly absorbing the atmospheric oxygen), the red oxide of
copper returns to the condition of blue cupric oxide, which is redissolved and
colors the liquor blue.

Now, to determine *the quantity of sugar contained in a litre of urine*, calculation must be made according to the following formula:

$$S = \frac{t \times 1,000}{n} \times \frac{V}{v}.$$

S. Quantity of sugar per litre.

t. Standard of the blue liquor, determined beforehand.

1,000. Number of cubic centimetres in a litre.

n. Number of cubic centimetres of the burette required to decolorize the
blue test fluid.

v. Original volume of the urine tested.

V. Volume of the urine after being diluted with distilled water.

Suppose the liquor of Fehling possesses the exact standard 20 c. c. = 1 decigramme of glucose, and that the original volume of urine 10 c. c. has been
diluted so as to measure 100 c. c. Likewise, 94 divisions of the burette holding this diluted urine are required to completely decolorize the 20 c. c. of the
blue test liquor. What amount of sugar is contained in a litre of this urine?
We have t $= 0$ grm .1; n $= 9$ cc .4; V $= 100$; $v = 10$; which gives us

$$S = \frac{0.1 \times 1,000}{9.4} \times \frac{100}{10} = 106 \text{ grms .38.}$$

The operation may be made with 10 c. c. of Fehling's solution, that is, one
volume of the blue liquor corresponding to 0.05 centigramme of glucose.
The results will be the same, and the operation more rapid.

Fermentation. Roberts has given a simple method for quantitative analysis
by fermentation; although less exact than those named above, it may be performed by those who have not the means ready for conducting the other
methods. It is as follows:—Into a 12-ounce vial place 4 fluid ounces of
urine, and a piece of German yeast of the size of a small walnut. Cover
this with a piece of glass, or cork it loosely, so that the carbonic acid gas
may escape, and set it in a warm place to ferment. Beside this, place a com-
panion vial (4 ounces) filled with the same urine, but without any yeast, and
corked tightly. In about 18 or 24 hours, according to the degree of warmth,
fermentation is completed. Ascertain .the specific gravity or density of the
urine in each vial, always at the same temperature; the difference between
which shows the "degrees of density lost," each degree lost indicating the
presence of one grain of sugar in one fluid ounce of the urine. Thus, if the

sp. gr. of the unfermented urine be 1.053, and that of the urine after fermentation be 1.052, the difference between the two sp. gr. is 1, consequently each fluid ounce of urine contains 1 grain of sugar. But if the sp. gr. after fermentation be 1.004, the difference between the two densities is 49, indicating 49 grains of sugar to each ounce of the urine. Ordinary yeast may be used when the German can not be procured. The patient, or some member of his family, may make this analysis every day, after having had a lesson or two, and report the density lost to the attending physician. German yeast can be had of the parties referred to in the early part of this work.

Remarks. As diabetic urine is not always to be had, and as it is very advantageous to be experienced in the management of its tests, normal urine may be artificially sweetened with honey, figs, raisins, or other preserved fruits, cut into pieces and boiled with the urine; then filter, and the clear fluid is ready for the examination. Or, ordinary cane sugar may be converted into glucose by placing a small fragment in water acidulated with a strong acid (nitric or hydrochloric), then boiling it, and diluting the solution with normal urine. Care must be taken to neutralize it with carbonate of soda until effervescence is no longer produced.

Microscopic Examination of Diabetic Urine. Sediments are rarely observed in this urine, though phosphates, urates, or uric acid may be found in it. If a drop or two of urine rich in sugar be quickly evaporated to dryness on a glass slide, rhomboidal crystals, sometimes disposed in arborescent tufts, will be seen; these crystals are, very probably, a combination of sugar and chloride of sodium. Gibbs states that when a considerable amount of salts is present in diabetic urine, the sugar crystallizes in small circular masses, with minute crystals projecting from the surface, which, when examined on a dark ground, resemble lumps of barly sugar.—Diabetic urine when left to itself, instead of becoming ammoniacal, becomes very acid, from the presence of carbonic acid resulting from its fermentation.

If it be then examined under the microscope, small white corpuscles will be seen, which are the true globules of the ferment, and of others which differ very little from them, being the spores of another mycoderm, the *penicilium glaucum*, which is developed in many sour organic liquids, especially milk. These mycoderms of diabetic urine are in the form of oval and transparent cellules, with one or two nuclei, with or without nucleoli, sometimes constricted in their center, or presenting a prolongation like a glove finger. In form and development they greatly resemble those of beer yeast. (This yeast can be procured at any brewery; it always contains a certain amount of grains of fecula, which must not be confounded with the yeast globules; a drop of a solution of iodureted iodide of potassium will reveal them.) If the fermentation is advanced, entangled filaments will be observed, resulting from the development of the mycoderms.

Clinical Import.— Sugar in the urine interests the medical practitioner only when present in more or less considerable quantity, persisting for a long time and uninterruptedly,—in which case the diagnosis is glucosuria, or diabetes mellitus. If the presence of sugar in the urine be variable, and its amount small, the fact is of no especial importance. Recent investigators state that sugar in small amount is present in normal urine. Sugar may be present in urine as the result of decomposition of *uroxanthin;* it may also be found after the employment of chloral and other anæsthetics, probably from deficient pulmonary action during the anæsthetic condition, in which the saccharine formation is not prevented or destroyed; it has likewise been detected during convalescence from certain acute maladies, after injuries to the nerve centers, from excess of saccharine or amylaceous matters received into the system, etc.; but its presence in the urine becomes important only when it *persistently* occurs in an appreciable amount. The exclusive use of ordinary sugar as food does not usually occasion saccharine urine, as might be supposed. In all cases of glucosuric urine, an examination should be made of the chloride of sodium and other salts that may be present. The persistence of sugar in the urine in anthrax, and furunculous diseases, uterine ulcers, gangrenous affections, certain inflammatory disorders, etc., has been observed by several investigators. Glycohemia, with a minimum amount of sugar in the urine, is, by some, considered the first or early symptom of diabetes mellitus.

Sulphates. Sulphates exist in normal urine in small quantity, principally those of soda and potassa, very rarely, if at all, that of lime. About 3 grms .69 (57 grains) of mixed sulphates pass in 24 hours, or from 1 grm .62 (25 grains) to 2 grms .72 (45 grains), the soda salt usually preponderating. In the examination of urine for sulphates, the same precautions must be observed as in that for the detection of phosphates, always operating upon clear and *non-albuminous* urine. As the urinary sulphates are soluble, they do not precipitate spontaneously.

‡ Detection of Sulphates.

1.	2.	3.
Acidulate the urine quite strongly with *hydrochloric acid*, and boil.	Drop an excess of *solution of chloride of barium* into the boiling fluid.	A white *precipitate indicates sulphuric acid* of the alkaline sulphates in the form of sulphate of baryta, insoluble in acids.

Observations.—1. The acidulation (20 c. c. of urine with 4 to 6 drops of acid) with hydrochloric acid is to prevent the precipitation of the phosphates in the form of barium phosphate, which is soluble in acids, while the sulphate is not. It must be borne in mind, however, that concentrated acids and concentrated solutions of many salts diminish the sensitiveness of the reaction. *Fresenius.*—Unless the urine be boiled, the resulting sulphate of baryta will pass through the pores of the filter on filtering. *Beale.*—3. The soda and potassa of the urinary sulphates remain in the fluid in the form of chlorides; the phosphoric acid is displaced, and the phosphate of baryta dissolves by means of the hydrochloric acid. The sulphate of baryta precipitated will be white if the urine contains normal pigment; greenish, if it is rich in bilifuivin; and from a rose to a dark red color, according to the abnormal amount of urohematin present.

Upon filtering or decanting the urine after all the sulphate of baryta has been deposited, an approximative estimation of the amount of *phosphoric acid* present may be made, by adding ammonia, which throws down phosphate of baryta. In this case the contact of air must be avoided.—Act according to the method indicated for phosphates.

Clinical Import. See *Sulphuric Acid.*

Sulphur. When sulphur or sulphides are taken internally, they pass off by the urine, in part, as, sulphates, or sulphuric acid. The neutral sulphates likewise increase the amount of this acid in the urine. Sulphur occasions the urine to evolve sulphureted hydrogen. Sulphur exists in the albuminoid substances of the system, in blood, bile, saliva, gastric juice, albumen, cystine, etc. Some kinds of food are quite rich in it, as, peas, beans, lettuce, cabbage, eggs, milk, corn, cauliflower, asparagus, turnips, celery, rice, ginger, hops, white mustard, flesh, etc., and which increase the urinary sulphates, when taken internally. See *Sulphuric Acid.* When sulphur exists in organic matters in not very minute quantity, it may be detected by mixing some of the matters with carbonate of soda and starch; place the mixture on platinum foil, and heat it by the blowpipe. The fused mass is then conveyed in a watch glass, with a drop or two of water added. Now, if a small crystal of the nitroprusside of sodium be placed in the mixture, if sulphur be present, the fluid will assume a splendid purple color, from formation of sulphide of sodium; this fluid passes from purple to a deep azure blue, and ultimately loses all its color. *Dana.*

Sulphureted Hydrogen. *Hydrosulphuric,* or *Sulphydric Acid. Hydrothion.* This is a poisonous, colorless gas, having a fetid odor resembling that of rotten eggs; it is inflammable, burning with a blue flame and evolving a sulphurous odor. When present in laboratories, or other places, the inhalation of its vapor is deleterious, and this may be avoided by the diffusion of a little chlorine, or by spraying the atmosphere of the apartment with an aqueous solution of chlorine. Water dissolves about three times its volume of this gas; solution of ammonia, or of potassa, dissolves it entirely. Sulphureted hydrogen is sometimes evolved from the urine. Sulphureted hydrogen is used, either in the form of gas, or solution, for the detection of lead in fluids, or for the separation of lead from liquids used in, or resulting from, chemical analyses. It forms a dark brown or blackish precipitate in liquids containing lead. It may be readily procured by gently heating sulphide of iron in dilute sulphuric acid; by the action of dilute sulphuric acid on sulphide of potassium; or, by dissolving sulphide of lime half an ounce in distilled water a pint, and then adding dilute hydrochloric acid 2 fluidrachms. Liquid sulphureted hydrogen is made by passing the gas, as it is formed, into water to saturation. Another mode of preparing sulphureted hydrogen has been given:—introduce into a flask equal weights of granulated zinc and galena, in small fragments, and then pour upon them hydrochloric acid diluted to the twentieth (1 of acid to 20 of water), so as to cover the mixture. Sulphureted hydrogen is soon disengaged regularly and in large quantity, mixed with a small proportion of hydrogen and hydrochloric acid; this latter gas may be absorbed by means of a washing bottle.

"In organic analyses, we have almost always to eliminate some principles with the acetate of lead. To separate the lead, authors say in a few words 'to remove it by means of a current of sulphureted hydrogen.' This is easily said; but the operator is to be pitied, for there are few manipulations more lengthy, more tiresome, and more uncertain in their progress and their results than this; and he is very fortunate if the flask or matrass does not break and throw the acid over his clothes, and perhaps upon his face. The current of gas diminishing, he stirs up the fire; the contents of the matrass or retort swell up and pass over into the wash bottle. If the acid is in excess, there is a disengagement of hydrochloric acid gas not foreseen in the programme; if the sulphur is too dry, the apparatus breaks, etc. By the process hereafter given, these small dangers and inconveniences are avoided. This very simple method acts with the same efficacy in the preparation of *chlorine* by hydrochloric acid and binoxide of manganese, and in that of *oxygen* by sulphuric acid and binoxide of manganese.—Mix with powdered sulphuret of antimony one-third its volume of siliceous sand, or powdered stone-ware; we may now throw upon it from the first a large quantity of hydrochloric acid and heat it actively without any uneasiness as to the results. The gas is regularly and abundantly evolved to the end. The mass becomes but slightly heated, and does not distil over into the wash bottle." *Mehu.*

For urinary investigations the gas may be prepared as required, by means of a long bottle filled with dilute acid to one-third its capacity; a copper rod is passed tightly through the cork, which has attached to its lower extremity a perforated leaden basin, in which the sulphuret is placed. When this basin is lowered into the dilute acid by means of the copper wire, the gas is evolved; when enough has been procured, the basin can be drawn up out of the acid. A glass tube, also passed through the cork, allows the gas to escape into a washing tube, or as may be required. Babo's sulphureted hydrogen gas generator is a very cheap and safe apparatus for preparing and using this gas as required.

Sulphuret or Sulphide of Ammonium. *Ammonium Sulphide.* Is sometimes employed in analyses in the place of sulphureted hydrogen. It may be made as follows:—Pass sulphureted hydrogen gas through 3 parts of liquor ammonia until no further absorption takes place, then add 2 parts more of the solution of ammonia. Keep this in small, accurately well stoppered, green-glass bottles. At first this fluid is colorless, and deposits no sulphur on the addition of acids; ultimately, however, from atmospheric action, it acquires a yellow tint, and gradually deposits sulphur, until the fluid contains in solution nothing but ammonia. It is better to wash the gas before it is allowed to enter the ammonia, by having it pass through a small vial of distilled water. When properly prepared, a drop of it added to a little

saturated solution of sulphate of magnesia, gives no precipitate; if a precipitate should occur, free ammonia is still present, and the fluid must be again
subjected to the action of a stream of sulphuretted hydrogen gas.

Sulphuric Acid. *Oil of Vitriol.* Sulphuric acid is present in the urine
in the form of sulphates of potassa, and soda, in the following proportions:
3 grammes .50 in 1,000 grammes of sulphate of potassa, and 3 grammes in
1,000 grammes of sulphate of soda. The average amount of this acid passed
in 24 hours is 1.943 grammes (30 grains). *Free sulphuric acid in presence of
sulphates* may be detected as follows, when this investigation is required: Mix
the fluid under examination with a very little cane sugar, and evaporate the
mixture at 212° F. to dryness over a water bath, in a small porcelain dish.
If free sulphuric acid is present a black residue remains, or, if the acid
exist in minute quantity, a blackish green. The other free acids in the urine
do not give this reaction.

Quantitative Analysis.

In estimating the sulphuric acid by chloride of barium, we must be careful
not to add any more of this latter solution after a barely perceptible cloudiness is obtained (in a clear portion of the fluid under examination) by the
addition of another drop of the barium solution, as well as in another specimen by a drop of solution of sulphate of potassa of strength exactly equivalent to that of the chloride of barium, about 1 part of pure sulphate of
potassa to 12 parts of distilled water. Two or three specimens of the clear
fluid under examination should be taken and tested.

Place 100 c. c. of the urine under examination, in a beaker or Florence oil
flask, and acidulate it with 20 or 30 drops of hydrochloric or nitric acid;
place on a sand or water bath and heat to boiling. Now let fall into the
boiling urine from a graduated burette, 4 or 5 c. c. of the *strong chloride of
barium solution* (see *Barium*), and allow the fluid to rest until the precipitate
formed has all subsided to the bottom; then add another c. c. of the chloride
of barium solution, and again allow the fluid to rest before adding any more
of this barium solution, and proceed in this manner, keeping the urine at the
boiling point whenever this solution is added, and then removing it from the
heat that the precipitate may subside. From time to time, by means of a
pipette, 12 or 15 drops of the clear upper stratum of urine are heated and
filtered through a small filter into a small test tube, and tested by the chloride of barium solution, to ascertain whether a further precipitate occurs. If
it does, return the few drops in the test tube to the beaker, and again add a
little more of the test solution of barium. And so proceed until no further
cloudiness appears on the addition of this test solution.

In the trials with the 12 or 15 drops in the test tube, above referred to,
Neubauer advises the *dilute solution of chloride of barium* to be used. If this
gives a precipitate, pour the fluid in the test tube back again, and add more
of the dilute solution, until a new trial with the test tube yields no further
precipitate; then 1 c. c. of this dilute solution corresponds with 1 milli-

13

gramme of sulphuric acid. Add the quantities of the two test solutions together, putting the figures for the dilute solution one decimal further back, and then it may be taken into account as a strong solution. Let us suppose that we have used 11 c. c. altogether of the strong test solution, and 10 c. c. of the dilute, then

11.0 c. c. of the strong solution, and

1.0 c. c. of the mild solution,

Make 12.0 c. c. of strong test solution used.

The patient passes 800 c. c. of urine in 24 hours, how much sulphuric acid does it contain? 12 c. c. of the strong test solution have been required to remove all the sulphuric acid from 100 c. c. of urine; and as 1 c. c. of this solution corresponds to 10 milligrammes of sulphuric acid,—12 c. c. are equal to 120 milligrammes. Multiply the quantity of urine passed in 24 hours, by the ascertained amount of sulphuric acid in 100 c. c. of this urine, and divide this by the amount of urine tested, 100 c. c.; then

$$\frac{800 \times 120}{100} = 960 \text{ grammes of acid in the urine of 24 hours.}$$

In the preceding process it should always be ascertained that no excess of the test solution of chloride of barium has been added to the urine under examination; this may be done by adding a drop of the clear fluid to the *standard solution of sulphate of potassa* (see *Potassa*), in a test tube; if no precipitate occurs, more of the barium test solution must be added; if, however, merely a slight haziness or cloudiness takes place, the analysis is finished; but, if too much precipitate is produced, there is an excess of the test solution of chloride of barium, and the analysis must be repeated upon a new specimen of the urine.

An approximative estimation may be made by a process similar to that named under *Phosphates in General, Estimation of Phosphates*. If the chloride of barium solution at the $\frac{1}{10}$th be used, $\frac{1}{10}$th of a cubic centimetre of the precipitate formed in the graduated tube, corresponds to 0 grm .50 of sulphate of baryta. Multiply the ascertained weight of the sulphate of baryta by 0.343, and we have the weight of the anhydrous sulphuric acid entering into its composition. 100 parts of sulphate of baryta correspond to 60.85 of anhydrous sulphate of soda, or to 74.76 of anhydrous sulphate of potassa. *Mehu.*

Clinical Import. Albuminoid substances in becoming disintegrated, form at the same time, urea and alkaline sulphates; the sulphates being the products of destruction of the tissues, and having no part in nutrition. The excretion of sulphuric acid (or sulphates) attains its maximum shortly after a full meal, then diminishes up to the corresponding meal of the next day, when it again increases. As heretofore stated certain kinds of food augment its quantity as well as the ingestion of all bodies containing sulphur. Animal diet, physical or mental exercise, and sulphates that do not produce a purgative effect, increase the amount of alkaline sulphates; while vegetable diet (with some exceptions), or fasting, diminishes it. The other causes that may modify the secretion of sulphuric acid are still imperfectly known. In disease, the variations have been but little, if at all, investigated. It is evident, however, that in acute febrile affections, and whenever the patient takes little or no nourishment, an increase of sulphates in the urine will be the indication of an abnormal decomposition of the sulphureted elements of the body.

Standard Solution of Sulphuric Acid. Analysis of Ammonia. Cautiously dilute 14 grammes of hydrated sulphuric acid with 200 grammes of distilled water; and, when the mixture has cooled down to the ordinary atmospheric temperature, the amount of sulphuric acid contained in every 10 c. c. of it, is determined in two such parts of 10 c. c., through precipitation with chloride of barium. When both experiments nearly agree, the average of them may be taken as correct. Thus, if we find that 10 c. c. of the diluted acid contain 0 grm .505 of sulphuric acid, they will then be exactly neutralized by 0 grm .2146 of ammonia. Consequently 1 c. c. of the dilute acid corresponds to 0 grm .02146 of ammonia.

Sulphurous Acid. If this acid be present in urine it may be detected by half filling a test tube with this fluid, and slightly acidulating it with hydrochloric acid; if now a strip of starched paper, stained blue with a weak solution of iodine, be suspended in the tube just above the urine, the disengagement of the sulphurous acid gas will decolorize the paper.

Sympexions. *Sympexis.* Nitrogenized concretions almost always observed in the fluid of the vesiculæ seminales, have been termed *sympexions* by Ch. Robin, who has carefully studied them, but has not been able to ascertain their composition. They are in the form of small grains, very variable in size, of waxy consistence, forming a homogeneous mass, and break into fragments under pressure. Their chemical reactions prove that they are formed of nitrogeneous matter other than a simple concrete mucus, for acetic acid, instead of shrivelling and corrugating them, causes them to swell, become more transparent, and finally dissolves them. Not unfrequently they enclose spermzoons, blood globules, or debris of epithelium, etc. They have also been found in the fluids of other parts of the body.

T.

Tannin. The effect of tannin upon the urinary constituents is not known, though, in some instances, it appears to increase the amount of urine. When administered internally, it passes off in the urine as gallic and pyrogallic acid. A saccharine substance is also supposed to be present. Allantoin is found in the urine when considerable tannin has been taken. *Schottin.—Solution of Tannin.* Take of pure tannin 1 part, distilled water 2 parts, by weight, and, to preserve the solution, add ether 1 part. Keep tightly corked. and in a cool, dark place. See *Albumen.*

Tar. See *Carbolic Acid.*

Taurin. A peculiar animal substance that may be extracted from bile; being present in *cholinic acid*, with *cholalic* (or *choloidic*) *acid.* It may be prepared from fresh ox-bile, which is to be boiled for some time with hydrochloric acid. Filter, to remove the precipitate of dyslysine; then evaporate the filtrate, and add alcohol. The taurin precipitates, while chloride of sodium and hydrochlorate of glycocoll remain in solution. Taurin forms tasteless, inodorous, colorless, transparent, regular hexagonal prisms, terminating in 4 or 6 planes. It is a nitrogenous substance consisting of 25 per cent. of sulphur, soluble in hot water, but not in alcohol or ether. Strecker produced it artificially. See *Cholinic Acid.*

Taurochloric Acid. See *Cholinic Acid.* *Bile Acids.*

Taurylic Acid. *Hydrated Oxide of Tauryl.* An acid found in small quantity in cow's urine, and, according to Staedeler, in very minute quantity in human. It resembles phenylic acid, but has a higher boiling point. Its presence in human urine has not been satisfactorily established.

Tea. The effect of tea is but slight upon the urine; it diminishes to a very small extent the chlorine, uric acid, and sulphuric acid; with an insufficient diet, tea causes the body to lose its weight less rapidly than when it is not used. A large amount of infusion of tea, taken at a meal, increases the water of the urine, but does not sensibly affect the solids.

Fragments of tea leaves are occasionally met with in urine. Under the microscope they present the cellular portion of the leaf, the cells being dark and of rather large size, with a light peripheral band between the dark part and the marginal outline, while minute spiral threads or vessels are projected from one or more parts of the margins of the cells.

Test Fluid. *In detection of Quinia in Urine.* The formula for this preparation having been overlooked in process No. 3, under *Quinia,* page 164, it is given at this place. It is as follows:—Take of pure acetic acid 30 minims, rectified alcohol 10 minims, dilute sulphuric acid 1 drop; mix. Process No. 3 will detect the $\frac{1}{100000}$th of a grain of quinia in the urine. The process is that of Dr. Herapath.

Test Papers. Test papers should be made of thick, substantial, pure, non-bibulous paper, and the color should be placed on it in a thin layer. Too much color destroys the sensitiveness for minute amounts of acid or alkali in solutions. *For Acids.* Exhaust litmus with strong alcohol; then extract it with water, and brush strongly sized paper with it. Dry the paper, and then wash with distilled water to remove free alkali. When dry, cut into strips and keep in a dark, well closed bottle. Acids redden it; it is very sensitive.—*For Alkalies.*—1. Dip blue litmus paper, prepared as just stated, in sulphuric acid diluted with just enough distilled water to redden the litmus. Dry, and then wash with distilled water to remove excess of acid. When dry, cut into strips, and keep in closed, darkened bottles. Alkalies restore the original blue color. N. B.—If strips of blue litmus paper are ruled with dilute sulphuric acid, so that each strip contains one red and one blue thick line, one immersion in a fluid will show at once whether an acid or an alkali be present. After penciling the paper with the acid, it should be dried and washed, as above.—2. *Turmeric Paper.* Macerate turmeric roots in water until they are exhausted; dry them, bruise them finely, and exhaust with alcohol. Apply on filtering paper, dry, and keep in corked bottles in the dark. This paper should have a fine yellow tint, which is changed to a brown by alkalies. Boracic acid changes it to a brown red.—3. Make an acid infusion of the red petals of the rose, and apply to paper. This is a very delicate test.—*Sulphureted Hydrogen.* This substance in urine may be known by its rendering dark-brown or black, paper that has been dipped in solution of acetate of lead, and then dried.

Other test papers are employed in chemical processes, but the above are sufficient in urinary investigations.

Thionurate of Ammonia. To a cold, strong solution of *alloxan* add an aqueous solution of sulphurous acid, until, on agitation, there is no longer any sulphurous odor; then add carbonate of ammonia to supersaturation, and keep boiling for half an hour. Abundant crystals form as the liquor cools, which are sparingly soluble in cold, but freely soluble in warm water.

Thionuric Acid. Dissolve *thionurate of ammonia* in hot water, and add solution of acetate of lead as long as any precipitate occurs. Suspend the filtered precipitate in water, and add sulphureted hydrogen to effect decomposition. Separate the sulphuret by filtration, and, on evaporating the clear filtrate, crystals of thionuric acid form. These are permanent, freely soluble in water, though decomposed by boiling.

Titrated. When a solution or chemical agent is quantitatively tested for the amount of such or such substance contained in it by the aid of standard or titrated solutions, it has been " titrated."—If an agent or liquid is directed to be titrated (tested or analyzed), it is to be quantitatively tested, as above observed.—A titrated or standard solution, is one whose strength or chemical power has been accurately found by experiment.

Torula Cerevisiæ. *Torula Sacchari. Mycoderma Cerevisæ. (Champignon du Ferment.* French.) *Yeast* or *Sugar Fungus.* (See *Penicilium Glaucum,* and *Microscopic Examination of Diabetic Urine,* under *Sugar.*) This fungus, like the common yeast plant, consists of numerous spores (minute oval cells) arranged in bead-like rows, very much resembling the ordinary yeast fungus, though nearly one-third smaller in size. The size of the spores vary according to the age of the plant. The delicate, gelatinous cells found soon after the formation of the fungus just below the surface of the fluid, have a diameter of about 0 mm .00362. At a later period when they have a fawn color, are heavy, and sink, their diameter varies from 0 mm .02647 to 0 mm .0425. See *Fig.* 31, page 178.

This fungus is developed in urine upon standing for a certain time depending upon the character of this fluid; in diabetic urine, it begins to form in a few hours, and is often developed within 20 to 36 hours after the urine has been voided, and this rapid appearance may lead to a suspicion of the presence of sugar. During the sporule and thallus stage of development, it is very difficult to determine the torula from penicilium; but the yeast fungus, in its aerial fructification has a globular head, while the penicilium presents a tuft of branches. During the development of the torula, the urine ferments, any sugar in it is destroyed, carbonic acid is evolved, and several acids are produced. It has been supposed that the torula was developed only in saccharine urine, but this is not the case, as it has been found in this fluid when no sugar could be detected. See *Fungi. Saccharomyces Cerevisæ. Vegetable Organisms.*

Transparency of Urine. Healthy urine is always transparent when voided, some time being required subsequently, before it becomes turbid, and which turbidity is due to a small, light, greyish white cloud, formed of the mucus and epithelia of the urinary passages, and which cloud gradually settles towards the bottom of the vessel. But, a transparent urine does not always indicate a normal condition of this fluid.—The transparency of the urine may be destroyed by several causes, as, an increased amount of *mucus*, the presence of *urates*, of *earthy phosphates*, of *pus*, of *albumen*, etc.—At times, the turbidity of the urine may exist at the time of its emission, but more frequently at some subsequent period, from formation and deposition of its salts, from the action of heat, or of chemical reagents. Occasionally, *coloring matter* may be present to a degree capable of interfering with the transparency of the urine. When turbidity is present in this fluid, we have to determine its cause, by the use of processes and reagents, that have been named throughout this work.

Trichomonas Vaginalis. An animalcule discovered in the muco-purulent discharge of leucorrhea by Dr. A. Donné. Its body is often round, though it assumes various forms, being more ordinarily elliptical, and having a diameter of from 0 mm .008 to 0 mm .02. It is provided with a flagellum and vibratile cilia, the former having a length of from 0 mm .025 to 0 mm .035, and of extreme tenuity. It moves in mucus with considerable activity, but much less so when water is added to it. Sometimes it is found alone, but more frequently united with others in groups of 2, 4, 6, and even more, their anterior extremities being free and moving their filiform appendages in every direction; their posterior extremities being held together. When the mucus containing them becomes cool, these infusoria liquefy and disappear. They are sometimes observed in the urine of women having leucorrhea. It requires a magnifying power of 300 or 400 diameters, and a proper illumination to distinguish them from the inanimate spherical purulent globules in the midst of which they are moving. They can, by means of a viscid matter, fasten their posterior extremities upon the object glass, and then elongate their bodies. Donné's observations regarding this infusorium have been confirmed by Robin, Velpeau, Kolliker, Scanzoni, and others.

Trimethylamin. An artificial or spontaneous volatile product of the decomposition of certain organic substances, characterized by its strong fishy odor. It has been found in herring pickle, in alcoholic preparations of animal tissues, and in putrid urine. It is a fluid, and is isomeric with propylamin. When a large quantity of ammoniacal urine is distilled, there passes over carbonate of ammonia, trimethylamin, and other volatile bases in small quantities. This substance has been found useful in gout and rheumatism.

Trommer's Test. *For Sugar.* To the suspected urine, in a test tube, add a drop or two of solution of sulphate of copper, and then liquor potassa in excess. Boil the mixture; if sugar be present there will be a red precipitate of suboxide of copper. As this test has proven rather unsatisfactory in its operation, the reduction tests of *Pavy* and *Fehling* are more generally preferred. Trommer's test (and similar tests) is frequently interfered with by certain organic elements of the urine; this fluid, upon being submitted to boiling after the addition of the reagents, presents a yellowish or yellow-brown color, but without any deposition of the red suboxide of copper; to overcome this difficulty, the urine should be previously treated with an excess of animal charcoal, then filtered, and the filtered liquid may then be examined for sugar.—Dr. George Hay, of Alleghany City, has recently verified the truth of the above statement, and has remarked that Trommer's test as usually employed is fallacious and unreliable. From carefully conducted experiments he found that when the suboxide of copper is not precipitated from this test, as usually applied, it may be taken as a certain evidence that sugar is absent; but the occurrence of such a precipitate in the urine is no positive indication of its presence. The method he advises, is to evaporate the urine over a water bath, then treat the dry residuum three or four times with boiling absolute alcohol, filter, and evaporate the filtrate to dryness, keeping it at the boiling point long enough to drive off all the alcohol. The remaining mass is dissolved in distilled water, then filtered, and the aqueous filtrate subjected to Trommer's test which now gives a positive result. *Druggists' Circular*, 1877, page 177.

Tubal Nephritis. Bright's renal disease. Albuminaria.

Tube Casts. See Renal Tube Casts.

Tubercle Fragments in Urine. When tubercle is located in some part of the urinary passages, a deposit of small yellow, oval, cheese-like fragments may be found in the urine. These fragments, examined under the microscope, are seen to contain pus corpuscles, which do not present the usual nuclei under the action of acetic acid; also disintegrated cells, considerable amorphous granular debris, disorganized connective tissue, cholesterin, minute oil globules, etc. They are about the $\frac{1}{2000}$th of an inch in diameter, and become transparent when acetic acid is added to them. Glycerin diminishes their conspicuousness.

Tubuli Bellini. *Ducts of Bellini.* The small non-tortuous uriniferous tubules which converge from the cortex of the kidney to the apices of the Malpighian pyramids.

Turnip. This vegetable appears to give rise to an augmented quantity of oxalate of lime crystals; and, eaten in abundance, imparts a peculiar odor to the urine.

Turpentine. This substance imparts the odor of violets to urine. In some instances, it appears to increase the quantity of urine. In large doses, or taken for a long time continuously, it may occasion hematuria, and even albuminaria.

Tyrosin. Tyrosin is of the same class of substances as *leucin*, which see. It is generally present with leucin, and is supposed to be produced in the liver, in which organ it has been detected by many investigators; it occurs in many of the animal tissues, is always present during the decomposition of albuminous substances, and has been found, by Liebig, as one of the products of decomposition of old cheese. Voit and Bauer think it very probable that leucin and tyrosin are among the first products of the decomposition of albumen. The mode of obtaining tyrosin from urine, when present in it, has already been stated under *Leucin*. In acute yellow atrophy of the liver, the urine, according to Neubauer, sometimes gives a spontaneous greenish-yellow crystalline sediment, composed of tyrosin and leucin, both of which will be present when the sediment is evaporated on an object glass; the color is due to the presence of bile pigment. Beale has found tyrosin in urine containing much uric acid, which had been allowed to remain for many weeks in a warm place.

Microscopical Characters. Unlike leucin, tyrosin is crystallizable. Sometimes it presents in globular masses, but more commonly in brush-like or sheaf-like clusters, constricted at their convergent points, and not unfrequently crossing each other, forming a cross of four brush-like arms, and narrowed at their point of union at the center. These masses consist of numerous long, fine, white (or yellowish when colored by bile pigment), shining, silky, acicular prisms; in the globular masses, which are readily broken down when compressed by the thin glass cover, these needles present a jagged appearance, in consequence of small spear-shaped crystals projecting from them.

Chemical Characters. Tyrosin has neither taste nor odor, but, when burned, emits a disagreeable smell, resembling that of burnt horn; it does not undergo sublimation. It is very soluble in boiling water, mineral acids, and alkalies; difficultly so in cold water, and acetic acid; and insoluble in alcohol or ether. Dissolved in ammonia and then spontaneously evaporated, it remains unchanged, but yields larger crystals. Acids precipitate it from its alkaline solutions.—Perhaps the easiest chemical test for its detection is that of Scherer's, viz.: Place a few crystals, or some of the urinary deposit, upon a platinum spatula, moisten them with chemically pure nitric acid, and slowly evaporate to dryness; a rich yellow substance is formed, together with oxalic acid. This yellow substance consists of the crystals of nitrate of nitro-tyrosin, $C_{18} H_{11} N_3 O_8$, which are reddened by hydrochloric acid; and when evaporated, a deep, brownish-black residue is left.—Hoffman's test is said to be a very delicate one : Place some crystals of tyrosin in a test tube, add a little water and boil; when the tyrosin is dissolved, add a few drops of solution of nitrate of protoxide of mercury to the hot liquid, which instantly assumes a rosy-red color, while a flocculent red precipitate forms after a short time.—

Piria's test, although very sensitive, is not of such easy performance as that of Scherer's, besides being unreliable when leucin is present; it is as follows: Place a small quantity of the crystals of tyrosin, or of the suspected urine from which any leucin present has been separated, in a watch glass, and add a drop or two of chemically pure sulphuric acid, which gives a red color. Cover the glass and allow it to stand for half an hour in a warm place; then dilute with distilled water. Saturate the acid of the mixture with carbonate of baryta; boil to decompose the bicarbonate of baryta; filter, and to the filtrate add a few drops of neutral dilute solution of perchloride of iron, when the presence of tyrosin, gives a beautiful violet color; the result of the action of the chloride of iron on the sulphate of tyrosin.—It is always proper, before investigating for tyrosin, to separate it from the *leucin* in the urine by the method described under this latter substance.

Clinical Import. When leucin and tyrosin are present in urine there is almost always a diminution of urea. They are never found in urine during health, but only when certain maladies exist, as, typhus, variola, acute yellow atrophy of the liver, and probably in any hepatic disease in which the parenchyma of the liver undergoes rapid disorganization. Tyrosin in the urine, is, according to Harley, "an almost certain sign of a rapidly approaching fatal termination."

U.

Uræmia. See *Uremia.*

Uramile. *Dialuramid. Murexan.* Heat a cold saturated solution of thionurate of ammonia, to the boiling point, then add hydrochloric acid, and boil for a few minutes; on cooling, a crystalline precipitate of hard shining needles occurs. These are soluble in cold concentrated sulphuric acid, and in aqueous solutions of the fixed alkalies. The ammoniacal solution, on exposure to the air, assumes a rose-red color, and if it be evaporated in the air, murexide is formed. Uramile is insoluble in water, alcohol, or ether, and in acetic, tartaric or citric acids. Boiling water, dilute phosphoric, sulphuric, and hydrochloric acids, dissolve a minute portion of it.

Urana. A term applied to the ureter.

Urane. A vessel for receiving the urine as it is voided.

Uranium, Standard Solution of Acetate of. See *Phosphoric Acid.*

Urapostema. A urinary abscess.

Urates. Urates are met with in urinary deposits more frequently than any other substance, and are generally colored a delicate rose, brick-red, or brown red; white urate sediments are rare. The more or less red, and occasionally white adhesive coating observed upon the sides or walls of urinals, consists of urates. Uric acid, being a bibasic acid, forms both acid and neutral salts, the latter, being much more soluble than the former which enter largely into the urate precipitate occurring in urine. Urate sediments are soluble by heating the urine, but fall again on cooling; the addition of an alkaline solution also dissolves them; and, like uric acid, they give the *muraxide* reaction. If, on heating in a test tube, a urine containing urates, they do not become thoroughly dissolved, it will be owing to the presence of insoluble salts mixed with them, as earthy phosphates, or of organized elements, pus, blood, etc. In this case the boiling urine should be filtered, and the filtrate, on cooling, will yield a deposit of urates free from foreign substances. There has been considerable difference of views among urologists concerning the composition of these amorphous deposits, some considering urate of ammonia, and others, urate of soda, as the principal salt in them; analysis has proven the presence of soda and ammonia, and sometimes traces of lime and magnesia in them. And it is generally admitted at this day by the best investigators that the sodium urate exists in the largest quantity. The specific gravity of urine containing an excess of urates varies considerably; thus a *pale urine*, becoming opaque on cooling, has sp. gr., at the time of its emission, of about 1.012; a *pale amber-colored urine*, giving a copious fawn-colored deposit on cooling, has a sp. gr. of about 1.018; a *high-colored urine*, giving a reddish-brown, or brick-red sediment, has a sp. gr. of about 1.025. More generally, however, the urine is high-colored, turbid, dense, and slightly acid; sometimes it may be neutral, or alkaline. If hydrochloric or acetic acid be added to urine containing urates, the uric acid is separated in crystalline forms. See *Figs.* 35, 36.

Under the microscope, if the sediment be recently deposited, the urates are presented in the form of small, irregular, amorphous, granules, often clustered in patches, or in ramified rows like sprigs of moss. If the sediment is old, other forms will be observed, either brown or yellow globules with dark outlines, isolated or grouped, or stellated masses, or spherules furnished with bristly points like chestnut burr. See *Fig.* 33.

Fig. 33.

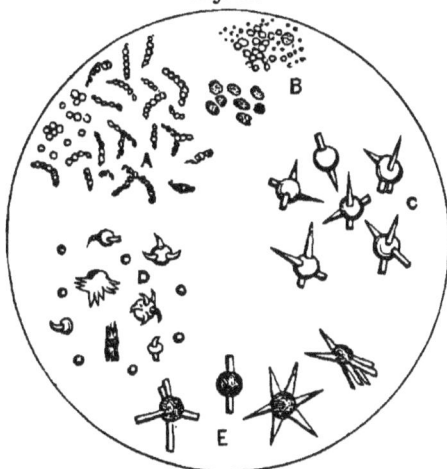

A. Ordinary appearance of urate of ammonia.
B. Urate of ammonia, less common than the preceding.
C. Rare forms of urate of ammonia; the spicula being, probably, uric acid; occasionally observed in albuminous urine, and occurring in dropsy after scarlatina.
D. Urate of soda, usual forms in urinary deposits.
E. Urate of soda, rare forms; differs from similar form of urate of ammonia, in the extremities of the needles of the latter being obtuse, while in the former they are acute.

The following are the urates met with in the urine:—

1. **Urate of Ammonia, Acid.** *Acid Ammonic*, or *Ammonium Urate*. This is more especially met with in alkaline urine, mixed with earthy phosphates. *Under the microscope*, it appears in opaque globular masses, which are beset with fine points similar to the spines of a hedgehog; sometimes in roundish or dumb bell-like masses, which polarize light; but more generally in a dark, granular, perfectly amorphous sediment, varying in color, according to the quantity of urohematin present, from a white to a pink, brick red, purple, or brownish red. See *Fig.* 33. This urate is dissolved when the urine is heated, reprecipitating as the fluid cools; it is soluble in liquor ammonia, and in liquor potassa, evolving ammonia when treated with the latter liquor, and from which solution uric acid is precipitated on the addition of acetic acid. Hydrochloric, or, acetic acid, added to urine containing urate of ammonia, occasions the gradual formation of uric acid, while the ammonia forms a soluble salt with the acid. Place any of the different forms of urate of ammonia in a watch glass, and add some liquor potassa; the urate will become decomposed, evolving ammonia which will restore the blue color to dampened and reddened litmus paper. If urate of ammonia be heated in a platinum capsule, it will be wholly dissipated. If a drop of acetic acid be added to a little of the deposit, plates of uric acid will gradually form. With nitric acid and ammonia, the deposit will give the *murexide* reaction. Urate of ammonia is less soluble in water than urate of soda, requiring a temperature of about 200° F., for its solution.

2. **Urate of Lime, Acid.** *Acid Calcic* or *Calcium Urate*. This salt exists in very small quantity in the urine, and is rarely met with. It forms a white amorphous powder, which, when exposed to a red heat, leaves carbonate of lime. It is sparingly soluble in water, though the presence of chloride of potassium increases its solubility; it parts with uric acid when treated with acetic acid. It has been found in calculi, and in gouty swellings around the joints.—If some amorphous, or, crystalline, urate of lime be placed upon a platinum spatula and calcined, a white gritty powder, carbonate of lime, will be left, which is soluble in hydrochloric acid with disengagement of carbonic acid gas. Acetic acid added to urate of lime decomposes it, and uric acid crystals are formed, which may be verified by the *murexide test.* See *Benzoate of Lithia.*

3. **Urate of Magnesia, Acid.** *Acid Magnesic*, or *Magnesian Urate.* This salt is very common in urinary calculi, though it has not been detected in human urine; it is probable, however, that a little of it is present, perhaps, occasionally. It crystallizes in small rectangular plates, of various lengths, and sometimes so delicate as to represent needles. In crystalline deposits, the elongated, ribbon-like plates are more abundant. The plates and needles are colorless, transparent, and usually grouped together parallelly, or in fan form; sometimes they form very minute stelliform masses, and again, spherical groups with a dark center. The crystals polarize light. According to Bensch, they are more soluble in cold than in boiling water. The dried

crystals are very efflorescent.—To analyze these crystals, calcine some of them on a platinum spatula, when a white residue of carbonate of magnesia will be left. Place some of this residue on a glass slide, and dissolve it by adding a small drop of hydrochloric acid; to this solution add a drop of solution of phosphate of soda, and then a drop of ammonia; the arborescent crystals of ammonio-magnesian phosphate will appear under the microscope. The presence of uric acid may be verified by subjecting the urate of magnesia to the *murexide test.*

Bigelow describes a *Biurate Hydrate of Magnesia,* which forms in large, quadrilateral crystals, having regular angles, transparent, of straw color, insoluble in water, and not decomposed by acetic, or ordinary hydrochloric acid, agents which promptly decompose the preceding form of urate of magnesia. Pure concentrated hydrochloric acid gradually decomposes them with formation of uric acid. Exposed to the air these crystals become opaque and silvery white; when heated, they break with a noise, violently scattering the fragments around; hence, in calcining them, the heat should be low at first to drive off their water of crystallization.

4. **Urate of Potassa, Acid.** *Acid Potassic,* or *Potassium Urate.* This salt has been found in urinary sediments, and in urinary calculi, in very minute quantity. It has been found in almost pure uric acid calculi. The remarks on urate of soda, as to its amorphous and crystalline forms, and its general constituent formation, will likewise apply to the urate of potassa. When treated with bichloride of platinum, similar to the method named under urate of soda, its behavior will be very different, regular octohedral prisms are formed, which are only slightly soluble in water, and do not polarize light.

5. **Urate of Soda, Acid.** *Acid Sodic,* or *Sodium Urate. Biurate of Soda.* This salt generally appears in the form of very small, irregular, amorphous granules. It may be prepared artificially by dissolving uric acid in a warm solution of ordinary phosphate of soda; it then appears in microscopic, hexagonal, prismatic crystals, usually united in stellar groups. Similar forms are met with in urine undergoing the process of alkaline fermentation. Microscopic examination sometimes detects very complicated forms in this transition period of the fermentation; the uric acid crystals, separated during the acid fermentation, have commenced re-solution in greater or lesser quantity, and become studded with elegant groups of prismatic crystals of urate of soda; at the same time, concentrically striated spherules, consisting, probably, of urate of ammonia, may be seen placed alongside of the prismatic crystals. Litmus is still feebly reddened by this urine. As the fermentation progresses, and when the urine has acquired a neutral reaction, prismatic groups of acid urate of soda may sometimes be seen, being now accompanied with fine large crystals of ammonio-magnesian phosphates. Solution of urate of soda allowed to evaporate spontaneously, gives a deposit of this urate in simple spherical masses and granules.

Urate of soda is soluble in water of about 100° F., and forms a less flocculent precipitate than that of urate of ammonia. It presents three forms under the microscope, viz., a colorless amorphous powder; regular globules with yellow center (*neutral*), and semi-transparent prismatic crystals that faintly polarize light. These crystals, whether met with in calculi or in sediment, may be determined by placing some of them on a platinum capsule, and calcining. A white residue, carbonate of soda remains; this fuses at an elevated heat; is dissolved in a few drops of water; and restores the blue color to reddened litmus paper. If a drop of this aqueous solution be placed on a glass slide, and a drop of solution of bichloride of platinum be added to it, there will form, when observed under the microscope, large, indefinitely long, transparent prisms, possessing, in a high degree, the power of polarizing light; these prisms are very soluble in water.—Urine containing urate of soda, becomes clear, from solution of the urate under heat; or, on the addition of a little liquor potassa; if to the urine, rendered alkaline by the potassa, an excess of acetic acid be added, crystals of uric acid will be deposited in from 6 to 12 hours.

6. **Urate of Soda, Neutral.** *Neutral Sodic,* or *Sodium Urate.* This is a pulverulent deposit, more commonly met with in the urine of persons laboring under severe fevers, and is generally associated with a little urate of lime, and traces of urate of ammonia. Having once appeared in the urine of men, or of carnivora, it is very apt to remain throughout the remaining part of the individual's life. When pure, it crystallizes in isolated mammillæ, or in mammillated groups. The granules of this urate in the urine are ovoid or spheroidal, sometimes one extremity more enlarged than the other; their periphery is distinctly marked, black or brownish, and their center is yellowish-brown, or, a more or less intense reddish color. When urine contains neutral urate of soda in solution, and an acid be added to it, the urine will become more or less opaque from formation of acid urate of soda, which will ultimately be deposited in fine granules. This may sometimes mislead the practitioner, who, having a urine of this character to examine, and observing an opacity produced in it by the addition of nitric acid, may mistake it for albumen; but, on the application of heat, the deposit will be redissolved and thus correct his error.

When these urates occur singly they present but little difficulty in analysis, but when they are mixed, or are united with uric acid, the analysis becomes more complicated. In such cases, which occur frequently, the practitioner will have to pursue the several processes heretofore named, on different portions of the sediment or urine, in order to detect the various bases, as, soda, ammonia, lime, etc., always observing the general aspect of the precipitate, and the characters of the crystals.

‡ *Microchemical Diagnosis and Separation of Urates.*—1. Place a drop of the deposit upon a glass slide; add a drop of dilute nitric acid to it; apply heat; as soon as the fluid evaporates, a reddish circle will appear. When the slide becomes cold add a drop of ammonia to the circle on the slide, if it be a urate or uric acid, the circle will change from a reddish to a purplish-red or

violet color (murexide reaction).—2. Place a drop of the deposit on a glass slide, and add a drop of acetic or hydrochloric acid; cover with thin glass, and examine under the microscope, crystals of uric acid will appear. See *Figs.* 35 and 36.

Treat a drop of urate deposit, freed from any foreign elements it may contain, with a drop of hydrochloric acid, and allow it to spontaneously evaporate on the glass slide under a bell glass. When the evaporation has terminated, upon examining the residue under the microscope, the following crystals will be found to have formed:—*a.* Cubes or octohedra of common salt, and of chloride of potassium, if the sediment contained urates of soda and of potassa. See *Figs.* 13 and 34.—*b.* If the sediment contained urate of ammonia, elegant arborizations of hydrochlorate of ammonia will be seen. These arborizations may be thoroughly observed on allowing a drop of limpid saliva to spontaneously evaporate on a glass slide.—*c.* Crystals of uric acid may be seen. See *Figs.* 35 and 36. The two preceding named salts are soluble in a drop or two of water added to the preparation; the uric acid crystals will remain intact. (For the separation of urates and other sediments, see *Microchemical Diagnosis of Earthy Phosphates,* etc., under *Phosphates.*)

Clinical Import. The sediments of urates appear under various influences. When the urine is concentrated or deficient in its fluid element, as, from excessive perspiration, watery evacuations at stool, abstaining from aqueous draughts, or from other causes, urates will be deposited as the urine cools. An excess of water in the urinary secretion, will hold the urates in solution for some time after cooling; the lower the temperature, the more quickly are they deposited. A slight departure in diet, an abnormal fatigue, any nervous shock, will suffice to determine a deposit of this kind. But in these cases there is no indication of disease, as the deposits are temporary. A deposit in urine is always an indication of some abnormal condition, more especially when it persistently remains for days and weeks, and should never be disregarded. The appearance of urates in the urine may be an indication of commencing gravel or calculus, especially when it persists, and no febrile or inflammatory symptoms are present; in most cases of this kind, disease of one or more of the abdominal organs is apt to exist. Urates may be present in phthisis, fever, acute inflammation, acute rheumatism, and in all cases where the tissues are destroyed more rapidly than their supply by the nitrogenized elements. The sudden appearance of urates in pneumonia and pleurisy when resolution and absorption commence, in gout and rheumatism, and in active inflammatory or febrile conditions, indicates a change, more generally, of a favorable character. An excessive use of animal food, or where the ordinary amount of nutrition continues the same but without much physical exercise, will occasion the appearance of urates in the urine there being, in such cases, a greater provision of nitrogen than is required for the reparation and replacement of the tissues. Again, the nitrogenous elements may not be in excess, but can not be assimilated by the digestive apparatus, as, in dyspepsia. When not due to any of the preceding conditions, nor to diminution of temperature, the presence of urates may indicate some derangement of the cutaneous functions, in which the kidneys are called upon to compensate for the defect in these functions. Blows over the region of the kidneys, fatigue of these organs, or congestion of them due to local causes, are very apt to be attended with deposits of urates; and when these are in such excess as to be deposited while the urine is in the bladder, the case assumes a serious aspect. The excess of urates in the urine of gouty persons, explains their presence in crystalline form in the articular surfaces. According to Charcot and Cornil, they may even be the cause of the albuminous nephritis so often supervening in these cases, as their presence in the kidneys would seem to demonstrate. See *Uric Acid.*

Urea. Urea (Ur) is the most abundant and the most important principle of urine; it is found in the urine of birds, serpents, and animals, but is more abundant in that of the mammalia. It is a normal organic constituent of urine, a great excess or deficiency of which, in this fluid, indicates some pathological condition. It is the chief final product of the metamorphosis of the albuminous or nitrogenized tissues of the organism which are eliminated in order to give place to a new substance the materials of which are furnished by the food. Urea represents five-sixths of the nitrogen absorbed with the food; it appears to be formed in or taken up by the blood, and is eliminated through the kidneys. Claude Bernard states that when the renal organs are removed, the urea is excreted by the gastro-intestinal mucous membrane, in the state of carbonate of ammonia. Urea has been found in the saliva, tears, milk, bile, liquor amnii, fluids of the eye, perspiration, etc., in small quantity. From 25 to 40 grammes of urea are passed daily, on an average; but this amount is subject to great variations, determined by the weight, age, sex, occupation, and diet of the person. When urea is present *in excess,* the urine is apt to have a high sp. gr. 1.020 to 1.030 or more, is usually clear, and free from sediments; when it is *deficient,* the sp. gr. may be reduced to 1.001 to 1.008, with an increase in the quantity of urine. When free from coloring matter and pure, urea is a white, semi-transparent, crystalline body, the crystals being four-sided or acicular, and having a bitter, saltpetre-like taste, and no odor. It is soluble in water and alcohol, hardly soluble in ether, insoluble in oil of turpentine. It is readily decomposed by heat, hydrated alkalies, strong mineral acids, putrescent organic matter, mucus, pus, and yeast. At 248° F., it fuses. Its presence in the urine causes chloride of sodium which usually crystallizes in cubes, to form octohedral crystals, crosslets, and daggers, see *Fig.* 13; while, should chloride of ammonium, which crystallizes in octohedra, be present in the urine, urea causes it to crystallize in cubes. Urea is prepared artificially, by various methods which it is not in the province of this work to explain.

According to M. Grimaux (*Chimie Organique Elementaire.* Paris, 1872), urea may be considered an amide, that is, an ammoniacal salt *minus* water: now, urea corresponds to neutral carbonate of ammonia *minus* two atoms of water; it is then the amide of carbonic acid, or carbamide. We know that urea was obtained, complete in all its parts, by Wiehler, in 1829; since which, this synthesis has been reproduced by various processes. It should likewise be understood that the absorption of water by urea converts this into carbonate of ammonia; this is precisely what occurs in the fermentation of urine, and hence its peculiar ammoniacal odor.—It is difficult for the medical practitioner to study the variations of urea in human urine and draw satisfactory practical indications from them. There does not yet exist any simple and sufficiently exact process of examination. The volumetric methods, which are considered the only ones, and of which most physicians can avail themselves, are very complicated and inexact.

Qualitative Analysis.

Urea is too soluble in water to form a spontaneous deposit in the urine. When it exists in excess, or in normal urine artificially concentrated, its presence may be readily demonstrated by combining it with nitric or oxalic acid. If albumen be present in the urine, it must be removed, by the process hereafter stated.—‡1. Place a small quantity of the (non-albuminous) urine, to be examined, in a watch glass, or on a glass slide, and evaporate it to the consistence of syrup; to this, when cold, having previously filtered it, while hot, to remove any urates or phosphates that may be deposited, add about one-third its volume of pure nitric acid. Immediately (or after a short time, depending upon the amount of urea present), a yellowish crystalline mass (yellowish from the coloring matter of the urine) of agglomerated pearly glistening scales or plates will be formed, which are crystals of nitrate of urea; should the crystallization occur slowly, fine rhombic prisms will be formed. (See *Fig.* 34).

Fig. 34.

A. Evaporated residue of healthy urine.
B. Oxalate of urea.
C. Nitrate of urea.

To study this salt under the microscope, a drop of the concentrated urine may be placed on a glass slide, in which we immerse a strand of cotton thread, and then cover with a thin glass, allowing about one-half of the thread to pass beyond the glass cover. Now, upon the protruding extremity of the thread place a drop of nitric acid; this reagent, by capillarity, penetrates the thread under the glass cover and acts upon the urea in the drop of urine;

after a variable time, crystals of nitrate of urea will appear in the fluid around the thread.

Pure *nitrate of urea* (*urea* or *ureal nitrate*), is soluble in water and alcohol, not very soluble in nitric acid, is not acted upon by atmospheric influence, explodes when quickly heated on platinum foil to a high temperature, but, at 284° F., it is decomposed into urea, suboxide of nitrogen, nitrate of ammonia, and carbonic acid.—When the crystals form rapidly in urine, the quantity and sp. gr. of the whole urine passed in 24 hours must be ascertained, because a single specimen of urine may appear richer in urea, from a diminished amount of water.

2. ‡ As nitric acid vapors injure the metallic parts of the microscope, a concentrated solution of oxalic acid may be added to the non-albuminous urine, and which will give crystals of oxalate of urea. *Oxalate of urea* (*urea oxalate*), crystallizes in long, thin, hexagonal plates or prisms, of rhomboidal form, the angles being less acute than those of the nitrate; sometimes these crystals are separate, at others adhering in groups. They are occasionally, but very rarely, observed in concentrated human urine. This salt is soluble in 23 parts of cold water, but freely soluble in boiling water. See *Fig.* 34.

3. Mix 20 c. c. of urine with 10 c. c. of baryta solution, and remove the precipitated sulphates and phosphates by filtration; neutralize the filtrate with nitric acid and evaporate to dryness over a water bath. Extract the residue with alcohol; evaporate; again exhaust the residuum with absolute alcohol, which will yield on spontaneous evaporation, very pure, colorless needles of urea. By filtration, drying the crystals and weighing them, we can form a pretty accurate estimation of the amount of urea passed at each urination. Thus, if, in the above process, we find that 1.5 grammes of urea are present, and the amount of the urine passed at the time was 240 c. c., we calculate — $\frac{240}{20} = 12 \times 1.5 = 18$ grammes of urea. By examining the urine of each urination in this manner, the amount of urea passed in 24 hours may be determined sufficiently correct for practical purposes. *Liebig.* —4. Evaporate 15 c. c. of non-albuminous urine to one-fourth its volume, and treat it with one volume of nitric acid; white nitrate of urea is precipitated. Treat this precipitate with solution of carbonate of baryta, carbonic acid is produced, with nitrate of baryta and urea. Separate the urea from the insoluble nitrate of baryta by alcohol; filter, and evaporate the filtrate to obtain the urea.—5. Take urine in the state of full alkaline fermentation, pour it upon filtering paper in a filter; after a time the filtration slackens, owing to the pores of the paper becoming filled with globules of a certain ferment. Now wash the paper with distilled water until the washings no longer give an alkaline reaction, and carefully dry the paper at 95° to 104° F. (This paper will convert even a very dilute solution of urea into carbonate of ammonia.) After it is dried, color it yellow with turmeric, dry it, cut it into strips, and preserve in dark, well closed bottles. Any solution, containing urea even to the $\frac{1}{10000}$th part, will cause this paper, when soaked

14

in it, to become covered with brown spots. The liquid to be tested must, in all cases, be neutralized to decompose alkaline carbonates and bicarbonates which interfere with the action of the test; neutral alkaline salts do not. *Musculus.*—See *Specific Gravity.*

N. B. As *albumen* interferes with the determination of urea in the urine by any of the qualitative or quantitative processes, it must first be removed. For this purpose, after having removed the sulphates and phosphates by baryta solution, as named above (No. 3), add, to the filtrate, absolute alcohol, which having a stronger affinity for urea than the albumen has, takes up the urea and renders the albumen insoluble. Filter to remove the albumen, and evaporate the filtrate to the bulk of urine employed, and proceed according to the process employed for the determination of urea. *Thudicum.*—If urine containing albumen be passed through a filter filled with crystals of sulphate of soda, these crystals will arrest the albumen and prevent it from entering into the filtrate.—De Luca removes albumen from the urine by ammoniated cupric nitrate, filtering and using the filtrate for the analysis of urea.— Mucus, pus, or any other deposit in the urine, must be removed by filtration or decantation before commencing an analysis.—The coloring matter of bile may be precipitated from urine by adding solution of acetate of lead to it.— Any nitrogenized matters in the urine may be precipitated by the addition of a sufficient quantity of solution of acetate of lead, acidulated with a few drops of acetic acid; then filter, remove excess of lead by a current of sulphureted hydrogen; again filter, and use the urine for detection of urea.

The following method has been recommended as sufficiently accurate for practical purposes in determining the (physiological or pathological) amount of urea present in a given specimen of urine: —

‡ Upon a glass slide or platinum foil, at 60° or 65° F., place a drop or two of the urine, and add to it an equal quantity of concentrated nitric acid, if crystals of nitrate of urea form at once, or within five minutes, there is an *excess of urea*, the proportion of which will depend upon the rapidity and extensiveness of the process. If the crystallization occurs within from 5 to 10 minutes, —or, if the urine be slowly evaporated to one-half its bulk, and then an equal quantity of the acid be added, followed by immediate crystallization,— the urine contains a *normal or average per cent. of urea.* If crystallization does not occur within 10 minutes,—or immediately, with the evaporated urine,— then there is a *deficiency of urea*, to a greater or lesser extent.

Quantitative Analysis.

Several methods have been made known for the quantitative determination of urea in the urine, as, Liebig's, Bunsen's, Davy's, Leconte's, Ragsky's, Knopp's, Huffner's, Yvon's, etc. Of these the following only will be given: 1. *Liebig's.* Into a beaker glass place 20 c. c. of the non-albuminous urine to be examined, and add to it 20 c. c. of the *solution of baryta* (for analysis of urea); this will precipitate all the sulphates, carbonates, and phosphates. Agitate the mixture thoroughly with a glass rod, and allow it to rest. Pass the mixture through a filter, previously moistened with distilled water, and refilter if the liquid is not clear. Of this filtrate, take 20 c. c., equal to 10 c. c. of the urine, and add to it a few drops of *solution of nitrate of silver*, in order to precipitate all the soluble chloride present, which interferes with the test, preventing the urea from being precipitated by the solution of protonitrate of mercury. Make sure that the liquid is alkaline, adding a drop or two of solution of carbonate of soda to make it so, if necessary. Now, having placed the standard solution of protonitrate of mercury into a burette or graduated pipette; filling it up to the level of the first graduation (zero), gradually allow this solution to drop into the above prepared urine, until there is added a slight excess of the standard solution, which may be known by a drop of the mixture producing a yellow stain when placed on prepared soda paper.* This color produced, the analysis is finished. Read off the amount of standard mercuric solution, on the graduated burette, required to produce this yellow stain, which will show the amount of urea in 10 c. c. of urine. Suppose 50 c. c. of this standard solution were required in the preceding process, then, as 1 c. c. of it corresponds to 0.01 gramme of urea, 50 c. c. will correspond to 0.5 gramme of urea. The patient having passed 800 c. c. of urine in the 24 hours, we multiply the 800 by the 0.5 gramme of urea found in 10 c. c. of the urine, and divide the result by the amount of urine tested, 10 c. c. Thus,

$$\frac{800 \times 0.5}{10} = 40 \text{ grammes.}$$

The above process of Liebig's, although giving very accurate results, requires considerable time and patience in its performance, and is better adapted for the chemist, than for the practitioner, who, after having once experimented with either of the following three processes referred to under section 2, will be enabled to readily and satisfactorily employ it in any case in which the quantitative estimation of urea is desired. Besides, there are certain modifications and corrections to be made in Liebig's method, not required in the others; thus, if the urine contains more than 2 per cent. of urea, it would require to be diluted to a certain extent; if more than 1 per cent. of chloride of sodium be present, we must either remove the chlorine by a graduated solution of nitrate of silver before undertaking the operation, or, subtract 2 c. c. from the total c. c. of mercurial solution used to determine the number of milligrammes of urea contained in 10 c. c. of the urine. If the urea sink to 1 per cent. or less, we must for every 5 c. c. of mercurial solution used less than 30 c. c., subtract 0.1 c. c. from the sum of c. c. actually used. Other modifications are also required for the presence of ammonia, albumen, etc., which it is unnecessary to name here, as but few physicians would ever undertake the process, as exact as it may be, when others, nearly as exact and of more ready performance, can be resorted to.

* The *prepared soda paper* is made by soaking a sheet of white filtering paper in a saturated solution of pure carbonate of soda. When dry it may be cut

into strips, and be kept in a well-closed wide-mouthed bottle. The above is Harley's modification of Liebig's process, and will be found useful.

‡ 2.—*Yvon's Process.* This is a modification of Knopp's and Huffner's, and, with the necessary articles on hand, the analysis can be made in 5 or 6 minutes at the bedside of the patient. The instruments consist of a long, wide bell-mouthed beaker or mercury jar, and a glass tube, ureometer, 40 centimetres long, with a calibre of 1 centimetre in diameter. This tube is divided into two compartments by a glass stop-cock placed towards its upper quarter, which allows a communication between the two compartments to be established or suppressed at will. Commencing at this stop-cock, the tube is graduated, above and below it, into cubic centimetres and tenths of the same. The lower part of the tube is the longest, from its extremity to the stop-cock, and is designed to be placed into a mercury jar, as hereafter described. —The *standard solution* employed consists of solution of caustic soda, 36°, 30 grammes, dissolved in distilled water 125 grammes, to which add, when cold, bromine 5 grammes, and agitate strongly. Allow it to rest for some time. and, after decantation, a fine yellow, clear, transparent fluid remains, which has to be frequently renewed, as it loses considerable of its strength in a very short time. By keeping the solution of caustic soda, separately, this preparation may be made in small quantity, whenever required, by adding to any given amount of it the proper proportion of bromine. As a *test liquor of* this standard solution, 1 gramme of pulverized urea, well dried by a prolonged sojourn under a bell glass with sulphuric acid, is dissolved in distilled water so as to have 5 c. c. of the solution equal to 1 centigramme of urea. After the action of the standard solution of hypobromite of soda, upon this solution of urea, 37 divisions of the tube, or $\frac{37}{40}$ths of c. c. of gas, will correspond to 1 centigramme of urea.

Ten cubic centimetres of the non-albuminous urine are diluted with distilled water to make a volume of 50 c. c., and of this diluted urine from 1 to 5 c. c. are operated upon, according to its richness in urea. Polyuric urine requires no dilution. We must always endeavor to have no more than 40 divisions of gas, in order not to too greatly increase the volume of the standard solution to be employed, and the dimensions of the column of liquid. The method of proceeding is as follows:—Open the stop-cock and plunge the inferior extremity into the jar filled with pure mercury, until this has arisen to a level with the lower part of the superior compartment; then close the stop-cock, raise the tube (but not entirely out of the mercury), the inferior compartment of which is filled with the mercury, and maintain it in this position by means of a strap or collar, which forms part of the complete apparatus. Into the upper apartment place from 1 to 5 c. c. of the diluted urine, open the stop-cock, and carefully make the fluid flow into the inferior compartment, closing the stop-cock before any air can pass along with it. Wash the receptacle at the superior part of the tube with several drops of distilled water, and carefully pass this into the lower apartment, by means of the stop-cock, as in the previous instance. This done, place into the upper receptacle 5 or 6 c. c. of the standard solution, and, as in the first instance, make it pass into the lower apartment, raising the tube a little should this be necessary. Agitate by an up and down motion, but without removing the tube entirely out of the mercury, having the stop-cock closed, of course. Decomposition occurs immediately, the urea is separated into water, nitrogen, and carbonic acid gas; the last being absorbed by the excess of alkali, nitrogen alone remains in the tube. To render the mixture more exact, raise the tube, close the lower orifice, while this is under the mercury, with a finger, and shake it; then carefully replace it in the mercury. The liquid soon becomes limpid when all the gas has collected together, and the amount of gas may be read off on the graduated divisions at the level of the fluid remaining in the lower compartment. The measure of the gas is made in a beaker filled with water, being careful to properly place the surface of the liquid in the tube on a level with that of the fluid in the beaker.

For instance, 5 c. c. of the diluted urine have been operated upon, and we find 22 divisions of gas produced, and as 5 c. c. of the standard solution represents 1 centigramme of urea, we have $\frac{22}{37}$ of a centigramme of urea. But the urine has been diluted to 5 times its volume with water, from which we can readily conclude that 1 c. c. of the original undiluted urine contains $\frac{22}{37}$ of a centigramme of urea. If 800 c. c. of the urine are passed in 24 hours we have $800 \times 22 \div 37 = 4$ grms .754 of urea in 800 c. c. of urine; or, 5 grms .945 of urea in 1,000 c. c. of urine.—Uric acid, creatine, creatinine, and the greater part of the nitrogenous matters of the urine are decomposed by the standard bromated solution. M. Leconte advises to deduct 4.5 parts for 1,000 of the nitrogen found. But, in pathological cases, can this rule be considered as exact? Should the urine be partly putrescent, or a part of the urea be converted into carbonate of ammonia, M. Yvon's process gives equally a good result. (The complete apparatus, with special instructions, is sold by M. Alvergniat, passage Sorbonne, Paris,).

More recently several similar processes have been presented to the profession, by various parties; it is unnecessary to name them all. Messrs. Russell & Watson, of England, have adopted one. Their proportions slightly differ from the preceding, thus: 20 grammes of caustic soda are dissolved in 50 c. c. of water, to which solution is added 5 c. c. of bromine. Their instrument costs 8s. 6d. sterling, and is manufactured by Messrs. Cetti, Brook street, Holborn,—full directions accompanying.—R. Apjohn has likewise devised a simple and readily worked apparatus, being in principle somewhat similar to the preceding. *Chemical News*, 1875, page 36. See *Esbach's Method.*—3. Dr. Geo. B Fowler, of New York, has advised a new and simple method for the quantitative estimation of urea, somewhat similar to that of Dr. Roberts' for determining the amount of sugar. Into a glass cylindrical jar, about 7 inches in depth and 1 inch in diameter, pour some of the urine to be examined and take its sp. gr. Pour out this urine into some convenient vessel, and carefully cleanse the jar. Now pour into it a portion of Squibb's hypochlorite of sodium and take its sp. gr. These two sp. gr.

are to be added together and their sum divided by 8, which will give the sp.
gr. of the following mixture previous to decomposition :—Into another glass
vessel, about 9 inches in depth and 1½ inches in diameter, pour 15 c. c. of the
urine that was contained in the smaller jar previous to cleansing it, and add
to it seven times its quantity of the hypochlorite (105 c. c). Active effervescence at once commences, and when it has ceased, agitate the mixture
from time to time until the urea is effectively decomposed. Then, take the
sp. gr. of the remaining fluid, and deduct this from the sp. gr. previous to
decomposition. Every degree of sp. gr. lost will indicate 7.791 milligrammes
of urea in every c. c. of the urine. Neither albumen nor sugar exert an
injurious influence upon the exactness of this process. For more minute
particulars see *New York Medical Journal*, June, 1877.

Clinical Import. Urea is derived from disintegration of the nitrogenous
tissues of the body, and from decomposition of nitrogenized food. Its secretion is greater during the waking hours than during sleep, and its presence
in urine is augmented by active physical or mental exercise, by animal diet,
and by drinking excessively of water; a purely animal diet increases the
proportion of urea in the urine nearly one-third, while vegetable diet only
will diminish it in about the same proportion; a diet consisting of non-ni-
trogenous articles will lessen its amount one-half or more. The weight,
occupation, and food of the person appear to regulate the quantity of urea;
the better the diet, and the more active the occupation, the greater the physiological amount of urea. (See *Agents*.) Its presence indicates, to a certain
extent, the wear and tear of the system during health, when the various
factors above named are taken into consideration, at the same time. The
average amount of urea in the urine has already been stated; when it greatly
exceeds, or falls below, this average, persistently remaining so for some time,
it indicates a pathological condition. As heretofore observed, when a urine,
not concentrated, to which an equal volume of nitric acid is added, promptly
yields a number of nitrate of urea crystals, urea is said to be "in excess."
When, even after concentration, but few crystals present, or form slowly,
there may then be a " deficiency of urea."

A persistent excess of urea (sp. gr. 1.030 or more), and especially when at
the same time a diminished amount of nitrogenized food is taken, indicates
a diseased condition in which there is an excessive disintegration of the tissues. This appears to be the case in the early periods of acute diseases, as,
fevers (excepting yellow fever), pneumonia, meningitis, exanthemata, etc.
This excess may likewise be present in certain deranged conditions of the
digestive, and the cutaneous, functions, diabetes insipidus, during an attack
of epilepsy, as well as in many other affections. As the acute malady reaches
its height, the abnormal increase of tissue-metamorphosis diminishes to an
extent that occasions a diminution of urea below the normal amount, and
when convalescence is about to occur, the quantity of urea gradually rises to
its normal figure.

Diminished or deficient urea in the urine (sp. gr. 1.001 to 1.008, with profuse urination), may be due to decrease in tissue metamorphosis, the result of a want of nutrition, or of a deficiency of oxygen; it may also be due to renal disease, in which, though urea be formed in normal quantity, it is not excreted by the kidneys. Deficiency of urea, when owing to the latter cause, is always of a serious nature, because the retained urea, acting as a poison, occasions uremia and death. (Ammonemia is entirely different in its symptoms from uremia, and is more frequently amenable to treatment. It is the result of reabsorption of the ammonia of the urine in the bladder, which may occur from a retention of urine, and especially when the urine becomes ammoniacal from decomposition of its urea.) Deficient urea indicates a serious disease of the cortical portion of the kidneys; this deficiency is met with in Bright's disease, and in any malady in which the kidneys fail to properly eliminate. It is also met with in minimum amount in dropsy, paralysis, anemia, yellow fever, acute yellow atrophy of the liver, cirrhosis, certain affections of the respiratory organs, and in many chronic affections.

In acute or inflammatory conditions, when the excess of urea continues, and even augments after the acme has passed, the prognosis is unfavorable. In albuminaria the prognosis is much more favorable when the urea remains normal, than where it is deficient. And in all conditions of the system, where deficient urea is gradually or rapidly followed by an increase in its quantity, with corresponding improvement in the patient's general condition, we may augur favorably.

In determining the excess or diminution of urea in the urine, the practitioner must not allow himself to be misled by judging of it from its proportion to a certain amount of fluid, because, during the febrile condition, when the urine is scant and high colored, or concentrated, it will then present an apparent increase or excess, and during convalescence, when this fluid is passed in large amount or profusely, the same proportion of urea in it would show an apparent great diminution or deficiency. The determination should be made upon the whole amount of the urea passed in 24 hours; by this method the decrease of this substance in acute diseases will be found much less than is generally supposed; indeed, it will frequently be reduced to a very minimum amount.

According to M. P. Brouardel there is a direct relation existing between the functional activity of the liver and the excretion of urea, so that, in all probability, a great part of the urea encountered in the urine is formed in the liver, this formation being in direct proportion to the size of the liver. From his investigations, he infers that the quantity of urea depends upon the integrity of the hepatic cells, and the greater or lesser activity of the hepatic circulation.

Ureal. Of or belonging to urea.

Urecchysis. Infiltration of urine into the cellular membrane.

Urema. See *Urine.*

Uremia. A generally fatal malady, due to the poisoning of the blood by urea, which has not been eliminated from this fluid by the kidneys. See *Urea (Clinical Import)*.

Ureoritrine. See *Uroerythrin*.

Ureorrhea. A flow of urine, as diabetes.

Uresis. The voiding of urine.

Ureter. The long membranous tube connecting the urinary bladder with the kidney, and through which the urine passes to be emptied into the bladder. There is one for each kidney.

Urethra. The canal through which urine is discharged from the bladder. In women it is an inch or an inch and a half long; in man, its length varies from 7 to 11 inches, and is divided into the *prostatic portion*, quite wide and dilatable, and about 16 lines in length ; the *membranous portion*, the narrowest portion of the canal (excepting the meatus), about 6 lines in length inferiorly, and 9 lines superiorly, concave on its upper surface and convex on its lower ; the *spongy portion*, about 6 to 8 inches long; and the *meatus urinarius*, or external orifice, the most contracted part of the canal.

Urethrorrhagia. Bleeding from the walls of the urethra.

Urethrorrhea. A continuous flow of any fluid from the urethral wall.

Urethroscope. An optical instrument designed for visual examination of the urethral walls, and so arranged as to permit the use of certain instruments during the inspection. The instrument consists of a tube to pass into the urethra, and a concave mirror to throw light through the tube into this canal, somewhat like that employed for ophthalmoscopic, or laryngoscopic examinations.

Uric Acid. *Lithic Acid. Urylic Acid. Bezoardic Acid.* Uric acid is found in the urine of omnivorous and carnivorous animals, but not in that of the herbivorous ; it is likewise present in the urine of birds, serpents, and insects. In human urine it is in combination with some base, as, soda, ammonia, lime, potassa, from which it may be separated in crystalline form by the addition of acetic, hydrochloric, or nitric acid. When met with, in urine just voided, in the free state, but which is rarely if ever the case, its presence is due to accidental or pathological causes; although it is frequently observed in gravel and calculi, sometimes forming the principal substance of a calculus. When urine drawn off from the bladder by a catheter, presents a deposit of more or less uric acid crystals, there will be great reason for fearing the formation of uric acid calculi. The amount of uric acid in health passed with the urine daily varies according to the character of the sex, the diet, climate, and exercise, from 0 grm .340 to 1 grm .478. When this acid is in excess, the urine will have a sp. gr. of 1.020 to 1.025, or more.—Uric acid is the product of the disintegration of the nitrogenized tissues or of the albuminous food, the metamorphosis being of a lower grade than that which eliminates urea, and which is due to less perfect oxidation. A more perfect oxidation converts uric acid into urea.

Pure uric acid is a white, light, crystalline powder, inodorous, tasteless, and feebly acid; it is insoluble in alcohol, ether, and acetic acid; nearly insoluble in nitric and hydrochloric acids; soluble in 11,000 parts of cold water, and in 2,000 parts of boiling; readily soluble in liquor potassa, and in concentrated sulphuric acid from which water precipitates it; and sparingly soluble in dilute solution of carbonate of potassa. Insolubility in ammonia, will distinguish it from cystine which is soluble in this alkaline fluid. Exposed to a red heat, uric acid is decomposed without being fused; by dry distillation, a sublimate of ring-form is obtained, consisting of urea and cyanuric acid; hydrocyanic acid and carbonate of ammonia likewise escape, and a porous coal containing nitrogen remains behind. One part of uric acid gradually added to four parts of concentrated nitric acid dissolves with effervescence, and is converted into a crystalline pulp containing alloxan. The uric acid is decomposed into alloxan and urea, and the nitric acid into nitrous, which immediately decomposes the urea into carbonic acid and nitrogen. It must not be forgotten, however, that urates and carbonates, as well as calcined ammonio-magnesian phosphate, equally dissolve with effervescence in nitric acid. Uric acid may also be determined by the murexide test (see *Purpurate of Ammonia*), and by Schiff's test (see *Carbonate of Silver*).

XIII. Table of
Chemical Characters of Uric Acid.

1. Calcination at a red heat. — No residue. A light black residue (a porous coal containing nitrogen. *Neubauer*.)

2. Solubility.
 a. Insoluble in cold water; slightly soluble in boiling.
 b. Insoluble in dilute hydrochloric acid.
 c. Soluble in potassa from which, if a slight excess of acid be added, crystals are precipitated.
 d. Soluble in alkalies.

3. Action of alkalies. — Triturated with caustic alkalies, unctuous compounds are formed, and *ammonia is not set free*.

4. Action of concentrated nitric acid. — Dissolves with *effervescence* and forms a crystalline pulp.
 Explanation of the reaction. Uric acid is decomposed into — Alloxan, which forms the crystalline mass. Urea. Nitric acid, do.....Nitrous acid } which give { Carbonic acid. Nitrogen.
 Causes of errors. Urates. Carbonates. — Dissolve equally with effervescence, as well as calcined ammonio-magnesian phosphates. *Beale*.

5. Action of dilute nitric acid. *Murexide*. — Upon heating and slowly evaporating to dryness, a *red residue* remains, which, treated by a few drops of ammonia, becomes purple or violet-red (murexide, *purpurate of ammonia*). If testing is for the detection of traces of uric acid, an excess of ammonia must be avoided; moisten a glass rod with this alkali, and blow the vapors from it upon the residue, the rod being held close to it without contact.—Caffein gives the same reaction.

6. Action of potassa. — The red residue, referred to above, when treated by potassa, gives a violet color, which disappears under heat.

7. Reaction of nitrate of silver. — Dissolve traces of uric acid in carbonate of soda. With this solution touch, lightly, a paper upon which a drop of solution of nitrate of silver has been allowed to spread. A dark spot is produced (reduction of the nitrate of silver.) Will detect 1-1,000 to 1-500,000 of uric acid.

Reaction of the Murexide.—According to Hardy's recent investigations, the characteristic color of uric acid is principally due to its modified anhydrous alloxan, then after the addition of ammonia to the isolloxalate of ammonium. *Chimie Biologique*, p. 455.

‡ Qualitative Analysis.

Add to the urine a little acetic, nitric, or hydrochloric acid, after a longer or shorter time, depending upon the amount of uric acid present, and the degree of concentration of the urine, crystals of this acid will be precipitated, which can be examined under the microscope. Or, under the microscope, a drop of a sediment of urates may be treated with a drop of acetic acid when the peculiar crystals will appear. The uric acid is displaced from its combination with the alkaline bases, and presents itself in more or less voluminous brown or yellow crystals. If these crystals be separated from the urine by a pipette, and placed in a test tube, the addition of liquor potassa will dissolve them ; upon adding acetic or nitric acid to this solution, the crystals will reappear, of different forms, and nearly colorless. Should any doubt exist as to their being uric acid, some of the preceding named chemical tests may be used.

Quantitative Analysis.

To 100 c. c. of urine add 10 c. c. of chemically pure nitric acid; agitate the mixture, and then set it aside for 24 hours in some cool place. Collect the uric acid crystals, that have been precipitated, upon a weighed filter, and wash thoroughly with distilled water until the addition of a solution of nitrate of silver to a little of the washings, no longer gives a precipitate. Dry the filter and crystals at a temperature of 212° F., and then weigh ; the weight of the two, minus the weight of the filter, will give the amount of uric acid contained in 100 c. c. of urine. Now, if 1,000 c. c. of urine be passed in 24 hours (or whatever may be the amount), and 5 grammes of uric acid be found in 100 c. c. of the urine, we calculate the amount of uric acid passed in 24 hours, thus : $\frac{5 \times 1000}{100} = 50$ grammes of uric acid in the 24 hours urine.—According to Neubauer, any error in this process, relative to the solubility of the uric acid, may be corrected by adding 0.0045 grm. to the amount of uric acid found in each and every 100 c. c. of the urine and the water used in the washing. Should the urine be considerably diluted, as in polyuria, it should be evaporated to ⅓th its volume before adding the nitric acid. If *albumen* be present, it should be removed by acetic acid or heat, and filtration.

2. To 100 c. c. of the urine add enough sodium carbonate to render it strongly alkaline; filter to remove the earthy phosphates, and to the filtered fluid, add 100 c. c. of saturated solution of chloride of ammonium, and allow the mixture to rest for several hours without agitation. The urate of ammonium which is deposited upon the sides and bottom of the containing vessel is collected upon a weighed filter ; the beak of the funnel is then corked, and hydrochloric acid at the 10th added, for the purpose of converting the urate into uric acid, which is washed, allowed to dry, and then weighed with the filter holding it. To correct for uric acid lost add 14 milligrammes for each 100 c. c. of urine. This process is based upon the slight solubility of the acid

urate of ammonium, which requires 1.600 parts of cold water to dissolve; though not exempt from errors, it gives tolerably accurate results. *Fokker.*

‡ *Microscopic Examination.* Pure uric acid is generally in oblong, smooth, square plates; but in the urine, many different forms are met with, being derivations of quadrangular rhombic tablets, or of hexangular plates, the angles of which are usually more or less obtuse. See *Figs.* 35, 36. Sometimes they are spindle-shaped, at others, short cylinders, cask-shaped; again, they may be in long needles variously grouped, in plates, or in prisms. Not' unfrequently they are in fan-shape, in coarse or fine rosettes, in comb-shape, dentated like a saw, in dumb-bell form, etc. With some very rare exceptions they are always colored yellow, red, or brown, and, sometimes, in abnormal urine blue, black, or yellow; the paler the urine, the less color they present, and vice versa. The colorless crystals are deposited from their solution in liquor potassa.

Fig. 35.

A. Crystals of uric acid with serrated edges, deposited from very acid urine.
B. Crystals do., rhomboid form, generally flat.
C. Crystals do., squares or cubes, in long continued deposits,—in calculous disease.
D. Crystals do., rare forms, accidental varieties of B and C.
E. Crystals do., often found mixed with urate of ammonia, or oxalate of lime. N. B.—All the crystalline forms of uric acid appear to be modifications of the rhombic prism; they vary in color from the palest fawn, to the deepest amber, or orange red,

These crystals are sometimes so large as to be recognized by the naked eye; those in plates or tablets present, under the microscope, granulations and lines, as if small fissures or cracked places abounded in them. The more transparent plates exhibit most splendid colors under polarized light. The various crystalline forms of uric acid are not always met with in any single specimen of urine; they may all be referred to some modification of the rhombus, square, or rectangle. They are very recognizable by their color, and soon undergo a change by uniting with the ammonia liberated by decomposition, and have a tendency to become amorphous. As the rule, though not invariably, a spontaneous crystallization of uric acid, within 24 hours or so after the emission of the urine, indicates an excess of this acid; it may also be

Fig. 36.

Varieties of Uric Acid Crystals.

deposited when there is a diminution in the proportion of alkaline base. Any doubts concerning the crystals present in urine, may be removed, by the solubility of uric acid in liquor potassa, and their subsequent reappearance on the addition of acetic or nitric acid. Urates and phosphates disappear when acetic or hydrochloric acid is added to them; uric acid remains undissolved.

Clinical Import. The presence of uric acid in the urine appears to be due to imperfect oxygenization of the blood, which augments the quantity of this acid and lessens that of urea. Thus, it is found in excess in febrile, inflammatory, and exanthematous diseases, in affections of the respiratory organs, the heart, spleen, and liver, in leucocythemia, derangement of the cutaneous functions, in certain cutaneous diseases, etc. It is common among intemperate persons, and in all those conditions of the system which interfere with perfect oxidation of the nitrogenized débris of the tissues. In cancer of the liver it is said to exist in great excess, being spontaneously deposited in the urine; this spontaneous deposition has also been observed in other diseases of this organ, as well as in dyspepsia, splenic, and cardiac affections. Uric acid may be precipitated from the urine, in the urinary passages, from an increased amount of the acid phosphate of soda, in this fluid, or, perhaps, from acid fermentation; hence, uric acid gravel in the kidneys or ureters, so frequently observed among certain patients. See *Urea* and *Urates* (*Clinical Import.*) Solvents to hold uric acid in solution in the urine are, phosphate of soda, phosphate of ammonia, benzoic acid, benzoate of lithia, and alkaline tartrates, citrates or carbonates.—Uric acid is stated to be lessened in quantity in the urine, in yellow and remittent fevers, in diabetes, in cholera, albuminuria, anemia, chlorosis, hysteria, and in certain stages of gout and rheumatism. Dr. Garrod has described a method for ascertaining the presence of *uric acid in the blood* of gouty persons; the "thread process." Place from 4 to 6 grammes of the serum of blood in a large watch glass, or flattened glass capsule, and add from .369 to .554 millilitres (6 to 9 minims) of acetic acid of ordinary strength, which addition is generally accompanied with the escape of a little gas in bubbles. Mix the fluids thoroughly, and immerse into the mixture two or three strands or fibres of thread drawn from a piece of muslin or linen. Allow the whole to rest in a dry, moderately warm place for 36 or 48 hours, when upon removing the threads and examining them under the microscope, crystals of uric acid will be seen deposited on them. This process will detect the $\frac{1}{85000}$th of uric acid in the blood. If the blood is not at the examiner's disposal he may substitute the serum from a blister, provided this be not applied over a point attacked by the gouty inflammation, because this inflammatory action causes the uric acid to disappear from points where it previously existed. Fresh fluid should always be used for these testings.—the presence of uric acid in the blood is an excellent means of diagnosis, as this acid is not found in acute articular rheumatism, nor in chronic rheumatism. But it may be found in other diseases than gout, as lead colic, Bright's disease, etc.

Uric Oxide. See *Xanthine.*

Uricemia. An excess of uric acid in the blood.

Urinal. A vessel for receiving the urine as it is voided. There are several kinds of these vessels, for males as well as females, adapted to the peculiar occasions for which they are required. A urinal in which urine is passed for examination, should hold at least 2,000 c. c., and be provided with a tight fitting cover to keep out foreign substances; if it be graduated, the amount of urine passed per 24 hours can be read off very quickly and without the extra trouble of measuring in a separate vessel.

Urinary. Of, or pertaining to the urine.

Urinary Casts. See *Renal Tube Casts.*

Urinary Sediments. The terms *urinary sediment, urinary deposit,* are given to the various substances precipitated from urine in which they are held in solution or suspension, when this fluid is at rest for a longer or shorter time. The terms *nebula, encœorema* (cloudiness or turbidity) are applied to a light-greyish flocculent cloud, formed by the urinary mucus holding various corpuscles in suspension, and appearing upon the cooling of the urine, the transparency of which it diminishes. See *Epithelium, Mucus.*—Sedimentary urine may be turbid at the very moment of its emission (organized sediments, *pus, blood, and phosphatic deposits*), or, it may issue perfectly clear and become turbid shortly afterwards (*urates*). In either instance, the urine should be passed into a conical graduated vessel, if this can be had; but any other vessel will answer provided it be *clean* and not too large. Indeed, if the vessel has not been carefully cleansed, the various foreign bodies it may contain will mingle with the urinary deposit and be so many causes of error. As stated heretofore, it is always advisable to accustom one's self, by previous examinations, to recognize under the microscope various kinds of dust, animal hairs, the textile fibres of our clothing and linen, grains of sand, oil globules, etc. In hospitals, grains of starch are very often found mingled with the sediments, and not unfrequently in large amount; they are apt to proceed from the starch powder so frequently employed as a topical application. We must expect to find them in the urine of women. They can be determined with certainty by means of *iodine water* which colors them blue or deep violet. See *Iodine;* also see *Preliminary Remarks* at the opening pages of the work.

It is an absolute illusion to attempt to class urinary sediments according to their color, their solubility by heat or in certain reagents, or, to the acid or alkaline reaction of the urine, and from these different characters to draw means of distinguishing between them. To determine the composition of a sediment from its color, would be to expose one's self to daily errors. This is why (whatever may be the authority and the competency of certain authors) we do not consider as practically useful tables given for the diagnosis of sediments, which suppose that we have under examination a type-sediment of urates, phosphates, or oxalates, etc.—In the usual conditions of practice, the practitioner by means of a small pipette takes up some of the urinary sediment, then allows a drop of it to fall upon the center of a glass slide, which he covers with thin glass, and, thus prepared, examines it under a microscope; that which he may see is given below in the table, under *General Microchemical Analysis of a Sediment*, in which are indicated the elementary physical and chemical characters that may enable him to form a provisional diagnosis concerning the elements under observation. This diagnosis he can subsequently confirm by referring to the detailed characters of each body, separately explained throughout the various pages of this work.

But this is not all. For instance, a drop of acetic acid is added to the object to ascertain the solubility of ammonio-magnesian phosphatic crystals. If these crystals *only* exist in the deposit, the reaction is very distinct; they all disappear. But matters do not usually occur in this manner. These crystals are mixed with other crystalline, amorphous, or organic elements, which the acetic acid dissolves, modifies, or respects. Then the reaction, instead of enlightening us, increases our uncertainty; and if the action, which this reagent exerts upon all bodies that may be met with in a sediment, is not perfectly present to the mind, we will be obliged to undertake long, wearisome, and frequently fruitless testings, upon the chemical properties of each body taken by itself. To obviate this difficulty, the reality of which we have too often recognized, we have summed up, in the form of a table, easy to consult, the action of the ordinary reagents upon the elements of which the sediments are usually composed. And that this table may preserve its practical character, those substances only accidentally met with in deposits, as well as rare bodies, are excluded.

Previous to this table, however, it is deemed best to give, for the information of the reader, the composition of the sediments after the reaction of urine, and the more simple chemical characters of several of them.

Probable Composition of a Sediment after the Reaction of the Urine.

Acid Urine. Sediments generally colored more or less deep red.—Urates; uric acid, always in small quantity; oxalate of lime, ditto; phosphate of lime, ditto (often crystalline or in small dark spherules, grouped or isolated);

hippuric acid, very rare; cystin, ditto; tyrosin, ditto (in urine containing biliary pigments).—Various organized bodies.

Alkaline Urine. Sediments generally colored more or less dirty white.—*Ammonio-magnesian phosphate; phosphate of lime;* oxalate of lime, in small amount; urate of soda, and of ammonia, in dark globules bristling with sharp spikes.— Mucus; pus; various organized bodies.

The deposits of urates and of uric acid are the only ones soluble by heat, or by the addition of an alkaline solution, soda or potassa.—The deposits of phospates and of carbonates are the only ones soluble in acetic acid, and insoluble by heat, or by the addition of an alkaline solution. The carbonates dissolve with effervescence. Observe that this effervescence may be due to the carbonate of ammonia of alkaline urine, which is more commonly the case; if it be due to carbonate of lime, by repeating the reaction under the microscope, bubbles of gas will be seen oozing from the surface of the small blackish bodies.

Complete Chemical Analysis of a Sediment. Place the deposit in a test tube with foot, and wash it with a small quantity of cold distilled water, and repeat it several times until the decanted supernatant liquid is colorless. Then treat the deposit the same as if it were a pulverized calculus. See *Calculi (Table of Qualitative Analysis.)*

‡ General Microchemical Analysis of a Sediment.

With a pipette, take up some of the sediment which it is desired to examine, and place a drop of it on a glass slide; it is always advantageous to dilute it with a little distilled water or some of the clear urine, so as to render the preparation more transparent; gently mix the additional fluid with the drop of the deposit, by means of a needle point, or a small point of wood, and then place a thin glass cover upon it. (This cover will not be necessary if the operator uses a chemical microscope.) The drop of fluid should not spread beyond the margins of the thin glass cover, an accident which will certainly happen with beginners who, in microscopic examinations, have a marked inclination to place an excess of the substance to be investigated upon the glass slide. In such a case, quickly dry the slide, and place another, but smaller, drop of the sediment upon it. Bear in mind, that the smaller the amount dropped upon the slide, the better it is for the examination; and by sufficiently diluting the small drop, the different bodies become separated, and the examination becomes much more easy and distinct.—In the investigation, a magnifying power of 300 diameters is at first employed, and then, to better observe certain minute elements, one of 500 diameters; in the table below, the magnifying power to use, in certain particular cases, has been indicated.

In applying reagents, two processes may be employed; the pointed extremity of some blotting or filtering paper may be insinuated between the slide and thin glass cover (or it may be closely applied against the latter) while a drop of the reagent is placed upon the other extremity of the paper, at a short distance from the thin cover. The paper, absorbing the fluid by capillarity, establishes a current passing from the point upon which the reagent was deposited, through the paper, to the thin glass. What occurs may then be observed, and the chemical reaction which ensues be investigated. This is a delicate process, and requires considerable experience in order to properly execute it; but, in certain doubtful cases, it is truly valuable. It has the inconvenience of displacing the various elements composing the preparation, but we may follow them by gradually moving the slide.

Another much more expeditious process consists in treating a small drop of the sediment upon a glass slide, by an excess of the reagent, then cover the whole with the thin glass, and examine the result under the microscope. When, by a preliminary observation of the original drop of the sediment, its contents have been well noted so that their forms, etc., can be correctly remembered, it can be seen whether, under the influence of the reagent, they disappear, or become changed.—Nitric acid should never be employed under the microscope, only exceptionally, and hydrochloric acid as seldom as possible; the vapors from these acids rapidly deteriorate the metallic mounting of the objective. When they are employed, the objective should be promptly and carefully wiped with a piece of fine old linen, or, a soft piece of glove leather.—When coloring reagents are used, a few seconds should be allowed that the various elements may be impregnated with them.

XIV. Table.

Non-organized Bodies.

Distinctly crystalline bodies.	Very voluminous crystals, generally isolated, transparent, with sharp edges. Typical form a coffin-lid.	Soluble in acetic acid.	Ammonio-magnesian Phosphates *See.*
	Large crystals, but generally grouped, always colored in yellow or brown; surface often fissured; outlines very dark.	Insoluble in acetic acid.	Uric Acid. *See.*
	Very small crystals, much smaller than a leucocyte, isolated, very transparent and very refractive, with sharp edges, octohedral form, sometimes letter envelope. Requiring 400 diameters.	Insoluble in acetic acid.	Oxalate of Lime. *See.*

Amorphous bodies.	Granules roundish or oval, with dark blackish outlines, isolated, or 3 or 4 united in a star-like form, in beads, etc. Granules very pale, much smaller, very transparent, and difficult to perceive; always united by irregular punctated patches (the most common aspect.)	Soluble in acetic acid.	Phosphate of Lime. *See.*
	Grains roundish, isolated, with concentric or radiating striæ (sometimes both together), more or less opaque, and blackish.	Soluble in acetic acid but with disengagement of gas bubbles, which issue from their surface.	Carbonate of Lime. *See.*
	Granules small, yellowish, of variable size, sometimes very small and disposed in branching series, like twigs of moss (recent sediments); sometimes larger, in the form of globules, with dark outlines and yellow center, united in a mass like frog's eggs, or else isolated and bristling with points (old sediments).	Slowly soluble in acetic acid, and, after a short time with the appearance of colorless tablets of uric acid.	Urates. *See.*
	Granulations very fine, isolated, agitated with a dancing, revolving movement (Brunonian movement).	Insoluble in acetic acid.	Molecular Granulations. *See Algæ, Bacteria, Vegetable Growths, Vibrios.*

15

Organized Bodies.

	Globules always round, with smooth or jagged outlines, without nucleus, more often presenting a central depression, yellowish, isolated, or united in coin-like piles, or intervolved in filaments of fibrin, or, of mucus.	Swell with weak *acetic acid,* or shrink up and present a raspberry aspect; not colored by carmine.	**Blood Globules.** *See.*
Cellular form, roundish, or oval.	Globules round or oval with slightly defined outline and greyish-white contents, granular or nucleolated, isolated or united in masses and then polygonal; often intervolved in mucus and elongated or expanded.	Rendered pale by *acetic acid* which causes 2 or 3 (generally 3) nucleoli to appear within them. Colored by *carmine.*	**Leucocytes.** *See.*
	Globules round or oval, very small, very refractive, sometimes presenting 1 or 2 brilliant nucleoli or verrucous expansions upon their margins, isolated or united in bead-form. Requiring 500 diameters.	Not changed by *acetic acid,* nor colored by carmine. The nucleoli or the interior of the cellule is colored brownish-yellow by *iodine water.*	**Spores.** *See Algæ, Fungi, Vegetable Organisms.*
	Corpuscles very small, oval, refractive, hyaline, furnished with a very long, delicate filament. 500 diam.	Unchanged by the reagents.	**Spermatozoids.** *See.*
Form variable, size more considerable than the preceding.	Roundish, cylindrical, fusiform, or polygonal, with very granular contents, or, more generally furnished with one or several nuclei.	Rendered pale by *acetic acid* which causes it or the nuclei to appear distinctly, distorting them. Colored by *carmine,* especially the nuclei.	**Epithelium.** *See.*
Cylindrical.	Voluminous, of greater or less length, with variable aspect, sometimes twisted or undulated. Requiring 120 diameters.		**Renal Tube Casts.** *See.*
	Very short and very small, transparent, more often agitated with oscillatory movements. Requiring 500 diameters.	Unchanged by *acetic acid* which retards or arrests the movements.	**Bacteria. Vibrios.** *See.*

Filaments, or Fibrillary.	Very thin, more or less ramified or inter-crossing.	Unchanged by *acetic acid*.	**Algæ. Fungi.** *See.*
		Rendered pale by *acetic acid;* the fibrillary aspect disappears and gives place to a swollen, transparent, amorphous mass.	**Fibrin.** *See.*
		Rendered more evident by *acetic acid,* which gives them a punctated or striated appearance.	**Mucus.** *See.*

N. B. The various filamentous substances just referred to above must not be confounded with the tubes coming from the textile fibres of linen and clothing, whether partitioned or not, and which are more or less frequently met with in the urine under examination.—*Starch grains* are in the form of roundish, oval, or polygonal, very refracting bodies, with concentric striæ, or with central punctiform depression, linear or stellated, etc., and almost always colored dark blue by *iodine water.*—Fat, under the form of highly refracting globules of variable dimensions, is accidentally, but very rarely, met with in urinary sediments. See *Starch. Fat.*

‡ XV. Table.

Action of Acetic Acid.

Add to the drop of urine on the glass slide a small drop of acetic acid, which must always be in excess; cover with thin glass, and examine under the microscope. If the sediment

Disappears.	Dissolving quickly.	Ammonio-magne-sian phosphates.
	Dissolving, but more slowly than the preceding.	Phosphate of lime.
	Dissolving, with disengagement of gas bubbles which issue from the surface of the deposit. Do not confound it with the disengagement of gas which often occurs, on treating, under the microscope, an ammoniacal urine with acetic acid, in which the bubbles arise in the midst of the fluid, and are quite large.	Carbonates.
	Dissolving slowly, but soon replaced by tablets of uric acid.	Urates.
Is modified.	Becomes pale; the nuclei, when they exist are more evident but deformed.	Epithelia.
	Becomes pale, certain urinary cylinders.	Epithelia. Those covered with urates.
	Becomes pale; swells, the fibrillary aspect disappears.	Fibrin.
	Becomes pale, with the appearance of 2 or 3 nuclei.	Leucocytes.
	Becomes pale and shrunken; sometimes, however, swollen.	Blood globules.
Remains without change.	....	Uric acid.
	....	Oxalate of lime.
	....	Spores, algæ, vegetable filaments, spermatozoids, vibrios, bacteria, molecular granulations.
There may appear.	Proceeding from the urates,—crystals in colorless, transparent, tablets, often disposed in longitudinal series.	Uric acid.
	Filaments, striated or punctated.	Mucus.

XVI. Table.

Action of Potassa at the Tenth.

Add to the (fresh) drop of urine on the glass slide, by the paper process (named above for applying reagents), an excess of a solution of potassa (1 part potassa to 10 parts of distilled water). One of the following changes may occur in the sediment:

Bodies disappear.	Non-organized bodies.	*Urates;* the older the urates the slower the action. *Uric acid;* dissolves slowly, so that its progress may be watched.
	Organized bodies.	*Blood globules;* will be seen to rupture, and instantly dissolve. *Leucocytes;* turn pale, and dissolve rapidly. *Nuclei of epithelia;* turn pale, and dissolve rapidly. *Renal tube casts;* turn pale, and dissolve rapidly. In the granular casts, the granulations become disassociated, and float in the fluid of the preparation. *Fibrin, mucus;* turn pale and dissolve.
Bodies are modified.	Epithelia.	*The nuclei disappear;* at the same time the cellule grows pale, swells, and becomes vesicular, its outlines are then so indistinct that they can be seen only by the aid of oblique light.—Pavement epithelia resist the action of the potassa the longest.
Bodies remain unchanged.	Non-organized bodies.	*Ammonio-magnesian phosphate.* *Phosphate of lime.* *Carbonate of lime.* *Oxalate of lime.*
	Organized bodies.	*Spores, vibrios, bacteria* (movements are arrested). *Spermatozoids.* Vegetable filaments. Molecular granulations.

For coloring agents, see *Ammonia, Carmine, Iodine,* etc.

Urine. Urine is a liquid in which is eliminated from the system a large proportion of its fluid and solid effete matters, and the quantity of these matters vary according to the variations occurring in the vital processes of the body. Healthy urine is composed of animal and saline substances, which have been separated from the blood by the kidneys; it is of a clear amber color, transparent, acid, reddening litmus paper, has a mawkish, aromatic, violet-like odor, frequently changed, however, by articles of diet, a saline, bitter, and rather disagreeable taste, a temperature on being passed from 92° to 100° F., and a specific gravity from 1.002 to 1.030, depending somewhat upon the amount of fluids taken. See *Specific Gravity.* One thousand parts of urine, contain the following constituents, the proportions of which, however, vary, according to the circumstances and influences to which the person is subjected: Water 938; urea 30; creatine 1.25; creatinine 1.50; phosphate of soda, acid phosphate of soda, phosphate of potassa, phosphate of lime, phosphate of magnesia, 12.45; coloring matter, mucus, .30; chloride of sodium, chloride of potassium, 7.80; urate of soda, urate of potassa, urate of ammonia, 1.80; sulphate of soda, sulphate of potassa, 6 90. From 600 c. c. to 1800 c. c. of urine are passed in 24 hours, holding in solution 38 grms .87 to 45 grms .35 of solid matters, but which amount of fluid may be increased by the ingestion of large quantities of fluid. The urine of infants is mostly colorless, inodorous, of low sp. gr. and having but slight reaction on litmus paper; on standing, it acquires an odor resembling that of veal broth.—On standing, urine at first evolves a urinous odor, succeeded by one resembling sour milk, and then one of a fetid, ammoniaco-alkaline character. Its color, odor, quantity, transparency, etc., may be greatly altered by diet, medicines, exercise, disease, etc. Acidity of the urine is common to carnivorous animals, that of the herbivorous is alkaline. A vegetable diet, renders the urine of man less acid, neutral, or alkaline, and likewise diminishes its proportion of urea. A visible deposit in recently passed urine is abnormal.

In the examination of urine. nothing definite can be determined unless the whole of the urine passed in every 24 hours is collected, preserved, and tested; in addition to which, the different specimens passed at various periods of the day should be tested, and examined as to their color, reaction, transparency, etc., immediately after their discharge from the bladder, as well as after the urine has stood for 10 or 12 hours. The urine of birds, serpents, and insects is *solid,* and, in all animals possessing distinct urinary bladders, it is *fluid.*— The temperature of the urine is regulated by that of the body from which it derives its heat; scalding of the urine does not depend upon a temperature of this fluid exceeding that of the body, but upon an acrid condition of the urine, or an inflamed condition of the canal through which it is passed. In disease, the temperature will be found to vary very much, and this should invariably be ascertained from day to day in all serious maladies; thus, in idiopathic tetanus it may rise to 112°.5 F., while in mania previous to death it will fall to 77° F.

When urine is quite clear, but slightly colored, and of but little sp. gr., it is termed Thin or Crude Urine (*U. tenue* or *crue,* Fr.); the urine following large draughts of aqueous or stimulating fluids, is termed the Urine of drink (*Urina pôtûs, U. de la boisson, U. aqueuse,* Fr.). That following two or three hours after a meal, is the Urine of digestion or cocted Urine (*Urine cuite, U. de chyle,* Fr.); that passed early in the morning, and which may be considered the normal type of urine, Urine of the blood (*Urina sanguinis,* Fr.). And according to the aspect and character of the urine, it has received the names; *bilious* or *icteral,* when of a deep yellow color with greenish reflections; *chylous,* when following digestion; *fatty* or *oily; febrile* or *lateritious,* when small in quantity and high colored; *flocculent,* when turbid from the presence of flocculi; *jumentous; lactescent* or *milky,* when white and turbid; *phosphorescent* or *luminous; sanguinolent* or *bloody; thick* or *mucilaginous,* from excess of mucus; *turbid* or *troubled,* etc. However these distinctions possess no great practical value, further than to indicate the necessity for an analysis, and perhaps to give a direction as to the course to be at first pursued in the chemical microscopical investigation of the urine under consideration.

The quantity of urine is increased in diabetes, and is diminished in all acute febrile and inflammatory diseases, in renal affections, and in dropsy; a persistent diminution is an unfavorable indication. More urine is passed in cold weather than during warm seasons, from diminished perspiration in the former instance, and increased, in the latter.—The color of the urine also affords certain indications; a pale urine from nearly colorless to straw yellow, with the exception of diabetes, contains less urea and solid matters, and indicates the absence of acute disease, while a highly colored urine contains a large amount of solids and is more or less unfavorable according to its persistency and the character and intensity of its color. A greenish yellow or greenish brown urine, or yellowish epithelial cells in this fluid, is generally due to the presence of bile pigment. A smoky tint is almost an absolute diagnostic of the presence of blood.—The froth on urine that is healthy, readily disappears; but if it persistently remains, albumen or bile matters may be present. Renal casts in urine would indicate the probability of albumen in this fluid; so would the presence of blood, or pus. When there is an excess or deficiency of urinary coloring matter, of phosphates of urea, or, of uric acid; when urine contains albumen, bile, blood, fatty matters, leucin, oxalate of lime, or sugar, some pathological condition of the system should always be suspected.—If to an abnormal urine, be added a concentrated solution of tannin, boiling it will give a white precipitate if albumen be present; a black, if hematin.—When bloody urine is dark red, ammonia will cause it to assume a clearer tint, hyacinth rose. In chlorosis, anemia, and certain nervous affections, there is an increase of water in the urine, and this fluid is limpid, nearly colorless; there is a diminution of water in profuse perspiration, diarrhea, cholera, etc., in which diseases the solid elements increase; in the first instance, should sugar be present, the sp. gr. would lead to a suspicion of its presence,—in the latter, a sthenic condition would be known by the accompanying acute symptoms.

Urine Ferment. *Kidney Ferment.* A substance to which M. Bechamp has given the name of "nephrozymasis." It is, according to him, the albuminoid ferment material of urine, which is obtained from various specimens mixed with calcareous, and magnesian, earthy phosphates. Its presence has been demonstrated by the property it possesses of fluidifying and saccharifying starch, the same as diastase and sialazymasis. It exists in the normal urine of both sexes in variable quantity, up to about 9 Troy grains for each quart of urine, being more abundant in infancy; under the influence of animal diet; and also, of exercise. In man, its average amount is about 8½ grains in 24 hours; in woman, about 6 grains. It has also been met with, but in smaller quantity, in the urine during pregnancy, in certain pathological conditions, acute as well as chronic, and even in albuminuria. M. L. Leblond has also observed it in affections of the nerve centers. That it is a different substance from albumen and albuminose is evident from its presence in normal urine during the healthiest condition of the body, and less frequently during pathological conditions. Again, while urine containing a large amount of albumen exerts no action upon starch, yet, if, in such a urine, this ferment is present, it will dissolve starch placed in contact with it, in a short time. The extreme variation in amount of this proteic substance, both during the state of health and that of disease, has not yet been satisfactorily accounted for. *Montpellier. Med.*

Urinometer. The urinometer is an instrument designed for determining the specific gravity of urine, and although not strictly accurate, it is sufficiently so for practical purposes. The urinometers met with in commerce are sold at very low prices, but, as the rule, they are inexact and not fit for use, giving no reliable data whatever. (See *Fig.* 37.) The best urinometers are those of Dr. Pile, Philadelphia, Penn.; Mr. Ackland, London; M. Bouchardat, Paris; and the German instruments, which have two spindles, one ranging from 1.000 to 1.020, the other from 1.020 to 1.040, by N. Niemann, of Alfeld. The 0 of the graduated scale gives the specific gravity of distilled water = 1,000. When the instrument is placed in the urine contained in a solution tube, or in any cylindrical vessel of sufficient length and diameter, it sinks to a certain number which will be found on a level with the surface of the fluid, and to which number 1,000 must be added to obtain the sp. gr. Thus, if the urinometer sinks until 30 of the graduation is on a level with the surface of the urine, its sp. gr. will be 1,030. As the sp. gr. of a fluid increases at a lower temperature, and decreases at a higher, these instruments are constructed for use at a certain temperature, 60° F., or 15° 5′ C.; hence the urine to be tested should always be artificially brought to a temperature of 60° F., when attempting to determine its specific gravity. A low sp. gr. indicates a maximum of water and a minimum of solids; while a high sp. gr. indicates a maximum or an increased amount of solids, as, urea, sugar, albumen, etc. In cases where the temperature of the urine is elevated, and can not be readily reduced to 60° F., the following table has been given by Mr. Ackland, for correcting the sp. gr. Thus, suppose with urine at 80° F., the urinometer gives a sp. gr. of 1.018; we add to this the amount found opposite 80° F., which is 1.90, and this will give a more correct sp. gr. of 1.019.9; and so on.

Fig. 37.

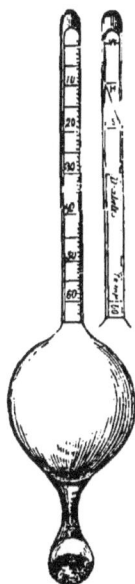

Urinometer.

XVII. Table.

Temperature.		Amount to be added.	Temperature.		Amount to be added.
F.	C.		F.	C.	
60	15.55	.00	73	22.78	1.20
61	16.11	.08	74	23.33	1.30
62	16.67	.16	75	23.89	1.40
63	17.22	.24	76	24.44	1.50
64	17.78	.32	77	25.00	1.60
65	18.33	.40	78	25.55	1.70
66	18.89	.50	79	26.11	1.80
67	19 44	.60	80	26.67	1.90
68	20.00	.70	81	27.22	2.00
69	20.55	.80	82	27.78	2.10
70	21.11	.90	83	28.33	2.20
71	21.67	1.00	84	28.98	2.30
72	22.22	1.11	85	29.44	2.40
			86	30.00	2.50

To use the urinometer, select a solution tube or cylindrical jar, in which the instrument will readily move, and place in it enough of the urine, at a known temperature determined by a thermometer. By pouring the urine carefully, froth will not be apt to form; but should any be present, it will attach itself to the stem of the urinometer, and prevent the exact degree marked from being distinctly seen, and may be removed either by blotting paper, or by completely filling the jar. letting it rest a few seconds, and blowing away the froth at the same time that we incline the vessel to decant the superfluous amount of urine. Now, introduce the urinometer into the urine and allow it to *gradually* sink into this fluid, until it will no longer descend spontaneously. To make certain that it has passed down to its proper level, gently press a finger upon the top of the stem, so as to cause it to sink still further about *one degree only*, no more, and then allow it to rise by removing the finger. If it be pressed downwards too far, its stem, becoming covered with urine, will be rendered heavier, and thus produce an error in the determination. The instrument must also be kept free from contact with the walls of the jar. The degree should be read off by looking at the graduated mark through the urine; that mark should be taken which is on a level with the surface of the fluid, and not that which is shown by an accumulation of the urine around the stem of the instrument, as this would give too high a sp. gr. In this reading the eye should be brought upon a line with the surface of the liquid.

It must be remembered that the specific gravity of urine varies according to the weight and age of the person, the quantity of the solids or fluids taken into the stomach, and the amount of exercise. A decreased amount of urine, with a slightly increased degree of sp. gr., does not necessarily indicate disease, and *vice versa.* The greater the amount of tissue destruction, the greater will be the quantity of solids in the urine (urea), and the higher will be the specific gravity. Dark-colored, scant urine is generally of high sp. gr.; while in pale urine, or when augmented in quantity, with the exception of diabetic urine, the sp. gr. is low. See *Pycnometer; Solids in Urine; Specific Gravity.*

Urobiline. A name given to the coloring matter of normal urine. Jaffe has obtained it by adding a large amount of ammonia to the urine, filtering, and adding chloride of zinc until a precipitate is no longer produced. This reddish precipitate is successively washed with cold water, then with warm, until the washings become no longer turbid by addition of nitrate of silver; then the precipitate is treated with boiling alcohol, and dried at a slightly elevated temperature. The pulverized mass is dissolved in ammonia, and the colored solution precipitated by acetate of lead. The red precipitate thus obtained washed with cold water, dried, and decomposed by alcohol containing sulphuric acid, yields the urobiline to this liquid.

Urochloralic Acid. The provisional name given by Musculus and De Merme to an acid found in the urine of persons who have taken considerable chloral hydrate; it appearing that chloral, similar to benzoic acid becomes materially changed in the organism, to a greater or lesser extent, being passed per urine in the form of a strong acid, which rotates polarized light to the left, decomposes carbonates with effervescence, reduces solutions of copper, silver, and bismuth salts, decolorizes sulphate of indigo, is soluble in water or alcohol, insoluble in ether, forms a precipitate of stellated crystals, insoluble in water, on the addition of basic lead acetate, and is stated to consist of 31.60 C, 4.36 H, 26.7 Cl. As much as 10 or 12 grammes of this acid have been found to exist in each litre of the urine.

Urochrome. A name given to a coloring matter of urine by Dr. Thudicum. According to Dr. T., it may be isolated in a state of purity, forming an amorphous substance, yellow, very soluble in water, less so in ether, and still less in alcohol. An increase of urochrome in the watery solution still occasions a yellow, not blackish color. It apparently has no immediate relation with the coloring matter of the blood, nor with that of bile, and is a derivative of the albuminous matters. Upon analysis it yields a red resin, principally consisting of uropittine ($C_6 H_{10} N_2 O_3$), and omicholic acid mixed with undetermined dark matters of uromelanin ($C_6 H_7 NO_2$), and other products. Probably by a simple process of oxidation, urochrome passes into the state of red coloring matter (*uroerythrin*) which sometimes colors the urine of patients without any trace of urates; this change is often effected after the micturition. This red coloring may likewise be due to omicholic acid, slightly soluble in the ammoniacal salts. The odor of acid or alkaline urine is due to the uropittine and omicholic acid, or, to the bodies derived from them. Carbonate of ammonia may increase them, but never originates them.—One of the first characters of uremia is retention of the urochrome, or of the uropittine and omicholic acid in the blood, which vitiates all the tissues, and may be found in the crusts of the teeth; their odor is also perceived

iu **expiration, and in the perspiration.** In this condition typhoid symptoms
are present. The treatment by acids promotes the retention of these toxic
substances in the circulation, and should be replaced by an alkaline treatment; the skin should also be bathed, and the cutaneous functions attended
to, until all the odor of uropittine has disappeared. See *Hæmochromogene.
Urohematin.*

Urocyanin. *Urocyanogen. Urocyanose.* See *Uroglaucin.*

Uroerythrin. *Urerythrin. Purpurin.* The names given to a red coloring
(acid) matter of the urine, which gives to this fluid and its sediment a dark
red, brick color. It is often found associated with urophein, and its presence
always indicates a morbid condition. It is met with in rheumatic and
periodic fevers, arachnitis, and in very acute inflammations of the meninges,
which enables us to diagnose this from typhus in which disease this pigment
is absent. It is, probably, only a modification of *urophein.*—To determine its
presence, into 10 c. c. of the urine let fall several drops of solution of acetate
of lead, which precipitates the urates, sulphates, and phosphates, decolorizing
the urine at the same time. If, upon standing for 15 minutes, the sediment
becomes white, uroerythrin is absent; but if the sediment becomes rose
colored, this pigment is present, and according to the greater or less intensity
of the red color will its abundance vary. See *Urohematin.*

Uroglaucin. *Urocyanogen. Indigo Blue. Cyanourin.* The names given
to a blue, or azure-green pigment, which may be obtained from uroxanthin.
Under the microscope the blue powder of uroglaucin is found to consist of
finely pointed needles, occasionally single, but more commonly in stellar
shape, or in groups having a radiated form. To obtain uroglaucin, mix 10
c. c. of urine with one-third its volume of ether, and another third of hydrochloric acid. Shake them together for several minutes; a red, blue or greenish-blue color (uroglaucin) is produced, which the ether carries with it upon
rising to the top of the fluid in the test tube. According to Kletzinsky, uroglaucin is identical with indigo blue. In Bright's disease, and in cystitis,
uroglaucin has been observed, being the product of the decomposition of
uroxanthin in the ammoniacal decomposition of the urine while in the bladder. It is an abnormal pigment, very probably formed from urohematin, and
to which no satisfactory clinical signification has yet been given, though its
presence has been supposed to be of an unfavorable character. See *Urohematin. Urozanthin.*

Urohematin. *Urophein.* The history of the urinary pigments is so complicated and obscure; that the most varied opinions have prevailed concerning
their nomenclature, as well as their origin, character, and semeiotic value.
Harley, from numerous and long continued researches, strongly maintains that
all the various urinary colors, are merely different degrees of the oxidation of
urohematin, the formation of which substance in the human body has been more
or less influenced and changed by the presence of disease. These colors are
not always present in the urine when passed, but become subsequently manifested from the influence of the atmosphere, or the action of certain chemical

substances. Urohematin proceeds from the decomposition of the red corpuscles or hematoglobulin of the blood, being more or less abundant in the urine according to the more or less rapid disintegration of these corpuscles. And his views appear to be rapidly gaining ground among medical men. Parkes states "That hematin is almost identical with, and can be changed into, normal urinary pigment and into bile pigment, is now generally admitted, and the close relationship of indigo to all three is just as certain." Again, "On the whole, it does not seem at all unlikely that indigo may be an occasional, but not an invariable, product of the metamorphosis of hematin, and that it arises from some slight perversion of metamorphosis, the nature of which is not yet known."

Harley gives two methods of determining this pigment in the urine, one for its presence in normal urine, and the other for its existence in the urine in excess. For the first method, the whole of the urine passed in 24 hours is mixed, and then diluted with distilled water until it measures 60 fluid ounces (1,800 c. c.) (should the 24 hours urine exceed 60 ounces, it must be concentrated to the required amount by evaporation). Of this prepared urine place 2 drachms (7.4 c. c.) in a test tube, and carefully add to it one-fourth its volume (1.85 c. c.) of chemically pure nitric acid, and then allow the mixture to stand for several minutes. If the urohematin is present in normal quantity, the tint of the urine will be but slightly modified; but if it be present in excess, the color will change to pink, red, crimson, or purple, according to the amount of this pigment   The different shades of coloration will be found referred to in *Vogel's Table of Colors.*

The change of color in this process may be hastened by heat, but it is better to employ no heat, and allow sufficient time for the change. The acid sets the coloring matter free; and in this way a very pale, clear urine may be found to contain an abundance, while, on the other hand, a high colored urine may contain but a small amount of urohematin. If it be required to collect the urohematin, a drachm of ether may be added to the above mixture; then strongly agitate, and set aside for 24 hours, when the ether will be at the top, and of a more or less red color. Decant the ether, evaporate to dryness, and treat the residue with chloroform.

When rapid disintegration of the blood globules is occurring, and an excess of urohematin is contained in the urine, the method is to boil 100 or 150 c. c. of the urine, and add 25 or 38 c. c. of nitric or hydrochloric acid to it. This liberates the coloring matter. When cool, place the urine thus colored into a half pint bottle, add to it ether 30 c. c., cork the bottle, agitate the mixture strongly, and allow it to stand for 24 hours, when the ether will present the appearance of a red, tremulous jelly, should the case be a very bad one. In the worst forms, the urine is neutral or alkaline. Sometimes nitric acid develops only a yellow color; hydrochloric, always gives the red.

Urohematin is a bright red, non-crystallizable, soft, sealing waxy, organic compound, becoming hard and brittle by age, and having a shining fracture. It is soluble in alcohol, chloroform, ether, fresh urine, and caustic ammonia, soda, and potassa ; insoluble in pure water, solution of chloride of sodium, or chloride of barium, and in hydrochloric, nitric, oxalic, sulphuric, and tartaric acids. A small quantity of it placed on a platinum spatula and burned until an ash remains, which ash is then dissolved in weak hydrochloric acid, —will give with a drop or two of sulphocyanide of potassium a fine red color, and, with a drop or two of ferrocyanide of potassium, a Prussian blue precipitate, thus demonstrating the presence of iron. The ash of any urine may be treated in the same way. See *Iron*. *Color of Urine*. *Hematin*.

Clinical Import. Urohematin may be present in abnormal quantity in pale urine, as well as in colored. It is present in excess in hysteria, chlorosis, dyspepsia, low fevers, pneumonia, diphtheria, and in nervous lesions; and the greater the amount above normal, the more serious is the patient's condition. In cerebral and spinal disease, it is apt to be present in the urine in great excess, associated with a saccharine or phosphatic condition of this fluid. The amount of urohematin in the urine will afford an idea of the extent and rapidity of the daily disintegration of the blood globules of the body; bearing in mind, however, that certain vegetable foods give rise to its presence, and that when it assumes a dark color, it may be due to a local destruction of the blood corpuscles, in the liver, the kidney, or in both combined. Urohematin may exist in a free state coloring the urine red, but this fluid, unlike that of hematuria, will be clear and transparent, and devoid of blood corpuscles, unless, indeed, the conditions of disorganization of blood corpuscles, and local hemorrhage, exist together. As before stated, the various colors of the urine, are chiefly due to the degree of oxidation of the urohematin, and, as the rule, the darker the color, the more serious the patient's condition. The remedies are pyrophosphate of iron, manganese, nux vomica, and nerve tonics.—The dark yellow urine from rhubarb and santonine assumes a blood-red color when treated with ammonia or potassa. The greyish yellow from resin turns blue with perchloride of iron. Campeachy wood, and aloes, impart a deep-red color to the urine, and madder, blackberries, raspberries, and fruit of cactus opuntia, a pure red, etc. Again, according to its degree of oxidation in the urine, urohematin becomes colored red, blue, green, yellow, or violet, with nitric, hydrochloric, and sulphuric acids.

Uromancy. *Uroscopy.* The pretended, charlatanic determination of diseases by simple inspection of the urine.

Uromelanin. See *Urochrome*. *Melanin*.

Urophein. *Urophain. Red Pigment of Urine.* See *Urohematin*. (Before Harley's determination of the coloring matter of urine, Heller had named it urophein, and gave the following method of detecting it:—Into a small beaker glass pour 2 c. c. of colorless concentrated sulphuric acid, and then, from a height of 3 or 4 inches, gradually pour a very fine stream of the urine to be tested, until 4 c. c. have been used. When the fluids are well mingled the mixture assumes a more or less dark brown or black color, according to the amount of this pigment present. Sugar, blood, bile pigment, and uroerythrin, have the same reaction, hence, they must be removed previous to the experiment. Ziegler states that the most intense coloration is present in cirrhosis of the liver. An abundance of urophein with a concentrated urine, occasions great acidity of the urine, with a sense of burning in urinating. Pale and abundant urine is apt to give a neutral reaction. Urophein is absent in diabetic urine. It predominates after a full meal, in profuse perspiration, in the decline of acute diseases, and when the urine is quite dense and rich in nitrogenized and extractive matters. It is in small quantity in chlorosis, anemia, and at the termination of long and serious diseases. It abounds in inflammatory fevers. Chronic liver disease, fatty, as well as amyloid degeneration, and cirrhosis, may occasion enough urophein in the urine to color it brown like Malaga wine). Isolated urophein is brown, acid, of urinous odor, and, in contact with the atmosphere becomes rapidly converted into carbonate of ammonia. When uroxanthin is present in typhus, it is stated that the reappearance of urophein is a favorable indication.

Urosacine. *Urrosacin.* A name given to the yellowish-red coloring matter of urine; probably a modification of *urohematin.* This coloring matter is more or less marked according to the greater or lesser amount of the fluid element in the urine.

Uroscopy. See *Uromancy.*

Uroses. Maladies pertaining to the urinary apparatus.

Urostealith. This is a constituent of urinary calculi first observed by Heller. When recent, it is soft and elastic like caoutchouc, becoming hard, brittle, and wax-like when dried. Calculi have been found consisting of pure urostealith, of various sizes, generally about the size of a pea, sometimes smaller, at others, a little larger, or, of this substance enveloped by crystals of ammonio-magnesian phosphate. When a fragment is heated on platinum foil, it remains solid for a time, then fuses, swells, and evolves a pungent odor, resembling that of shell-lac and benzoin. In this state, if a flame be applied to it, it takes fire, burning with a clear yellow flame, leaving a voluminous charcoal, which, when thoroughly burned, leaves a very minute alkaline ash, consisting principally of lime. It softens but does not dissolve in boiling water; is difficultly soluble in warm alcohol, from which it is recovered in an amorphous state on evaporation. Ether dissolves it pretty freely; it is soluble in hot liquor potassa forming a brown soap, and from which it may be separated by treatment with an acid. Heated with nitric acid a colorless solution is obtained, a slight quantity of gas being evolved; if this solution be evaporated, and the residue treated with ammonia or potassa, a dark-yellow color is given to it. As this substance has only been met with two or three times, a thorough analysis of it has not been made; consequently it has not been positively ascertained whether it is a fat or a resin, nor is its elementary composition known.

Uroxanthin. *Indican. Indigose.* This substance was obtained from urine by Heller. It is a normal coloring matter, yellow, existing in small quantity in normal urine, but which may appear in abundance under the influence of disease. Urine is not colored by it, except from its decomposition. Under treatment with acids, Heller divided it into a blue coloring substance, *uroglaucin*, and a red one, *urrhodin*. According to Harley it is one of the pigments derived from urohematin. See *Color of Urine; Urohematin.*—To detect uroxanthin, Heller advises to place 4 or 5 c. c. of pure fuming hydrochloric acid into a small beaker glass, and while constantly stirring it with a glass rod, add to it drop by drop, not exceeding 10 or 20 drops of the urine. If an abnormal amount of uroxanthin be present, it is decomposed, and the fluid passes from a violet-red up to a blue color. This discoloration of the fluid occurs more rapidly as the indican is more abundant; if from the small amount of uroxanthin in the urine, the reaction is slow or indistinct, it may be rendered more rapid and marked by the addition of 2 or 3 drops of pure nitric acid. If bile pigment be also present in the urine, it must be removed by precipitation with solution of acetate of lead, and filtration, previous to employing the above-named test.—It may also be detected by heating a small quantity of urine diluted with a few drops of hydrochloric acid. As soon as the mixture becomes heated it will give a violet-red, and then a deep blue color. When cold, the coloration occurs very slowly and is not very intense. The coloring matter obtained by the action of acids and heat, precipitates, when allowed to stand for 24 hours, in a fine powder which, examined under the microscope ($\times$ 550), consists of very fine roundish, uniform, transparent granules, which are agitated with Brunonian movements when they are separate or united in irregular, dark-blue patches.

Prof Senator, of Berlin, adopts a very ready method for the determination of indican in the urine, as follows: To 5 or 10 c. c. of the urine, add an equal quantity of fuming muriatic acid, if indican be present, a dark-blue cloud is produced, frequently at once; this cloud becomes more distinct, if a saturated solution of chloride of lime be carefully added, drop by drop. An excess of the lime solution decolorizes the blue cloud. Chloroform added to the blue urine, combines with the indigo, and precipitates, with it, to the bottom of the vessel. The amount of indigo present can be approximately estimated from the degree of the coloration it imparts to the chloroform. Prof. S. does not believe that decomposition of food in the intestine occasions the presence of indican, as he has observed it to be present in greater quantity in cases where but little or no food was contained in the intestine. He has found this substance present in the urine, in cases of ileus, peritonitis, gastric cancer and other malignant abdominal affections, gastric ulcer, pernicious anemia and leucemia, phthisis, atrophy of the kidney, and acute fevers.

Uroxanthin or indican, like indigo, is colorless, and may be separated from urine, forming a clear brown syrupy fluid, readily soluble in water, ether, and alcohol; urine containing it in excess presents no external character that would lead to a suspicion of its presence. By decomposition, under the influence of boiling acids, as well as of ferments, it yields indigo blue, indigo red, and sugar (indigo glucin) as the principal products. It is found in excess in the urine of persons afflicted with diabetes, albuminuria, typhus and other malignant fevers, cholera, pyemia, variola, after coition and sexual excesses, and especially in maladies of the spinal cord, and in cancer of the liver. The gravity of which diseases is in direct ratio with the abundance of indican. It is also found in the urine of the dog.

Urrhodin. *Indigo Red. Indirubin.* According to Heller, this is one of the products of oxidation (decomposition) of uroxanthin, being, however, less oxidized than uroglaucin, and generally found in larger quantity. To obtain it from pathological urine, add solution of acetate of lead to the urine as long as any precipitate occurs, and then filter; remove the lead from the filtered liquor by hydrosulphuric acid; filter; and boil to free the filtrate from the acid. Now, drop this urine, into an equal volume of pure, fuming hydrochloric acid, constantly stirring, until the mixture becomes of a dark-blue color. (If only a violet or red color is obtained, no deposit of indigo will occur.) Let the mixture stand for 12 hours, then add an equal volume of cold distilled water, and set aside for 24 hours. Separate the precipitate by filtration, wash it with boiling water until the washings have a neutral reaction; then wash with a little dilute alcohol, and dry, on the filter, in a drying oven, over sulphuric acid. Wash the precipitate and the perfectly dry filter with pure ether, as long as a red-colored filtrate is obtained. This ethereal liquor has an acid reaction, and, when evaporated, leaves an amorphous brownish-red resin, which is insoluble in water, but soluble in cold alcohol and ether to which it imparts a beautiful carmine-red color (urrhodin). From the washed residue on the filter, uroglaucin is obtained, by boiling it in alcohol (several successive quantities) as long as a blue colored fluid is obtained; these colored liquors are united, filtered while boiling hot, then evaporated to one-half their bulk, and set aside to crystallize.—Méhu, instead of acting upon the urine with strong mineral acids, agitates the crude urine with ether or chloroform, which removes the coloring matter, especially the red. The blue matter, more exclusively in suspension, upon filtering remains on the filter, but aided by the red matter it dissolves a little in the ether or the chloroform, whence the violet coloration of these liquids.—See *Uroglaucin.* For clinical import, see *Uroxanthin.*

Urylic Acid. See *Uric Acid.*

V.

Valerian. Taken internally in large doses, communicates its odor to the urine. Valerianic acid is one of the products of the metamorphosis of *leucin.*

Vegetable Coloring Matters. When urine, containing a vegetable coloring matter is rendered alkaline by the addition of ammonia, it more generally becomes red or crimson, according to the amount of such coloring matter present; if it now be acidulated by adding an acid to it, the color becomes paler or yellow; upon again restoring it to an alkaline condition, the red color reappears. By this method vegetable coloring matters in urine may usually be determined from the coloring principle of blood, and of uroery-thrin, which do not give these reactions.

Vegetable Oils. Many vegetable oils impart their own, or a peculiar, odor, to urine, as, oils of turpentine, valerian, juniper, garlic, etc. The changes they undergo in the system, or the influence they exert upon the character of the urine, is yet undetermined.

Vegetable Organisms. Besides the various organized and crystalline elements composing urinary sediments, various vegetable (or animal) organ-isms in process of development, are met with. Upon examining an alkaline urine under the microscope, especially during summer, very fine, slender filaments will be observed crossing each other in an inextricable manner, and myriads of specks, or of small rods which move in every direction. In other specimens of urine, divers spores (fructifying or reproductory organs) will be found, either isolated, grouped, or arranged in lines like strings of beads, and belonging to undetermined species of fungi (mucedines). These spores are generally roundish or ovoid, very refracting, present a homogeneous, amor-phous composition, with or without nucleoli. When these latter are present, they are to the number of 1 or 2, rarely more, and are remarkable for their brightness. They vary in diameter from the $\frac{1}{3500}$th to the $\frac{1}{12000}$th of an inch.

All these bodies are of external origin, and have neither interest or signifi-cation for the practitioner. But it is important that he should know them, to avoid confounding them with other more or less apparently similar ele-ments eliminated by the organism. Whenever they are encountered it may be concluded that the fluid, under examination, containing them, is under-going a change. It is a mistake to suppose that vibrios and bacteria, which are vegetable and not animal, may normally exist in the different humors of the economy, notwithstanding the fact of their presence in pus, and other fluids affected or produced by diseased conditions. Ammonia arrests the movements of spirillum, vibrios, and bacteria, without affecting them, while it dissolves the substance of infusoria, of which two kinds only are represented in human urine.

Some very minute granulations are daily met with in the urine under microscopic examination, which have an uninterrupted dancing or gyratory movement, termed the Brunonian molecular movement. These granula-tions, frequently called "proteinous molecular granulations," are either isolated and mobile, or united in groups or granular masses. According to the recent researches of M. Bechamp, these granules, which he terms *micro-zymas*, are only one of the phases of development of vibrios and bacteria, which results from their attachment and union end to end, in thread-like chains. Others regard them as the reproductive organs of the algæ or fungi. M. Hoffman considers them to be products of disaggregation, of organic detritus, and not of new organs. Spirillum, vibrios, and bacteria represent the various phases of development of filamentous algæ (leptothrix and others). —The various bodies just noticed are not colored by *solution of carmine*, and

16

acetic acid makes but little impression upon them. Iodine is the proper agent to render them conspicuous (See *Solution of Iodine*, No. 1). It colors their tissue yellow, arrests their movements, and colors the contents of the tubes of leptothrix, without acting upon its external envelop.—Salicylic acid, as well as thymol, prevents the growth of these minute organisms, and checks fermentation and putrefaction, in urine. See Tables XIV, XV, and XVI.

Vesical Oxide. See *Cystine*.

Vibriones. *Vibrios.* Filiform bodies, more or less distinctly jointed from imperfect division, and having an undulatory movement like that of a serpent. Their length varies from 0 mm .001 to 0 mm .003. See *Vegetable Organisms*. Vibriones are met with in the pale urine of cachectic and debilitated persons, and in the urine of persons extremely prostrated by phthisis, mesenteric and syphilitic diseases, etc.

Vitali's Method for Quinia. Dr. D. Vitali states that the following process for the detection of quinia in urine is both satisfactory and readily practiced :—Into a test tube place 8 or 10 c. c. of the urine, add to it 5 or 6 c. c. of ether, and then 8 to 10 drops of ammonia.* Agitate the mixture thoroughly and set aside. As soon as all the ether forms an upper layer upon the fluid, remove it by means of a small pipette, place it in a small capsule, add a drop of diluted pure hydrochloric acid to it, and evaporate by a gentle heat. When cool, place a drop or two of saturated chlorine water in the capsule, and by means of a glass rod rub up all the hardly visible residue on the sides and bottom of the vessel, and then add a drop of ammonia. The characteristic green color will be produced if the urine contains $\frac{1}{10000}$th part of the alkaloid. If a drop of yellow prussiate of potassa be added to the residue left by the evaporation, and then a drop or two of chlorine water, followed by a trace of ammonia, a reddish-purple color will be produced.

* Dr. Vitali prefers, at this stage of the process, the employment of a solution of caustic soda 1 part to 60 parts of distilled water,—instead of the ammonia.

Vogel's Table of Colors. This is a very ingenious and valuable mode of determining, approximatively, the quantity of coloring matter in the urine, prepared by Dr. Vogel after having made a large number of comparative observations; it is well adapted for the purposes of the medical man, enabling him to ascertain the amount of urohematin or coloring matter present in any specimen of urine. These colors are divided into three groups, viz., yellowish, reddish, and brown urines, and although there may be intermediate colors, yet these nine will be found sufficient for bedside investigations. As the absolute quantity of urohematin required in a given bulk of urine, to produce any one of these colors, is not known. Vogel assumed, as a starting point, that in 1,000 c c. of pale yellow urine the pigment should be considered as a unit or 1, no matter how much its actual amount might be. The same volume, 1,000 c. c. of yellowish-red urine (V) will consequently contain 16 parts of pigment, and reddish-brown 128 parts. If, therefore, a person passes in 24 hours 1,000 c. c. of pale yellow urine, and another party passes 1,000 c. c. of reddish-brown urine, the former has discharged 1 part of urohematin and the latter 128 parts, in equal volumes of urine. Again, if a patient passes in 24 hours, 1,000 c. c. of yellow urine, and another, in the same time, passes 4,000 c. c. of pale-yellow urine, both have discharged an equal quantity of pigment in the same time. Suppose, in 24 hours, 1,800 c. c. of yellow urine were passed, how much pigment would this urine contain? Now 1,000 c. c. of this colored urine contain, as will be seen by the table, 4 parts of pigment, hence,

$$1,000 \; : \; 4 \; :: \; 1,800 \; : \; x.$$

$$x = 7.2 \text{ parts of coloring matter in the}$$
1,800 c. c. of yellow urine.

XIX. Table.

Vogel's Table of Colors.

I.	II.	III.	IV.	V.	VI.	VII.	VIII.	IX.		
1	2	4	8	16	32	64	128	256	Pale Yellow	= I.
	1	2	4	8	16	32	64	128	Light Yellow	= II.
		1	2	4	8	16	32	64	Yellow	= III.
			1	2	4	8	16	32	Reddish Yellow	= IV.
				1	2	4	8	16	Yellowish Red	= V.
					1	2	4	8	Red	= VI.
						1	2	4	Brownish Red	= VII.
							1	2	Reddish Brown	= VIII.
								1	Brownish Black	= IX.

To obtain uniform results, and render these comparable, it is necessary that the urine be absolutely clear, and which, when required, may be obtained by filtering it. A cylindrical glass jar, of not less than 4 or 5 inches in diameter, must be used in this investigation, and the color be ascertained by transmitted light.—Books of colored papers, already gummed, representing the colors named in the above table, can be had of dealers in chemical apparatus, and of many surgical instrument makers; they are useful for recording the color of specimens of urine passed at various times by an individual, by simply attaching a slice of the colored paper to the minutes which every physician should keep of his patients.

When the color of the urine depends upon the presence of bile pigment or other substances, the above table could not, of course, be depended upon; but these cases are rare, and the coloring matter may be detected in most instances by the means named in various parts of this work. Occasionally, although the color may be present, it will not give the proper amount of pigment corresponding with the scale; yet, according to Vogel, the maximum amount of error is one-third or one-fourth of the figures found, and the table will be found fully sufficient and satisfactory for ordinary clinical purposes.

In chlorosis, and anemia, the daily amount of pigment may even fall below 1; while in typhoid fever, scarlatinous nephritis, etc., it may amount to 16, 64, and even 256.

W.

Water Doctor. *Uroscopist.* A name given to a person who pretends to diagnose disease by merely inspecting the urine, without subjecting it to any chemical or microscopical examination.

Wayne's Analysis. For the determination of sugar in the urine, Prof. E. S. Wayne, of Cincinnati, Ohio, has proposed the following process:—1. Precipitate the coloring matter, etc., from the urine, by means of a solution of neutral acetate of lead (common sugar of lead), and filter. Again precipitate with a solution of subacetate of lead (basic acetate of lead), and filter. To the filtrate add ammonia as long as it occasions a precipitate. Collect this precipitate; decompose it with sulphureted hydrogen; filter to remove the sulphuret of lead, and evaporate the clear solution, in a water bath, to dryness, and sugar remains, if it were present in the urine.

White Blood Corpuscles. See *Blood; Leucocytes.*

Wood Fibre. Fibres of deal occasionally exist in urine, and may be confounded with renal casts. They may enter urine in uncovered vessels, during, or soon after, sweeping an uncarpeted floor, and in other ways. The pores of the woody fibres, in some respects, resemble epithelial cells, but they are more regular in their arrangement, have a less regular outline, and more or less refractive power. The practitioner should make himself thoroughly acquainted with the appearance of the various kinds of wood in ordinary use, when present in minute fragments.

Wool Fibre. Woollen hairs have an appearance of firm cylinders, with slight indentations along their margins, together with fine transverse markings. They are soluble in liquor potassa or soda, but not in ammoniacal solution of copper. Wool becomes colored brown under the action of plumbate of soda, in consequence of the sulphur it contains, while silk remains unaffected.

X.

Xanthin. *Xanthic Oxide. Urous Acid. Uric Oxide.* This substance was first observed by Marcet in a small calculus. It has only been observed three or four times in the form of a calculus, and very rarely in urinary sediments. It has been found, however, in several of the tissues of the system, and appears to be one of the intermediate products of the metamorphosis of protein substances. It possesses two atoms of oxygen less than uric acid, one atom of oxygen more than hypoxanthine, and, unlike cystine, contains no sulphur. Xanthin, when pure, is a white, non-crystalline, waxy organic substance, insoluble in cold water, ether, alcohol, and hydrochloric acid, slightly soluble in hot water and acetic acid, moderately soluble in hot hydrochloric acid, which on cooling precipitates crystals of the hydrochlorate of xanthin, and is also soluble in nitric acid without effervescence, the solution leaving, on evaporation, a bright yellow residue, which gives a dark purplish-red color when treated with liquor potassa. Alkalies dissolve it, from which solutions it is precipitated by acids, in a white powder. Under the blowpipe, it crumbles, blackens, gives out an odor like that of burnt horn, burns, and leaves a small amount of ashes. It may be obtained from urine by digesting the urine in a weak solution of carbonate of potassa, which separates the uric acid, and leaves the xanthin undissolved. In calculi, and in urinary sediments, it presents a yellow color. It has been artificially prepared from uric acid, and from guanine. Neubauer states that he has procured only about 15 grains of it from 600 pounds of healthy urine. The concretions termed "Oriental bezoards," contain a large quantity of this substance. Its clinical importance is unknown.

Xanthopsin. A poisonous substance into which santonin, under certain imperfectly ascertained circumstances, is transformed when it has remained in the system a certain length of time, and which is supposed to occasion the dangerous symptoms often met with from the administration of the former vermifuge. Xanthopsin is eliminated by the urine, to which it imparts a yellow color, resembling that of this fluid in jaundice. Caustic alkalies added to urine containing xanthopsin, cause this fluid to assume a red color. The addition of podophyllin, in a purgative dose, to santonin, when the latter is administered as a vermifuge, will prevent the formation of xanthopsin. *Falck. King.*

Z.

Zoosperms. See *Spermatozoids.*

Fig. 38.

Litre Bottle.

SUPPLEMENT.

Albumen. Dr. Wm. Roberts, of Manchester, Eng., has given a new process, observed by us since the preceding pages were printed, for estimating albumen in the urine, and which is effected by progressively adding water to the urine, and then observing the action of nitric acid upon each dilution, until the albuminous opacity gradually diminishes, and finally ceases to appear. Thus: Take a clear glass jar, capable of holding 2,000 or 3,000 c. c. of fluid; into this place 5 c. c. of the albuminous urine, and dilute it with 5 c. c. of clear water. This dilution may be termed the "first degree;" add to it some nitric acid, a few drops. If opalescence occurs, again add another 5 c. c. of water, and again, if necessary, test with nitric acid; this second dilution forms the "second degree." And continue in this manner until the acid occasions no reaction after the liquid has stood for 30 seconds, but occasions a faint opalescence at the 45th second; this is the zero point of the reaction. Now divide the number of the dilutions or degrees required, by the 5 c. c. of urine to which they were added, which will give "the degrees of albumen," each degree of which he has ascertained by calculation to be an indication of .0034 per cent. of albumen. This, multiplied by the degrees of albumen will give the per cent. of albumen in the urine tested. Suppose 1,200 c. c. of urine to be passed in the 24 hours, 5 c. c. of which required 1,250 dilutions before the zero reaction was attained; then $1,250 \div 5 = 250 \times .0034 = 85$ per cent. of albumen. Now as this albumen in 1,200 c. c. of urine is to be determined, we find that $\frac{1,200}{100} \times .85 = 10.2$ grammes of albumen in this urine of 24 hours. This method appears to be applicable to all albuminous urines, and requires less time and trouble than other quantitative processes. The nitric acid may not require to be added after each dilution, but only from time to time during the operation. *Medico-Chir. Trans.* Vol. XLI.

Uroxanthin. *Indican.* Prof. Senator, of Berlin, in a careful examination of the urine in more than 100 cases of various diseases, has found the presence of indican in this fluid, more frequently in chronic than in acute diseases, and especially in wasting diseases, as, in pththisis, innutrition, diffuse peritonitis, pneumonia, meningitis, pleurisy, typhoid fever, chlorosis, progressive pernicious anemia, leucæmia, multiple glandular swellings with children, rickets, and particularly in cancer of the stomach, multiple lymphomata, lympho-sarcomata, tabes mesenterica, amyloid degeneration of the liver, spleen, and kidneys, and granular kidney. In many of these cases, the presence of indican was accompanied by an increase of lime in the urine.

His process for detecting the presence of urine, was to free it from albumen, and then to carefully mix 10 or 15 c. c. of it, in quite a large glass, with an equal volume of fuming hydrochloric acid. Now, add gradually, drop by drop, a concentrated solution of chloride of calcium, until the indigo-blue color is fully developed, and agitate the whole with chloroform, which takes up the freshly precipitated indican, and sinks with it to the bottom of the liquid. Its amount can 'now be accurately estimated. Pale urine is the richest in this substance. Highly colored urine requires to be first decolorized by acetate of lead (avoiding an excess of this salt), previous to employing the above named tests. *Med. Times and Gaz.* 1877.

Xanthuria. A term applied to that peculiar condition of the system in which xanthin is present in the urine.

APPENDIX.

M. Bouchardat has given the following tables for correcting the specific gravity of urine at certain temperatures, and which are based upon the fact that urinometers are graduated at the temperature of 15° C. Thus a urine at the temperature 28° C., giving a specific gravity of 1.021, will require 2.5 to be added to it, making its sp. gr., 1.023.5; or if it be saccharine urine, it will require 3.1 to be added, making the sp. gr. 1024.1, and so on.

XX. Table.

Corrections for Non-saccharine Urine.

To Subtract from the Degree Obtained.			To Add to the Degree Obtained.		
TEMPERATURE.		SPECIFIC GRAVITY.	TEMPERATURE.		SPECIFIC GRAVITY.
Cent.	Fah.		Cent.	Fah.	
0	32	0.9	15	59	0.0
1	33.8	0.9	16	60.8	0.1
2	35.6	0.9	17	62.6	0.2
3	37.4	0.9	18	64.	0.3
4	39.2	0.9	19	66.2	0.5
5	41	0.9	20	68	0.7
6	42.8	0.8	21	69.8	0.9
7	44.6	0.8	22	71.6	1.1
8	46.4	0.7	23	73	1.3
9	48.2	0.6	24	75.2	1.5
10	50	0.5	25	77	1.7
11	51.8	0.4	26	78.8	2.0
12	53.6	0.3	27	80.6	2.3
13	55.4	0.2	28	82.4	2.5
14	57.2	0.1	29	84.2	2.7
15	59	0.0	30	86	3.0
			31	87.8	3.3
			32	89.6	3.6
			33	91.4	3.9
			34	93.2	4.2
			35	95	4.6

XXI. Table.

Correction for Saccharine Urine.

TO SUBTRACT FROM THE DEGREE OBTAINED.			TO ADD TO THE DEGREE OBTAINED.		
TEMPERATURE.		SPECIFIC GRAVITY.	TEMPERATURE.		SPECIFIC GRAVITY.
Cent.	Fah.		Cent.	Fah.	
0	32	1.3	15	59	0.0
1	33.8	1 3	16	60.8	0.2
2	35.6	1.3	17	62.6	0.4
3	37.4	1.3	18	64.4	0.6
4	39.2	1.3	19	66.2	0.8
5	41	1.3	20	68	1.0
6	42.8	1.2	21	69.8	1.2
7	44.6	1.1	22	71.6	1.4
8	46.4	1.0	23	73	1.6
9	48.2	0.9	24	75.2	1.9
10	50	0.8	25	77	2.2
11	51.8	0.7	26	78.8	2.5
12	53.6	0.6	27	80.6	2.8
13	55.4	0.4	28	82.4	3.1
14	57.2	0.2	29	84.2	3.4
15	59	0.0	30	86	3.7
			31	87.8	4.0
			32	89.6	4.3
			33	91.4	4.7
			34	93.2	5.1
			35	95	5.5

XXII. Table.

Diagnosis of Blood in Urine.

LOCATION, DISEASE, ETC.	BLOOD PASSED.	CHARACTER OF URINE.	PAIN, ETC.
From KIDNEYS.	Bloody urine, with elongated clots from ureters, is generally albuminous — usually tube casts present, and symptoms of renal disease. Blood brown colored, or like porter, and not as profuse as when from the bladder.	Urine smoky or blackish – brown, if acid; bright red, if alkaline. Forms a brownish-red pulverulent mass or deposit. Albumen is present, as well as renal casts.	Pain when moulds of clotted blood form in the ureters, and are discharged in the urine.
NEPHRITIC COLIC.	Frequently bloody urine from kidneys.	Constant desire to urinate. Sometimes containing bloodclots.	With pain in region of kidneys, and along the course of the ureters.
RENAL CANCER.	Frequent and profuse bloody urine.	Pus and encephaloid matter present, in the advanced stages.	Tumor found in the loins; and deep-seated pains.
From BLADDER.	Blood in small flaky clots, not mixed with the urine, but passing with it.	Urine ammoniacal; with tenacious mucus and phosphatic deposits in feeble persons. Urine alkaline.	Dull pain in region of bladder, and at its neck; apt to have frequent desire to urinate. Sometimes retention from a coagulum in the urethra.
ABRASION, or ULCERATION of the BLADDER.	Blood mixed with mucus or pus in the urine.	Frequent desire to urinate, and urine fetid, containing more or less muco-purulent matter.	Acute burning pain in pelvic cavity, with uneasiness.
MALIGNANT DISEASE.	Blood dark colored, with putrid offensive matters.	Urination often difficult, painful, and with frequent desires to void the urine.	More or less severe pain in vicinity of the disease.
From the URETHRA.	Blood coming without the urine, in drops or in a small stream. Sometimes small clots.	The first jet of urine only is bloody, the balance becoming clearer, and natural.	Perhaps soreness at the part from which the blood issues.
VESICAL CALCULUS.	Blood in urine after exercising; or a drop or two, with pain, in the last expulsive effort at urination, and with pain at the time.	Urine passed often during the day; apt to be of a florid color; and the desire to urinate caused by any movements or exercise.	Pain in penis or perineum felt after (and often before) urinating, especially when the pain is increased by exercise. Usually the pain is at the end of the penis.
Probably PROSTATIC HYPERTROPHY.	Blood intimately mixed with urine, dark colored, and not much altered by circumstances.	Urine frequent; especially during the night.	More or less constant irritation, at neck of bladder.
CHRONIC CYSTITIS, or CHR. INFLM. NECK of BLADDER.	Urine contains blood corpuscles, if any hemorrhage be present. Mucus in increased amount.	Urine frequent and in small amount during the day (frequently alkaline, fetid).	Pain low down in the belly. Slight pain in expelling the last drops of urine.

LOCATION, DISEASE, ETC.	BLOOD PASSED.	CHARACTER OF URINE.	PAIN, ETC.
CHRONIC PROSTATI-TIS.	Blood rarely, if ever, passed.	Urination unduly frequent; a small mu-co-purulent discharge from urethra; urine a little cloudy.	Diminished sexual desire. Pain at end of penis. Dull pains in perineum and vicinity.
DISTENDED MUCOUS MEMBRANE of BLADDER.		Urine difficult and incomplete; passes by drops involuntarily, but in full stream by catheterization.	Pain in penis or perineum, felt before urinating.
CHRONIC INFLM. of MUCOUS COAT of BLADDER.	Blood occasionally observed, especially when ulcerations or abcesses have formed.	Urine passed often during the day (over 5 or 6 times), in small quantities at a time; alkaline. Mucus increased.	Dull pain in region of bladder. Heaviness in perineum; weakness in back.

N. B.—Always have the urine passed into two vessels—say an ounce or two in the first, and the balance in the second. Examine only what is found in the second vessel.

In cold weather, urates are deposited in urine in which none would be seen in summer. If the thickness or turbidity of the urine be due to urates, the application of a little heat will clear it up—which is never the case when it is due to pus, mucus, or other organic matters.

When the urates do not appear habitually, it amounts to nothing; only when they are heavy and constant, in which case correct the patient's habits and digestion, and check indulgence in diet.

Examine for albumen, sugar, etc.; and in all cases pass a bougie, gently and carefully.

Weights and Measures.

To facilitate investigations, it has been deemed useful to give some of the French Metrical Weights and Measures, their conversion into American, and *vice versa*. It is an excellent plan for every investigator to make these reductions and calculations for himself, and to figure them on a pasteboard, of easy access. This can readily be done as follows:

I. One milligramme = .015434023453 Troy grain.

One centigramme = .15434 + Troy grain.

One gramme = 15.434 + Troy grains.

Now, to determine fractions of these weights, divide the Troy grains by the fraction; thus, $\frac{1}{45}$th of one milligramme is ascertained by dividing .015434 + by 45 = .000343 Troy grain.

To determine any increase of these weights, multiply by the number exceeding the unit; thus, 15 centigrammes = .15434 + — 15 = 2.315 + Troy grains. The same course may be pursued for the determination of the American measures in the following:

One millimetre = .0393707904 English inch.

One centimetre = .393707904 English inch.

One metre = 39.3707904 English inch.

One millilitre or c. c. =   16.2319 Minims.
One centilitre   =   162.3190 Minims.
One litre   = 16231.90   Minims, or
   2 pints, 1 fl. ounce, 6 fl. drachms, 6.32 minims.

N. B. A millilitre, a gramme, and a cubic centimetre, each, = .061028 cubic inch, 15.434 Troy grains of distilled water, and 16.2319 minims, Apothecaries measure.

    II. One Troy grain = 64.79895   milligrammes, or,
                    6.47989   centigrammes, or,
                     .06479   gramme.
  One English inch   = 25.3995408 millimetres.
  One cubic inch   = 16.3861758 millilitres, or, c. c.
  One Apoth. minim =   .0616052 millilitres.

The French Metrical value of any of these American measures may be found, as in the preceding case, by multiplying the metrical measures by the increased amount of American measures, or, dividing them by the fractions; thus, 10 inches = 253.996 millimetres; $\frac{1}{12}$th Troy grain = 5.3993 milligrammes.—The long array of figures given in the milli-conversions is rarely employed, though stated here for the benefit of those desirous of knowing them. For instance, the reduction of hundred-thousandths of an inch into millimetres, is sufficiently accurate for all ordinary purposes, by dividing the fractions of the inch into 25.399541 millimetres. Then $\frac{1}{400000}$ inch = .0000634 millimetres.

The metrical weights and measures, whether they be grammes, metres, or litres, etc., bear the following proportions to the unit (that is, to 1 gramme, 1 metre, 1 litre), according to their prefix:

Myria-	Kilo-	Hecto-	Deca-	Gramme, Metre, Litre.	Deci-	Centi-	Milli-
10000	1000	100	10	1	$\frac{1}{10}$	$\frac{1}{100}$	$\frac{1}{1000}$
10000.0	1000.0	100.0	10.0	1.	.1	.01	.001

A decigramme is $\frac{1}{10}$th of a gramme, a centigramme is $\frac{1}{100}$th of a gramme, a hectogramme is 100 grammes, and so on, whether grammes, litres, or metres. The lower line shows the method of dividing and reading the Metrical Weights and Measures, from the situation of the decimal dot. Thus, 1. grm, or, 1. grm .00, is read as 1 gramme (litre or metre); .01 grm is one centigramme, .001 grm is one milligramme, 10. grms is a decagramme. 1263. grms .845 may read, 1 kilogramme, 2 hectogrammes, 6 decagrammes, 3 grammes, 8 decigrammes, 4 centigrammes, 5 milligrammes; or, it may be read as 1263845 milligrammes (—litres, or—metres).

XXIII. Table.

To Reduce Metrical to American Measures.

	Cubic Centimetres, Grammes, or Millimetres, to			Litres, to			Millimetres, to
	Wine Measure.	Cubic Inches.	Troy Grains.	Wine Measure.	Cubic Inches.	Troy Ounces.	English inch.
	M			Pints.			
1	16.2	.061028	15.434	2.1135	61.028	32.104	.03937
2	32.4	.122056	30.868	4.2270	122.056	64.208	.07874
3	48.7	.183084	46.302	6.3405	183.084	96.312	.11813
	ℨ						
4	1. 4.9	.244112	61.736	8.4541	244.112	128.416	.15748
5	1.21.	.305140	77.170	10.5676	305.140	160.520	.19685
6	1.37.4	.366168	92.604	12.6811	366.168	192.624	.23622
7	1.53.6	.427196	108.038	14.7947	427.196	224.728	.27559
8	2. 9.8	.488224	123.472	16.9082	488.224	256.832	.31496
9	2.26.	.549252	138.906	19.0217	549.252	288.936	.35433
10	2.42.4	.610280	154.340	21.1353	610.28	321.04	.39371
20	5.24.6	1.22056	308.680	42.2706	1220.56	642.08	.78742
	ℨ						
30	1.0. 6.9	1.83084	463.020	63.4059	1830.84	963.12	1.18113
40	1.2.49.2	2.44112	617.361	84.5412	2441.12	1284.16	1.57484
50	1.5.31.6	3.05140	771.701	105.6765	3051.40	1605.20	1.96855
60	2.0.13.9	3.66168	926.041	126.8118	3661.68	1926.24	2.36226
70	2.2.56.2	4.27196	1080.381	147.9471	4271.96	2247.28	2.75597
80	2.5.38.5	4.88224	1234.721	169.0824	4882.24	2568.32	3.14968
90	3.0.20.8	5.49252	1389.062	190.2177	5492.52	2889.36	3.54339
100	3.3. 3.2	6.1028	1543.402	211.3530	6102.8	3210.4	3.93710
200	6.6. 6.4	12.2056	3086.804	422.706	12005.6	6420.8	7.87420
	Oct.						
500	1.0. 7.16	30.514	7717.011	1056.765	30514.0	16052.0	19.68550
1000	2.1. 6.32	61.028	15434.023	2113.530	61028.0	32104.0	39.37079

By the preceding table we may readily reduce metrical figures into American. Thus, 10 c. c. = 154.340 Troy grains, or .610.28 cubic inch. Four litres = 8.4541 pints, or 128.416 Troy ounces of distilled water. Ten millimetres = .39371 English inch. If we desire to ascertain the equivalent of 15 millimetres, we add together those of 10 and 5, as .39371 + .19685 = .5905. If the 30th part of a millimetre is required, divide 1 millimetre, .03937 by 30 = .00131 Tr. grn.; and so with the other measures. 10,000 c. c. requires to multiply the equivalent of 1,000 c. c. by 10, which will give, in wine measure, 22 pints, 2 ounces, 1 drachm, and 20 minims. If the American value of centimetres be required, say 20 centimetres,—as 10 millimetres equal one centimetre, we can multiply the 20 cm. by the mm. 200 mm. will equal 20 cm.; then 200 millimetres (or 20 centimetres) = 7.874 English inches.

APPENDIX.

XXIV. Table.

To Convert American to Metrical Measures, etc.

	Inch to Millimetre.	Grains to Milligramme.	Minims to Millilitres.	Cubic Inch to Cubic Centimetres.
$\frac{1}{10000}$	.00254	.00647	....	.001638
$\frac{1}{5000}$	.00507	.01295	....	.003279
$\frac{1}{1000}$	.02539	.06479	....	.01638
$\frac{1}{500}$	.05079	.12958	....	.03277
$\frac{1}{200}$	.1269	.32395	....	.08193
$\frac{1}{100}$	.2539	.64791	....	.1638
$\frac{1}{90}$	.2822	.7199	....	.1820
$\frac{1}{80}$	.3175	.8099	....	.2048
$\frac{1}{70}$	.3628	.9256	....	.2341
$\frac{1}{60}$	.4233	1.0799	....	.2731
$\frac{1}{50}$	.5079	1.2958	....	.3277
$\frac{1}{40}$	.6349	1.6198	.001	.4096
$\frac{1}{30}$	.8466	2.1597	.002	.5462
$\frac{1}{20}$	1.2699	3.2395	.003	.8193
$\frac{1}{10}$	2.5399	6.4791	.006	1.6386
$\frac{1}{5}$	5.0799	12.9583	.012	3.2771
$\frac{1}{4}$	6.3498	16.1979	.015	4.0964
$\frac{1}{2}$	12.6998	32.3959	.031	8.1929
	Centimetres.	Centigrammes.		
1	2.539	6.4792	.062	16.3861
2	5 079	12.9583	.123	32.7718
3	7.619	19.4375	.184	49.1577
4	10.159	25.9167	.246	65.5436
5	12.699	32.3959	.308	81.9295
6	15.239	38.8751	.370	98.3154
7	17.779	45.3543	.431	114.7013
8	20.319	51.8335	.493	131.0873
9	22.859	58.3137	.554	147.4731
10	25.399	64.7989	.616	163.861
20	50.799	129.5838	1.232	327.718
30	76.198	194.3757	1.848	491.577
40	101.598	259.1676	2.464	655 436
50	126.997	323 9595	3.080	819.295
60	152.397	388.7515	3.696	983.154
70	177.796	453.5434	4 312	1147.013
80	203.196	518.3353	4.928	1310 873
90	228.595	583.1272	5.544	1474.731
100	253.995	647. 192	6.160	1638.617
500	1269.977	3239.5959	30.804	8192 951
1000	2539.954	6479.8919	61.605	16386.1757

In the above table, $\frac{1}{50}$th of an inch equals .5079 millimetre; 8 grains equal 51.833 centigrammes; 5 minims equal .308 millilitre; 8 cubic inches equal 131.0873 cubic centimetres; and so on. How many millimetres are there in the $\frac{1}{3200}$th of an inch? In 3,200 there are 16 two hundreds; $\frac{1}{200}$th of an inch equals .12699 millimetre; then 16 two hundredths = $\cdot\frac{12699}{16}$ = .007937 milli-metre. Or, it may be done as follows:—$\frac{1}{2}$ inch = 12.6998 millimetres,—then 1 inch = 25.3996 millimetres. Therefore, 25.3996 ÷ 3,200 = .007937.

XXV. Table.

For the Conversion of Metrical Weights and Measures into American.

	Millimetres to Inches.	Centimetres to Inches.	Milligrammes to Tr. Grains.	Centigrammes to Tr. Grains	Cubic Centimetres; Grammes; or Millilitres— to Wine Measure.
					Minims.
.01	.00039	.0039	.00015	.0015	.16
.02	.00078	.0078	.00031	.0031	.32
.03	.00118	.0118	.00046	.0046	.48
.04	.00157	.0157	.00061	.0061	.65
.05	.00197	.0197	.00077	.0077	.81
.06	.00236	.0236	.00092	.0092	.97
.07	.00276	.0276	.00108	.0108	1.13
.08	.00315	.0315	.0012	.0123	1.29
.09	.00354	.0354	.0014	.0138	1.46
.1	.00394	.0393	.0015	.0154	1.62
.2	.00787	.0787	.0031	.0308	3.24
.3	.01181	.1181	.0046	.0463	4.87
.4	.01575	.1575	.0061	.0617	6.49
.5	.01969	.1969	.0077	.0771	8.11
.6	.02362	.2362	.0092	.0926	9.74
.7	.02756	.2756	.0108	.1080	11.36
.8	.03149	.3149	.0123	.1234	12.98
.9	.03543	.3543	.0139	.1389	14.61
1.	.03937	.3937	.0154	.1543	16.23
2.	.07874	.7874	.0308	.3086	32.46
3.	.11811	1.1811	.0463	.4630	48.69
4.	.15748	1.5748	.0617	.6173	fʒ 1. 4.92
5.	.19685	1.9685	.0771	.7717	1.21.16
6.	.23622	2.3622	.0926	.9260	1.37.39
7.	.27559	2.7559	.1080	1.0803	1.53.62
8.	.31496	3.1496	.1234	1.2347	2. 9.85
9.	.35433	3.5433	.1389	1.3890	2 26.08
10.	.39371	3.9371	.1543	1.5434	2.42.32
20.	.78742	7.8742	.3086	3.0868	5.24 63
30.	1.18113	11.8113	.4630	4.6302	fʒ 1.0. 6.95
40.	1.57484	15.7484	.6173	6.1736	1.2.49.27
50.	1.96855	19.6855	.7717	7.7170	1.5.31.59
60.	2.36226	23.6226	.9260	9.2604	2.0.13.91
70.	2.75597	27.5597	· 1.0803	10 8038	2.2.56.23
80.	3.14968	31.4968	1.2347	12.3472	2.5.38.55
90.	3.54339	35.4339	1 3890	13.8906	3.0.20 87
100.	3.9371 ·	39.3708	1.5434	15.4340	3.3. 3 19
500.	19.6855	196.8539	7.7170	77.1701	16.7.15.95
1000.	39.3710	393.7079	15.4340	154.3402	33.6.31.90

In the above table .09 millimetre equal .00354 inch; .3 centimetre equal .1181 inch; .8 milligramme equal .0123 Troy grain; 3. centigrammes equal .4630 Troy grain; 10 c. c., or 10 grammes equal 2 fluidrachms 42.32 minims and so on.

In England, and in this country, the inch is by common consent, taken as the standard of measurement; on the European continent, lines, and parts of a line, as well as millimetres are employed for a similar purpose. The following tables of comparative micrometrical measures are given, as they may be useful for reference:

Millimetre.	Paris Lines.	Vienna Lines.	Rhenish Lines.	English Inches.
1	0.443296	0.455550	0.458813	0.0393708
2.255829	1	1.027643	1.035003	0.0888138
2.195149	0.973101	1	1.0071625	0.0864248
2.179538	0.966181	0.992888	1	0.0858101
25.39954	11.25952	11.57076	11.65354	1.

The following are given merely to aid in refreshing the mind of the investigator:

To Determine the Magnifying Power of an Objective.

Multiply the size of the divisions of the stage micrometer by the numerator of the fraction of parts of an inch it is magnified, and divide this by the denominator. Thus, suppose each division of the stage micrometer equals $\frac{1}{500}$th of an inch, 1 of which is magnified to $\frac{7}{10}$ths of an inch; then $\frac{1}{500} \times 7 = \frac{7}{500} \div 10 = 350$ diameters. If 10 divisions are magnified to $\frac{7}{10}$th of an inch, then $\frac{1}{500} \times 7 = 3,500 \div 10 = 350$ diameters, or 10 divisions = 350 diameters; and 1 division = 3,500 diameters. Again, the stage micrometer divisions equal, each, $\frac{1}{500}$th of an inch, 1 of which is magnified to $1\frac{3}{8}$th inch, or $\frac{11}{8}$th inch; then $\frac{1}{500} \times 11 = 5,500 \div 8 = 687.5$ diameters.

To Measure an Object with the Eye-piece Micrometer.

In order to *measure an object*, the value of the spaces of the ocular micrometer must first be had; and which varies with each objective. This is obtained as follows: We find that with an inch objective, three of the divisions of our ocular micrometer occupy just one of the stage micrometer spaces, $\frac{1}{500}$th of an inch; then, when measuring with this objective, an object occupying three ocular micrometer divisions, is $\frac{1}{500}$th of an inch large, or, if it occupies only one of these divisions it is $\frac{1}{1500}$th of an inch. We now remove the inch objective, and employ the quarter inch, and find that twenty divisions of the ocular micrometer occupy just one of the stage micrometer spaces; then, when measuring with this objective, twenty eye-piece micrometer divisions are equal to $\frac{1}{500}$ of an inch, and one is equal to the one-twentieth of this, or the $\frac{1}{10000}$ of an inch. And so with all the other objectives. But, suppose the eye-piece micrometer gives with an objective five and a half divisions to one of the stage micrometer; we reduce the included divisions of the former to halves, which would make eleven, and we then multiply this by the measure of the one division of the stage micrometer (500), which would make $\frac{1}{5500}$ inch, equal the half of one ocular micrometer division, or $\frac{1}{2750}$ inch, the value of one space in this eye-piece. These measurements having been once made should be recorded, so as to be ready, for future investigations, without any loss of time in renewing them.

To Determine the Size or Diameters of an Object with any Combination of Objectives and Oculars, and without an Eye-piece Micrometer.

The microscope having been mounted with the required objective and eye-piece, attach the camera lucida, and place a stage micrometer on the stage of the instrument in the focus of the objective; then fix the compound body in the horizontal position. A sheet of plain paper, or of Bristol card, is now to be laid upon the table at the same distance from the center of the eye-glass of the ocular as it is from this center to the stage micrometer. Throw the image of the micrometer lines upon the card, by means of the camera, and carefully mark two, three, or four of them consecutively; divide each of the magnified spaces thus obtained between the micrometer lines, into fifths, tenths, or twentieths, and, in one corner of this card, record the divisions of the stage micrometer used, the focal length of the objective, and the number of the eye-piece. This card, with its divisions, will always serve to give the measure of any object observed with the same objective and eye-piece, when the compound body of the microscope is brought to the horizontal position (at the same distances), and the image of the object is, by means of the camera, directed upon the graduated lines of said card. If the diameters of the observed object are unequal, we may, after having obtained one diameter, move the card around as required, so as to obtain the others.—A series of these cards should be prepared for each objective with the different eye-pieces, and be preserved for future use whenever it is required to measure the size of one or more objects.

It does not matter whether the stage micrometer be ruled into hundredths, five hundredths, or thousandths, of an inch, nor what combination of objective and eye-piece is employed, the result of the measurement will always be the same for each combination, as shown by the following table:

17

XXVI. Table for Measuring Microscopic Objects.

Scale into 20's.	Scale into 10's.	Scale into 5's.	MAGNIFIED MICROMETER SPACE.			
			1-2000 inch.	1-1000 inch.	1-500 inch.	1-100 inch.
			80,000	40,000	20,000	4,000
1			40,000	20,000	10,000	2,000
			26,666+	13,333+	6,666+	1,333+
2	1		20,000	10,000	5,000	1,000
			16,000	8,000	4,000	800
3			13,333+	6,666+	3,333+	666+
			11,428+	5,714+	2,857+	571+
4	2	1	10,000	5,000	2.500	500
			8,888+	4,444+	2,222+	444+
5			8,000	4,000	2,000	400
			7,272+	3,636+	1,818+	363+
6	3		6,666+	3,333+	1,666+	333+
			6,154-	3,077-	1,538+	307+
7			5,714+	2,857+	1,428+	285+
			5,333+	2.666+	1,333+	266+
8	4	2	5,000	2.500	1,250	250
			4,706-	2,353-	1,176+	235+
9			4,444+	2,222+	1,111+	222+
			4,210+	2,105+	1,052+	210+
10	5		4,000	2,000	1,000	200
			3,809+	1,904	952+	190+
11			3,636+	1,818+	909+	181+
			3,478+	1,739+	869+	174-
12	6	3	3,333+	1.666+	833+	166+
			3,200	1.600	800	160
13			3.077-	1,538+	769+	153+
			2,963-	1,481+	740+	148+
14	7		2,857+	1,428+	714+	142+
			2,758+	1,379+	689+	138-
15			2,666+	1,333+	666+	133+
			2,380+	1,290+	645+	129+
16	8	4	2.500	1,250	625	125
			2,124+	1,212+	606+	121+
17			2,353-	1,176+	588+	117+
			2,285+	1,143-	571+	114+
18	9		2.222+	1,111+	555+	111+
			2,162+	1,081+	540+	108+
19			2,105+	1,052+	526+	105+
			2,051+	1,025+	512+	102+
20	10	5	2,000	1,000	500	100

SCALE OF DIVISIONS.

Thus, if a micrometer of $\frac{1}{2000}$th of an inch ruling be employed, it will be manifest that, if the magnified image of one of its spaces on the card be divided into twentieths, one of these divisions will invariably give the measurement of $\frac{1}{40000}$th of an inch; while twelve and a half of these divisions will give $\frac{1}{3200}$th of an inch; and so on. This will be found in the table, by referring to the column under the heading, "$\frac{1}{2000}$th inch," and that of "Scale into 20's."—If the micrometer space on the card be divided into tenths, by referring to the table, in the column under the heading, "Scale into 10's," we find one of these spaces, for a $\frac{1}{2000}$th inch micrometer, equal to $\frac{1}{20000}$th of an inch; while five and one-fourth spaces equal $\frac{1}{3809}+$th of an inch.—If the micrometer be ruled into $\frac{1}{500}$ths of an inch, and the magnified image of one of its spaces, on the card, be divided into twentieths, upon referring to the columns in the table, under the heading "$\frac{1}{500}$ inch" and "Scale into 20's," we find that one of these spaces equal $\frac{1}{10000}$th of an inch; and six and one-half of them equal $\frac{1}{1538}+$th of an inch. But if the spaces are divided into fifths, we must refer to the column under the heading "Scale into 5's," and then run along the horizontal line until we come to the column for "$\frac{1}{500}$ inch," in which we will find one of these spaces equal to $\frac{1}{2500}$th of an inch; and four and one-eighth of them equal to $\frac{1}{515}+$th of an inch.—If the image of $\frac{1}{100}$th inch micrometer have its spaces between two lines divided into fifths, by referring to the column under "Scale into 5's," and following out horizontally to the column for "$\frac{1}{100}$th inch," we find one space equal to $\frac{1}{500}$th of an inch; and four spaces equal to $\frac{1}{125}$th of an inch; and so on.—If the stage micrometer be ruled into $\frac{1}{200}$ths of an inch, we have simply to double the figures given for that of $\frac{1}{100}$ths of an inch.

Whenever the magnified space between any two micrometer lines is divided into twentieths, we must refer to the column in the table under the heading "Scale into 20's;" if into tenths, to that under "Scale into 10's;" and if into fifths, to that under "Scale into 5's;" ascertaining the amount of measurement of one or more of these divisions, in the columns under the headings $\frac{1}{20000}$th, $\frac{1}{10000}$th, $\frac{1}{500}$th, or $\frac{1}{100}$th inch, according to the ruling of the micrometer used. When — is placed after any figures, it means that the measurement is slightly less than given; and +, that it is slightly above; in either case to an almost inappreciable amount.—To reduce the fractional parts of an inch into millimetres, divide them into 25.399540871; then $\frac{1}{40000}$th of an inch equal to .00063498+ millimetre.

To Reduce Compound Fractions to Simple Ones. Multiply the numerators together for a new numerator, and the denominators for a new denominator.

What is the $\frac{19}{20}$ of $\frac{1}{2000}$?

$\frac{19}{20} \times \frac{1}{2000} = \frac{19}{40000} = \frac{1}{2105 \cdot 26} +.$

What is the $\frac{3}{5}$ of $\frac{1}{100}$?

$\frac{3}{5} \times \frac{1}{100} = \frac{3}{500} = \frac{1}{166 \cdot 66} +.$

To Reduce Decimals to Fractions. Place the denominator under the decimal, and reduce this fraction to its lowest term; thus—

Reduce .225 to fractions. $\frac{225}{1000} = \frac{9}{40}$.

Reduce .059375 to fractions. $\frac{059375}{1000000} = \frac{19}{320}$: or, $\frac{059375}{1000000} \div 25 = \frac{2375}{40000}$ $\div 25 = \frac{95}{1600} \div 5 = \frac{19}{320}$.

To Reduce Fractions to Decimals. Divide the numerator by the denominator, adding ciphers as long as may be necessary. Thus—

Reduce $\frac{7}{8}$ to a decimal. $\frac{7000}{8000} \div 8 = \frac{875}{1000} = .875$.

Reduce $\frac{5}{6}$ to a decimal. $\frac{50000}{60000} \div 6 = \frac{8333}{10000} + = .8333 +$.

Reduce $\frac{1}{100}$ to a decimal. $\frac{10000}{4000000} \div 400 = \frac{0025}{10000} = .0025$.

To Add Fractions. Reduce the fractions to a common denominator; add their numerators together, and place the sum over a common denominator; thus—

Add together $\frac{2}{3}$, $\frac{3}{4}$, and $\frac{5}{6}$. $\frac{8}{12} + \frac{9}{12} + \frac{10}{12} = \frac{27}{12} = 2\frac{1}{4}$.

To Subtract Fractions. Find the common denominator, as in the preceding instance, and place the difference of the numerators above it. Thus—

Subtract $\frac{1}{4}$ from $\frac{1}{3}$. $\frac{3}{12} - \frac{4}{12} = \frac{1}{12}$.

Subtract $\frac{1}{11}$ from $\frac{1}{7}$. $\frac{7}{77} - \frac{11}{77} = \frac{4}{77}$.

To Multiply Fractions. Multiply the numerators together for a new numerator, and the denominators for a new denominator. Thus—

Multiply $\frac{5}{13}$ by $\frac{11}{16}$. $\frac{88}{208} = \frac{11}{26}$.

Multiply together $\frac{7}{8}$, $\frac{3}{10}$, $\frac{8}{9}$, $\frac{5}{6}$, $\frac{2}{3}$, $\frac{5}{7}$. $\frac{10080}{90720} = \frac{1}{9}$.

To Multiply Fractions by a Whole Number. Multiply the numerator of the fraction by the whole number, and place the result above the denominator. Thus—

Multiply $\frac{5}{36}$ by 9. $\frac{5}{36} \times 9 = \frac{45}{36} = 1\frac{1}{4}$.

To Multiply a Whole Number by a Fraction. Multiply the whole number by the numerator of the fraction, and divide by its denominator. Thus—

Multiply 29 by $\frac{3}{4}$. $29 \times 3 = 87 \div 4 = 21\frac{1}{4}$.

To Divide Fractions. Invert the divisor; then form new numerators and denominators, by separately multiplying each of these together. Thus—

Divide $\frac{5}{6}$ by $\frac{2}{3}$. $\frac{5}{6} \times \frac{3}{2} = \frac{15}{12} = 1\frac{1}{4}$.

Divide $12\frac{5}{6}$ by $4\frac{1}{3}$. $\frac{77}{6} \times \frac{3}{13} = \frac{616}{198} = 3\frac{1}{3}$.

To Divide Fractions by a Whole Number. Multiply the denominator of the given fraction by the whole number, and place the result beneath the numerator. Thus—

Divide $\frac{8}{10}$ by 4. $\frac{8}{10} \div 4 = \frac{8}{40} = \frac{1}{5}$.

To Divide a Whole Number by a Fraction. Multiply the denominator by the whole number, and divide the result by the numerator. Thus—

Divide 45 by $\frac{3}{6}$. $45 \times 6 = 270 \div 3 = 90$.

Divide $10\frac{2}{3}$ by $3\frac{2}{5}$. $10\frac{2}{3} = \frac{32}{3}$; and $3\frac{2}{5} = \frac{17}{5}$. Therefore, $\frac{32}{3} \div \frac{17}{5}$ inverted to $\frac{32}{3} \div \frac{5}{17} = \frac{160}{51} = 3\frac{7}{51}$.

The above simply shows how many times one fraction or whole number is contained in another; for instance, $\frac{2}{3}$ of $\frac{5}{6}$ can not be $1\frac{1}{4}$, which is simply the number of times that $\frac{2}{3}$ can be contained in $\frac{5}{6}$. The $\frac{2}{3}$ of $\frac{5}{6}$ is ascertained by multiplying the numerators for a new numerator, and the denominators for a new denominator. Thus—

$\frac{2}{3} \times \frac{5}{6} = \frac{10}{18} = \frac{5}{9}$.

What is the $\frac{3}{5}$ of $\frac{7}{8}$? $\frac{3}{5} \times \frac{7}{8} = \frac{21}{40}$.

Proportions of Chemicals in Dilute Solution. A two per cent. solution of any salt, or other substance, means 2 parts of the salt to 100 of water or other fluid; one-half per cent., means 1 part of the chemical to 200 of fluid; one-tenth per cent., 1 part to 1,000 of fluid; one per cent., 1 part to 100 of fluid. The French prepare solutions as follows: Potassa at the tenth, means 1 part of potassa to 10 parts of water; acetic acid at the fifth, is 1 part of acetic acid to 5 parts of water, and so on.

XXVII. Table.

Of Symbols or Formulæ.

Those articles to which a star (*) is prefixed, have no trustworthy symbols, as several of them are found to vary, while others have not been formulated, except empirically. Margarin is, at present, believed to be a mixture of two substances.

Acetate of Lead, *neutral*.................................$(C_2 H_3 O_2)_2 Pb. 3Aq.$

" " *basic*........................$(C_2 H_3 O_2)_2 Pb, 2 Pb O.$

" Soda...............................$(C_2 H_3 O_2) Na.$

" Uranium.......................... $Ur_2 O_2 (C_2 H_3 O_2)_2 + 2 Aq.$

Acetic Acid, *anhydrous.* ....$C_2 H_4 O_2.$

Acetic Acid............................$H (C_2 H_3 O_2).$

Acetone................................$C_3 H_6 O.$

Aconitia...............................$C_{54} H_{40} N O_2.$

* Albumen; in varying proportions of.......$C_{400} H_{310} N_{50} O_{120} S_2 P.$

Alcohol..................................$C_2 H_5 (O H).$

* Alkapton.

Allantoin................................$C_4 N_4 H_6 O_3.$

Alloxan.................................$C_4 N_2 H_2 O_4.$

Alloxantine..............................$C_8 N_4 H_4 O_7, 3 Aq.$

Ammonia, *gas*.........................$N H_3.$

" *liquid*........................$N H_4^- HO.$

Ammonio-magnesian Phosphate..............$Mg N H_4 P O_4 + 6 Aq.$

* Ammonio-oxide of Copper................$N_2 H_6 Cu O_2(?).$

Ammonio-sodic Phosphate.................$Na_2 N H_3 P O_4.$

Amygdalin.............................$C_{20} H_{27}, NO_{11} 3 HO_2.$

Antimony................................Sb.

Arsenic.................................As.

Arseniureted Hydrogen............................As H_3.

Atropia.... ....$C_{17} H_{23} N O_3$.

Barium............ ....Ba.

Baryta........................ ....Ba O.

    "   hydrate, crystals....................Ba $H_2 O_2$ 8 Aq.

Benzamide.................... ....N $H_2 C_7 H_5$ O.

Benzoate of Lithia........... ....$C_7 H_5 O_2$ Li.

Benzoglycholic Acid..... ....$C_9 H_8 O_4$.

Benzoic Acid.................. ....$C_7 H_5$ O, H O.

Bicarbonate of Soda.........................Na H C O_3.

Bichloride of Platinum.................. Pt Cl_4.

Bichromate of Potassa......................$K_2 Cr_2 O_7$.

Bilirubin............ ....$C_{16} H_{18} N_2 O_3$.

* Biliverdin.............................. ....$C_8 H_9 N O_3$ (?).

Bimeta-antimoniate of Potassa....... ......$Sb_2 O_5 K_2 O 7H_2 O$.

Bisulphide of Carbon................ .... C S_2.

Bromide of Potassium.........................K Br.

Bromine..........,...............................Br.

Butyric Acid......... ....H $C_4 H_7 O_2$.

Calcium..........................,.. ......Ca.

Carbolic Acid..............................$C_6 H_6$ O.

Carbonate of Ammonia............... ....$(N H_4)_2 C O_3$.

    "   Barium.................. .... Ba C O_3.

    "   Lime.................................Ca C O_3.

    "   Magnesia, *anhydrous* .... Mg C O_3.

    "   "   *hydrated* ....$_4$C O_3 Mg, Mg $H_2 O_2$, 6 HO.

             (Composition of above not constant.)

    "   Silver....................... .... Ag_2 C O_3.

    "   Soda................... ....Na_2 C O_3.

Carbonic Acid, *gas*..............................C O_2.

* Casein, an albuminoid.

Chloral.............. ....C_2 H $Cl_3 O_2$

Chlorate of Potassa....................... ....K Cl O_3.

Chloride of Ammonium.......................N H_4 Cl.

    "   Barium .... Ba Cl_2.

    "   Gold and Soda...................Au Cl_3 Na Cl $2H_2$ O.

    "   Lime........... ....Ca Cl_2.

    "   Sodium.........................Na Cl.

    "   Tin ....Sn Cl_2 *(proto)* and Sn Cl_4 *(bi.)*

Chlorine.......................,........ Cl.

Chloroform................. ....C H Cl_3.

* Chlorophyll ....$C_9 H_9 N O_4$ (?).

Cholalic Acid............................ ....$C_{26} H_{40} N S O_7$.

Cholepyrrhine. See Bilirubin.

Cholesterin................................$C_{26} H_{44}$ O.

Cholic Acid...... $C_{24} H_{40} O_5$.
Cholinic Acid...... $C_{26} H_{45} S_2 N O_7$.
Choloidic Acid...... $C_{24} H_{38} O_4$.
Chromate of Potassa, *neutral*...... $K_2 Cr O_4$.
Cinnamic Acid...... $C_9 H_8 O_2$.
Copaivic Acid...... $C_{20} H_{30} O_2$.
Copper ...... Cu.
Creatine...... $C_4 H_9 N_3 O_2$ 2 Aq.
Creatinine...... $C_4 H_7 N_3 O$.
Creosote...... $C_8 H_9 (C_6 H_5) O_2 C_8 H_{10} O_2$.
Cryptophanic Acid...... $H_2 C_5 H_7 N O_5$.
Cuminic Acid...... $C_{10} H_{11} O_2$.
Cyanate of Ammonia, hydrated...... $N H_4 C N O$.
Cyanide of Mercury...... $Hg (C N)_2$.
Cystine...... $C_3 H_7 N S O_2$.
Damaluric Acid...... $C_7 H_{12} O_2$.
Deutoxide of Copper...... $Cu_2 O$.
Ether...... $(C_2 H_5)_2 O$.
Ferrocyanide of Potassium...... $K_4 Fe (C N)_6$.
* Fibrin, variable.
Fuchsin...... $C_{20} H_{19} N_3$.
Gallic Acid...... $C_7 H_6 O_5$.
Glycerin...... $C_3 H_8 O_3$.
Glycocoll...... $C_2 H_5 N O_2$.
Glucogene
Glucose $\Big\}$ ...... $C_6 H_{12} O_6$.
Grape Sugar
Guanine...... $C_5 H_5 N_5 O$.
* Hæmatin...... $C_{68} H_{70} N_8 Fe_2 O_{10}$.
* Hæmatocrystallin.
* Hæmatoidin...... $C_7 H_9 O, N O$ (?).
Hippuric Acid...... $C_9 H_9 N O_3$.
Hydrochlorate of Ammonia...... $N H_4 Cl$.
   "   "   Anilin...... $C_6 H_8 N Cl$.
Hydrochloric Acid...... $H Cl$.
Hypochlorite of Lime...... $Ca Cl_2 O_2$.
   "   "   Soda...... $Na Cl O$.
Hypoxanthin...... $C_5 H_4 N_4 O$.
Indigo-blue...... $C_8 H_5 N O$.
Indigo-carmine, or $\Big\}$ ...... $C_8 H_4 NO S O_2 OK$.
Sulphindigolate of Potassium
* Indigo-glucin...... $C_6 H_{10} O_6$ (?).
Inosite...... $C_6 H_{12} O_6$.
Iodine...... I.
Iron...... Fe.

Lactic Acid...C_3 H_6 O_3.
Lead...Pb.
Leucin...C_6 H_{13} N O_2.
Lithium..Li.
Lithia..Li_2 O.
Magnesia...Mg O.
Margaric Acid.......................................C_{17} H_{34} O_2.
*Margarin...C_{54} H_{104} O_6.
Mercury..Hg.
Molybdate of Ammonia.......................Mo O_4 (N H_4)$_2$.
Morphia...C_{17} H_{19} N O_3.
Murexan...C_4 H_5 N_3 O_3.
Murexide..C_6 H_6 N_5 O_4.
Nickel...Ni.
Nitrate of Silver..................................Ag N O_3.
Nitrate of Urea....................................C O H_4 N_2, H N O_3.
Nitric Acid..H N O_3.
Nitro-prusside of Sodium....................Na_2 N O, Fe 2 C N.
Nitroso-nitric Acid..............................H N O_3 + N_2 O_4.
Oleic Acid...C_{18} H_{34} O_2.
Olein........ ....(C_3 H_5) (C_{18} H_{33} O_2)$_3$.
Oxalate of Ammonia............................(N H_4)$_2$ C_2 O_4 2 Aq.
     "   Lime ....Ca C_2 O_4 4 Aq.
     "   Urea.................................(C O H_4 N_2)$_2$ C_2 H_2 O_4.
Oxalic Acid..C_2 H_2 O_4.
Parabanic Acid....................................C_3 H_2 N_2 O_3.
Perchlorate of Potassa........................K Cl O_4.
Perchloride of Iron..............................Fe_2 Cl_6.
Permanganate of Potassa....................K Mn O_4.
Pernitrate of Mercury....... ....2 (Hg (N O_3)$_2$) H_2 + O.
Phosphate of Lime, *acid*.....................H Ca P O_4.
     "   "   *basic* ....Ca_3 (P O_4)$_2$.
Phosphate of Magnesia........................H Mg P O_4 7 HO.
Phosphate of Potassa......... ....K_3 P O_4.
Phosphate of Soda, *acid*........ ....H_2 Na P O_4.
     "   "   *neutral*.....................H Na_2 P O_4.
     "   "   and Ammonia...........Na_2 N H_4 P O_4.
Phosphoric Acid. ....H_3 P O_4.
Phosphorous Acid................................H_3 P O_3.
Phosphorous..... ....P.
Picric Acid...C_6 H_3 N_3 O_7.
Picro-carminate of Ammonia.
Platinum...Pt.
Potassa.. ....K HO.
Protonitrate of Mercury.....................Hg_2 (N O_3)$_2$.

APPENDIX.

Protoxide of Copper .... $Cu\ O$.

Prussiate of Potassium, *red.* .... $K_3\ Fe\ (C\ N)_6$.

Purpurate of Ammonia................... $C_8\ N_6\ H_8\ O_6\ Aq$.

Purpuric Acid............................ $C_8\ N_5\ H_5\ O_6$.

Purpurin................................ $C_9\ H_6\ O_3\ H_2\ O$.

Pyrophosphate of Magnesia........ .... $Mg_2\ P_2\ O_7$.

Quinia.............................. $C_{10}\ H_{24}\ N_2\ O_2$.

Salicin.................................. $C_{13}\ H_{18}\ O_7$.

Salicylic Acid........................ $C_7\ H_6\ O_3$.

Santonin............................. $C_{15}\ H_{18}\ O_3$.

Silicate of Potassa............ .... $K_2\ Si\ O_3$.

Silver.................................. Ag.

Stearic Acid........................... $C_{18}\ H_{36}\ O_2$.

Stearin................................ $(C_3\ H_5)\ (C_{18}\ H_{35}\ O_2)_3$.

Strychnia.............................. $C_{21}\ H_{22}\ N_2\ O_2$.

Succinic Acid........................ $C_4\ H_6\ O_4$.

Sugar, *cane*......................... $C_{12}\ H_{22}\ O_{11}$.

"   *grape*................ $C_6\ H_{12}\ O_6 + H_2\ O$.

Sulphate of Copper, crystals.............. $Cu\ S\ O_4 + 5\ Aq$.

"   Potassa........................ $K_2\ S\ O_4$.

Sulphide of Ammonia....... .... $N\ H_4\ H_2\ S$.

"   Calcium......................... $Ca\ S$.

"   Carbon.................... $C\ S_2$.

Sulphomolybdate of Ammonia............... $(N\ H_4)_2\ S\ Mo\ S_3$.

Sulphur.............................. S.

Sulphureted Hydrogen................ $H_2\ S$.

Sulphuric Acid........................ $H_2\ S\ O_4$.

Sulphurous Acid, *gas*.............. $S\ O_2$.

"   "   *liquid*............. $H_2\ S\ O_3$.

Tannin............................ $C_{27}\ H_{22}\ O_{17}$.

Taurin.............................. $C_4\ H_7\ N\ S_2\ O_3$.

Taurylic Acid......................... $C_7\ H_8\ O$.

Thionurate of Ammonia.................. $N\ H_4\ (N_3\ H_4\ C_4\ S\ O_6)_2$.

Thionuric Acid.......................... $C_4\ H_5\ N_3\ S\ O_6$.

Trimethylamin........................ $C_3\ H_9\ N$.

Tyrosin............................. $C_9\ H_{11}\ N\ O_3$.

Uranium .... Ur.

Urate of Ammonia, *acid.*.............. $C_5\ H_3\ (N\ H_4)\ N_4\ O_3$.

"   Lime, *acid.*.............. $C_5\ H_2\ N_4\ Ca\ O_3$.

"   Magnesia, *acid.*.............. $C_5\ H_2\ N_4\ Mg\ O_3$.

"   Potassa, *acid*..... .... $C_5\ N_4\ H_3\ K\ O_3$.

"   Soda, *acid* .... $C_5\ H_3\ N_4\ Na\ O_3$.

"   "   *neutral.*.............. $C_5\ H_2\ N_4\ Na_2\ O_3$.

Urea.............................. $C\ H_4\ N_2\ O$.

Uric Acid..$C_5 H_4 N_4 O_3$.
*Urochloralic Acid.
* Uroglaucin.
* Urohematin.
* Urophein.
* Urosacine.
* Urostealith.
* Uroxanthin.
* Urrhodin.
Veratria...$C_{56} H_{86} N_2 O_{15}$.
Xanthine ....$C_5 N_4 H_4 O_2$.

Thermometric Conversions.

1. *To Convert any Degree of Centigrade above 0° to One of Fahrenheit.* (0° C. $= + 32°$ F.)

Multiply the centigrade degree by 9, then divide by 5, and to the result add 32. Thus : 11° C. $\times$ 9 $\div$ 5 $= 19.8 + 32 = + 51°$ 8 F.

2. *To Convert any Degree of Centigrade below 0° to One of Fahrenheit.* ($-17°$ 7778 C. $= 0°$ F.)

Multiply the centigrade degree by 9, then divide by 5, and subtract the result from 32. Thus: $-11°$ C. $\times$ 9 $\div$ 5 $\div$ 19.8 $- 32 = + 12°$ 2 F.

N. B. But if 32 has to be subtracted from the result, then the Fah. degree obtained is $-$ minus, or below 0°. Thus: $-21°$ C. $\times$ 9 $\div$ 5 $= 37.5 - 32 = -5°$ 8' Fah.

3. *To Convert any Degree of Fahrenheit above $+32°$ to One of Centigrade.* ($+ 32°$ F. $= 0°$ C.)

Subtract 32 from the Fah. degree, then multiply by 5, and divide the result by 9. Thus: 34° F. $- 32 = 2 \times 5 \div 9 = + 1°$ 1111 C.

4. *To Convert any Degree of Fahrenheit below $+32°$ to One of Centigrade.* ($- 0°$ F. $= -17°$ 7778° C. All degrees of Fah. below $+32$ form $-$ minus degrees of Cent.)

Subtract the Fah. degree from 32, then multiply by 5, and divide by 9. Thus: 30° F. $- 32 = 2 \times 5 \div 9 = -1°$ 1111 C.

N. B. But if the Fah. degree is minus or below 0°, then add 32 to it, multiply by 5 and divide by 9. Thus: $-6°$ F. $+ 32 = 38 \times 5 \div 9 = -21°$ 1111 C.

($-40°$ F. $+ 32 = 72 \times 5 \div 9 = -40°$ C.)

THE END.

BEACH—THE AMERICAN PRACTICE CONDENSED, OR THE FAMILY PHYSICIAN. Being the Scientific System of Medicine, on Vegetable Principles, designed for all classes. In nine parts. By Wooster Beach, M. D., Member of the Medical Society of the City and County of New York; of the Medical and Physiological Society of Wetterau, Germany; of Leipsic, Saxony, etc. Fifty-fifth edition, revised. Illustrated with nearly two hundred engravings. Octavo, 873 pages, sheep, ... $4 50

BEACH—THE FAMILY PHYSICIAN AND HOME GUIDE: For the Treatment of the Diseases of Men, Women, and Children, on Reform Principles. By Wooster Beach, M. D. With the Laws of Health, a Glossary of Medical Terms, etc. Octavo, 1,000 pages, morocco, gilt, ... 6 00

HOWE—ON DISLOCATIONS AND FRACTURES. Illustrated. By A. Jackson Howe, M. D., Professor of Anatomy in the Eclectic Medical Institute. Octavo, 424 pages, sheep, ... 4 00

HOWE—A MANUAL OF EYE SURGERY. By A. Jackson Howe, M. D. 8vo, 204 pages, cloth, ... 2 50

HOWE—ART AND SCIENCE OF SURGERY, Illustrated. By A. Jackson Howe, 8vo, 836 pages, sheep, ... 7 00

JONES—SCUDDER—GENERAL AND SPECIAL THERAPEUTICS. By L. E. Jones, M. D., and J. M. Scudder, M. D. 8vo. 312 pages, sheep, ... 3 00

KING—THE AMERICAN DISPENSATORY—By John King, M. D. Eighth edition, completely revised and largely rewritten. Royal octavo, 1,440 pages, sheep, ... 10 00

KING—THE CAUSES, SYMPTOMS, DIAGNOSIS, PATHOLOGY, AND TREATMENT OF CHRONIC DISEASES. By John King, M. D., Professor of Obstetrics in the Eclectic Medical Institute. This is the most complete work ever published. It embraces a life-time's investigation in this field by one of the best observers and writers on medicine. Every thing that is known of Chronic Diseases will be found here properly classified, and rendered as practicable as possible. Royal octavo, 1,607 pages, sheep, 15 00

KING—AMERICAN ECLECTIC OBSTETRICS—By John King, M. D., Professor of Obstetrics in the Eclectic Medica Institute of Cincinnati. Third edition, revised and enlarged. With seventy illustrations. Royal octavo, 739 pages, sheep, ... 6 50

KING—WOMAN, HER DISEASES AND THEIR TREATMENT. By John King, M. D., Professor of Obstetrics in the Eclectic Medical Institute. Octavo, 366 pages, sheep, 3 50

KING—CHART OF URINARY DEPOSITS: "Carefully arranged, and will prove useful as a reference to the practitioner to refresh his memory, and materially aid the student in getting a clear idea of the subject."—American Journal of Pharmacy, . . . 50

KING—UROLOGICAL DICTIONARY: Containing an Explanation of Numerous Technical Terms; the Qualitative and Quantitative Methods employed in Urinary Investigations; the Chemical Characters and Microscopical Appearances of the Normal and Abnormal Elements of Urine, and their Clinical Indications. By John King, M. D. With twenty-seven useful tables and thirty-nine wood cuts. 8vo, sheep, . . . 3 00

KOST—ELEMENTS OF MATERIA MEDICA AND THERAPEUTICS. Adapted to the American Eclectic or Reformed Practice. With numerous illustrations. By John Kost, M. D., late Professor of Materia Medica, Therapeutics, and Botany, in the American College, Cincinnati, etc. 8vo, 830 pages, sheep, ... 4 50

SCUDDER—THE ECLECTIC PRACTICE OF MEDICINE. By John M. Scudder, M. D., Professor of Pathology and Practice of Medicine in the Eclectic Medical College. Eighth edition, 8vo, 813 pages, sheep, ... 7 00

SCUDDER—THE ECLECTIC PRACTICE OF MEDICINE, FOR FAMILIES. By John M. Scudder, M. D. Illustrated. 8vo, 884 pages, sheep or half morocco, . 5 00

SCUDDER—PRACTICAL TREATISE ON DISEASES OF WOMEN. By John M. Scudder, M. D. Illustrated by colored plates and numerous wood engravings. 8vo, sheep, . . . 4 00

SCUDDER—SPECIFIC MEDICATION AND SPECIFIC MEDICINES. By John M. Scudder, M. D. 12mo, 394 pages, cloth 2 50

SCUDDER—SPECIFIC DIAGNOSIS. A Study of Diseases with Reference to the Administration of Remedies. By John M. Scudder, M. D. 12mo, 400 pages, cloth, . 2 50

SCUDDER—THE ECLECTIC PRACTICE IN THE DISEASES OF CHILDREN. By Professor John M. Scudder, M. D, Octavo, 456 pages, sheep, . . . 5 00

SCUDDER—THE PRINCIPLES OF MEDICINE. By Professor John M. Scudder, M. D. 8vo, 361 pages, sheep, . . . 4 00

SCUDDER—ON THE USE OF INHALATIONS, in the Treatment of Diseases of the Respiratory Organs. By John M. Scudder, M. D. 12mo, 94 pages, sheep, ... 1 00

SCUDDER—ON THE REPRODUCTIVE ORGANS AND THE VENEREAL. By John M. Scudder, M. D. Colored illustrations. 8vo, 394 pages, sheep, ... 5 00

SCUDDER—A FAMILIAR TREATISE ON MEDICINE. By John M. Scudder, M. D. Two vols. in one. 8vo, 840 pages, cloth, ... 3 00

SYME—NEWTON—PRINCIPLES AND PRACTICE OF SURGERY. By James Syme, Professor of Clinical Surgery, University of Edinburgh, Surgeon to the Queen, etc. Edited with illustrations, by Robt. S. Newton, M. D., Professor of Surgery. Third edition. 8vo, 908 pages sheep, ... 6 00

Either of the above sent by Mail, Postage Prepaid, on receipt of the Price Annexed.

The American Dispensatory,

By JOHN KING, M. D.,

PROFESSOR OF OBSTETRICS AND DISEASES OF WOMEN AND CHILDREN, IN THE ECLECTIC MEDICAL
INSTITUTE OF CINCINNATI; FORMERLY PROFESSOR OF THE SAME IN THE ECLECTIC COLLEGE
OF MEDICINE, CINCINNATI; OF MATERIA MEDICA, THERAPEUTICS AND MEDICAL
JURISPRUDENCE, IN THE MEMPHIS INSTITUTE; AUTHOR OF "AMERICAN
ECLECTIC OBSTETRICS," "WOMEN, THEIR DISEASES AND
TREATMENT," "CHRONIC DISEASES." ETC., ETC.

EIGHTH EDITION, Completely Revised and Largely Re-Written.

ONE VOLUME, ROYAL OCTAVO, 1440 PAGES, SHEEP, PRICE $10.00.

Although many valuable Dispensatories have been presented to the physicians and pharmaceutists of this country and Europe, they have all, except the former editions of this work, been confined to an account of those remedies only which have been recognized and employed by that class of physicians termed "Old School" or "Allopathic," ignoring all the recent discoveries and improvement of Medical Reformers, and have, therefore, only partially answered the purposes of the large number of progressive medical men found in these countries. In the present Dispensatory, not only are all the known medicinal plants described, as well as their numerous pharmaceutical compounds, alkalies, resinoids, oleo-resins, etc., but likewise all those poisonous mineral agents so strongly objected to by the New School physicians; thus forming a volume full and complete in itself. There is no other work, in Europe or America, containing such completeness of information. To render the work practically useful to the physician and pharmaceutist, and to bring it up to the discoveries and improvements in medical science of the present day, neither pains nor expense have been spared.

NOTICES OF FORMER EDITIONS.

GOOD OLD-SCHOOL AUTHORITY.—The "*American Journal of Pharmacy*" speaks of the work as follows: "We have taken some pains to give it a careful examination, although pressed for time. * * * The numerous plants which are brought forward as Electic Remedies, embrace many of undoubted value. * * * The work embodies a large number of facts of a Therapeutical character, which deserve to be studied. Many of these are capable of being adopted by physicians, especially by country physicians, who have the advantage of more easily getting the plants. * * * The attention which is now being given by the Eclectics, in classifying and arranging facts and observations relative to American plants, will certainly be attended with excellent results. "It would afford us much pleasure to extract a number of the articles from the *Eclectic Dispensatory*, but the length of this article admonishes to stop; yet we can not close without adjudging to Dr. KING the merit of giving perspicuity and order to the vast mass of material collected under the name of Botanical Medicine, and for his determination to oppose the wholesale quackery of Eclectic Chemical Institutes. The Eclectics have opened a wide field for the rational Therapeutist and the organic chemist; and we hope that physicians and apothecaries will not be repelled from reaping the harvest which will accrue to observation and experiment."

"The examination we have been able to give it, has convinced us that a great deal of labor has been bestowed upon the production, and that it contains an account of a larger number of the medical plants indigenous to our country, than any other work with which we are acquainted."—*Michigan Journal of Medicine.*

WILSTACH, BALDWIN & CO., PUBLISHERS,

141 and 143 Race Street, Cincinnati.

Sent by Mail, Postage Prepaid, on Receipt of the Price Annexed.